Medical Technology and Society

This book is published as part of an Alfred P. Sloan Foundation program.

Medical Technology and Society

This book is published as part of an Alfred P. Sloan Foundation program.

Medical Technology and Society: An Interdisciplinary Perspective

Joseph D. Bronzino,
Vincent H. Smith, and
Maurice L. Wade

The MIT Press
Cambridge, Massachusetts
London, England

342.1
B8Um

© 1990 Massachusetts Institute of Technology

All rights reserved. No part of this book may be reproduced in any form by any electronic or mechanical means (including photocopying, recording, or information storage and retrieval) without permission in writing from the publisher.

This book was printed and bound in the United States of America.

Library of Congress Cataloging-in-Publication Data

Bronzino, Joseph D., 1937–
    Medical technology and society : an interdisciplinary perspective
/ Joseph D. Bronzino, Vincent H. Smith, and Maurice L. Wade.
       p.   cm. — (New liberal arts series)
    ISBN 0-262-02300-8
    1. Medical technology—Social aspects. 2. Medical technology
—Moral and ethical aspects. 3. Social medicine. 4. Medical
economics. I. Smith, Vincent H. II. Wade, Maurice L. III. Title.
IV. Series.
    [DNLM: 1. Delivery of Health Care—trends—United States.
2. Technology Assessment, Biomedical.   W 84 AA1 B79m].
R855.3.B76 1990
362.1—dc20
DNLM/DLC
for Library of Congress   89-13410
                                    CIP

Contents

UNIVERSITY LIBRARIES
CARNEGIE MELLON UNIVERSITY
PITTSBURGH, PA   15213-3890

Series Foreword

The Alfred P. Sloan Foundation's New Liberal Arts (NLA) Program stems from the belief that a liberal education for our time should involve undergraduates in meaningful experiences with technology and with quantitative approaches to problem solving in a wide range of subjects and fields. Students should understand not only the fundamental concepts of technology and how structures and machines work but also the scientific and cultural settings within which engineers work and the impacts (positive and negative) of technology on individuals and society. They should be much more comfortable than they are with making calculations, reasoning with numbers and symbols, and applying mathematical and physical models. These methods of learning about nature are increasingly important in more and more fields. They also underlie the process by which engineers create the technologies that exercise such vast influences over all our lives.

The program is closely associated with the names of Stephen White and James D. Koerner, both vice presidents (retired) of the foundation. Mr. White wrote an internal memorandum in 1980 that led to the launching of the program two years later. In it he argued for quantitative reasoning and technology as "new" liberal arts, not as replacements for the liberal arts as customarily identified but as liberating modes of thought needed for understanding the technological world in which we now live. Mr. Koerner administered the program for the foundation, successfully leading it through its crucial first four years.

The foundation's grants to 36 undergraduate colleges and 12 universities have supported a large number of seminars, workshops, and symposia on topics in technology and applied mathematics. Many new courses have been developed and existing courses modified at these colleges. Some minors or concentrations in technology studies have been organized. A Resource Center for the NLA Program, located at Stony Brook, publishes and distributes a monthly newsletter, collects and disseminates syllabi, teaching modules, and other materials prepared at the colleges and universities taking part in the program, and serves in a variety of ways to bring news of NLA activities to all who express interest and request information.

As the program progressed, faculty members who had developed successful new liberal arts courses began to prepare textbooks. Also a number of the foundation's grants to universities were used to support writing projects of professors—often from engineering departments—who had taught well-attended courses in technology and applied mathematics that had been designed to be accessible to liberal arts undergraduates. It seemed appropriate not only to encourage the preparation of books for such courses but also to find a way to publish and thereby make available to the widest possible audience the best products of these teaching experiences and writing projects. This is the background with which the foundation approached The MIT Press and the McGraw-Hill Publishing Company about publishing a series of books on the new liberal arts. There enthusiastic response led to the launch of the New Liberal Arts Series.

The publishers and the Alfred P. Sloan Foundation express their appreciation to the members of the Editorial Advisory Board for the New Liberal Arts Series: John G. Truxal, Distinguished Teaching Professor, Department of Technology and Society, State University of New York, Stony Brook, Chairman; Joseph Bordogna, Alfred Fitler Moore Professor and Dean, School of Engineering and Applied Science, University of Pennsylvania; Robert W. Mann, Whitaker Professor of Biomedical Engineering, Massachusetts Institute of Technology; Merritt Roe Smith, Professor of the History of Technology, Massachusetts Institute of Technology; J. Ronald Spencer, Associate Academic Dean and Lecturer in History, Trinity College; and Allen B. Tucker, Jr., Professor of Computer Science, Bowdoin College. In developing this new publication program, The MIT Press has been represented by Frank P. Satlow and the McGraw-Hill Publishing Company by Eric M. Munson.

Samuel Goldberg
Program Officer, Alfred P. Sloan Foundation

Preface

Medicine is a special profession, and health care is a special commodity. No other group of professionals and no other industry has such an obvious and dramatic effect on our day-to-day welfare. When we are sick, we want to be cured right away or, at least, to begin the process of recovery. Even when we are not sick, we want to monitor the general state of our health and be warned about potential medical problems. Many of us also feel that we, our families, and our communities have a right to receive the best care the system can offer, even when we cannot pay for it ourselves.

Today medical care is provided through complex heirarchical delivery systems that involve complicated science, sophisticated machinery, and a multitude of highly specialized personnel. The contemporary American health care system is a far cry from the structures that provided care at the turn of the century, which often consisted of little more than a rural doctor with few surgical techniques and a drug cabinet almost devoid of effective remedies.

Since the turn of the century, the primary source of the changes that have taken place in health care has been technical innovation. In recent years the rate of innovation has been astonishing and created a medical world about which most of us know very little. Although we certainly appreciate its benefits—longer life, reduced morbidity, lower risks of adverse effects from surgery, reductions in the time spent recovering from illness, safer and more precise diagnostic techniques, and even the disappearance of some forms of disease—technical innovation has often been coupled with increased economic and societal costs. Many patients feel that the system has become inhumane. They are overwhelmed by the complexity of the technologies that are being used to help them and often feel that they are treated as products processed in a hospital assembly line. More are frightened by the prospect of the catastrophic hospital bills that may accompany heroic efforts to save their lives or the lives of their families and friends, whereas others are terrified of being sustained solely by mechanical support.

The ethical dilemmas posed for society as a whole by a technologically based health care system are also profound. We are no longer sure when efforts to sustain life should be continued. It is not clear how much of society's resources should be allocated to the provision of health care based on sophisticated technologies. Nor are we certain about how to fairly allocate the benefits of the system among potential recipients. In short, both as individuals and societies, many of us do not understand the technologies on which our health care system is based or the complex economic and ethical issues that confront us.

This book is intended to explain the technological bases of some of the most important innovations in medical technology and the economic and ethical issues associated with their development and use. It is written for the reader with little or no specialized background in either medicine, engineering, economics, or philosophy and is especially well suited for undergraduates pursuing a liberal arts education as well as those interested in becoming familiar with the social issues posed by the use of medical technology.

The book has three parts: The first part consists of three chapters that identify the major technological, economic and ethical issues associated with the changes that have taken place in the American health care system since the turn of the century. Chapter 1 provides a historical overview of the development of medical technology and its effect on the evolution of the American health care delivery system. Economic issues are examined in chapter 2, in which we describe the role played by changing economic conditions as well as federal and state programs in affecting the development and adoption of new medical technologies since 1950. This chapter also provides an introduction to the techniques of benefit-cost analysis, a set of procedures often used to assess the potential or actual social worth of new medical technologies. Chapter 3 presents an analysis of the types of moral arguments that surround debates concerning the use of new medical technologies.

The second part of the book consists of four chapters, each of which provides an in-depth examination of specific areas of medical technology. Chapter 4 explores cardiovascular medical technology, including cardiac pacemakers, defibrillators, cardiac assist devices, and the artificial heart. In this chapter benefit-cost analysis is used to assess the economic viability of these cardiac technologies and, in addition, ethical issues posed by human experimentation are examined with special attention given to the artificial heart. Chapter 5 covers critical care technologies, including respiratory therapy, patient monitoring and intensive care units. Economic issues involving the costliness and overuse of intensive care units in contemporary hospital settings are discussed. The ethical dilemmas posed by the life sustaining potential of modern resuscitative and support devices are examined, particularly the question of euthan-asia. Chapter 6 covers the utilization of computers in health care, especially in those applications relating to the clinical laboratory, patient medical records, and diagnostic support systems. The questions of whether the use of computers in health care has increased costs, affected the rate of technical change, and reduced employment of health care professionals are considered. The chapter concludes by defining legal liability, examining liability for harm caused by defective software, and discussing the impact of computers on confidentiality and the privacy of medical information. Chapter 7 covers the fundamental principles of several medical imaging modalities, including nuclear medicine, diagnostic ultrasound, computed tomography and magnetic resonance imaging. The economic issues associated with the optimal use of these technologies and their distribution throughout the health care system are discussed. This chapter also deals with the nature of technology assessment and explores the ethical concerns raised by its use.

The last part consists of chapter 8, which focuses on contemporary ethical and social concerns raised by the highly technological character of modern health care. Four issues are discussed: First, the question of whether too much of society's health care budget is devoted to acute care rescue medicine as opposed to preventive medicine, is addressed. Second, the concerns raised by the competition of health care resources between technology and basic research are examined. Third, the complaint that today's highly technological medicine results in the dehumanization of patients is investigated. Finally, the issue of equitable access to medical technology is discussed.

Acknowledgments

This study is the result of approximately three and a half years of collaborative research. Its genesis was the organization of a year-long public lecture series at Trinity College focusing on the effects on society of medical technological innovations. The series was funded by the Alfred P. Sloan Foundation under its program to support initiatives in the new liberal arts curriculum. We are deeply grateful to the Sloan Foundation for its continued support over the past three years. In particular we especially thank Sam Goldberg of the Sloan Foundation for his enthusiasm and encouragement and John Truxal of the Foundation's editorial review committee for his detailed insights and comments during a crucial stage of the book's development.

Trinity College has also provided considerable support. We especially appreciate of the efforts of Dean Ronald J. Spencer, who also serves on the Sloan Foundation editorial board. He has been enthusiastic about the project from the outset and has made important contributions through his comments and suggestions on various drafts. We are also deeply grateful to Jim English, former president of Trinity College, and to Borden Painter, former dean of Trinity College, for their support. In addition we thank Montana State University, and particularly Myles Watts, for providing resources and travel funds to facilitate the completion of the project.

Inevitably many of our colleagues have encouraged us and have provided helpful coments and insights that have improved the quality of this work. We thank especially David Ahlgren, Naomi Amos, Ruth Anderson, W. Miller Brown, Bob Battis, Ward Curran, Neville Doherty, Frank Egan, Betsy Frazao, Dale Hoover, John Lapp, Ken Mathews, Kim Steele, and Rick Stroup for their contributions. Also we express our appreciation to Frank Satlow and Laura Radin, our editors at The MIT Press. Laura was helpful in her efforts to tighten our writing. Our wives—Barbara, Janet, and Laura—deserve thanks for their tolerance, love, and patience.

Gerry Donovan, administrative assistant for the Department of Engineering and Computer Science at Trinity College, coordinated and organized preparation of the manuscript. From start to finish she managed to handle our crises and panics with kindness, humor, and common sense. We are very grateful to her for shouldering these burdens so well.

Finally, a special acknowledgment goes to Addison-Wesley, publisher of Bronzino's *Computer Applications for Patient Care* (1982), and to PWS, publisher of Bronzino's *Biomedical Engineering: Basic Concepts and Instrumentation* (1986), for permitting generous use of materials from these texts.

1

Evolution of the Modern Health Care System: The Impact of Technological Development

Since the beginning of the twentieth century, technological innovation in the fields of basic science and engineering has taken place at such a rapid and accelerating pace that sophisticated and complex technologies permeate almost every facet of contemporary life. The consequences of this process have been dramatically apparent in the realm of medical science and the delivery of health care services. Although the art of medicine has a long history, the evolution of a health care system capable of providing positive therapeutic treatments in the prevention and cure of illnesses is a decidedly new phenomenon. In particular the establishment of the modern hospital as the center of a technologically sophisticated health care system has occurred only during the past sixty years. In the process it has created the need for highly specialized health care professionals. The result has been the emergence of technologically sophisticated hospitals dominated by a technologically sophisticated staff.

Technology has always molded medical care; therefore, to appreciate the present state of affairs, it is necessary to briefly review the evolution of medical technology, in general, and the American health care system, in particular. Because the purpose of this chapter is to provide just such a review, let us go back to the beginning.

The Beginnings of the Evolutionary Process

From analysis of archeological evidence as well as myths and legends, it is clear that the treatment of health care problems in prehistoric times did not consist entirely of superstitious practices. Certainly early human societies used a variety of herbs and roots to alleviate the effects of disease and actively sought ways to relieve those in distress. Even crude surgery was attempted by early medical practitioners. For example, skulls with holes in them have been uncovered in various parts of Europe, Asia, and South America. The holes were cut out of the bone with flint instruments to gain access to the brain. Although the rationale for these early operations is unknown, one can speculate. Perhaps they were performed because of magical or religious beliefs, for example, to liberate malicious demons from the skull who were thought to be the cause of extreme pain (as in the case of migraine) or attacks of falling to the ground (as in epilepsy). That this procedure was carried out on living patients, some of whom actually survived, is evident from the rounded edges on the bone surrounding the hole, an indication that the bone had grown again after the operation. These survivors also achieved a special status of sanctity so that after their death pieces of their skull were used as amulets to ward off convulsive attacks. From these beginnings the practice of medicine has evolved to become an integral part of all human societies and cultures.

Early Medical Practice

As early civilization grew in sophistication, the status of medicine increased. In ancient Egypt, Imhotep, physician and builder of pyramids, was held in great esteem. Imhotep's name signified "He who cometh in peace" because he visited the sick to give them "peaceful sleep." This early physician practiced his art so well that he became deified in the culture of ancient Egypt as the god of healing. The concepts and practices of Imhotep and the medical cult he fostered were duly recorded on papyri and stored in ancient tombs. One scroll (c. 1500 B.C.), discovered in 1873, contains hundreds of remedies for numerous afflictions ranging from crocodile bite to constipation. A second famous papyrus (c. 1700 B.C.), discovered in 1862, is considered to be the most important and complete treatise on surgery of all antiquity. These writings outline proper diagnoses, prognoses, and treatment in a series of surgical cases. These two papyri are certainly among the outstanding writings in medical history.

As the influence of ancient Egypt spread, Imhotep was identified by the ancient Greeks with their own god of healing, Aesculapius. Inevitably mythology has become entangled with historical facts, and it is not certain whether Aesculapius was in fact an earthly physician like Imhotep. However, one thing is clear: by 1000 B.C. medicine was already a highly respected profession. In Greece the Aesculapia were temples of the healing cult and may be considered precursors of the first hospitals. By modern standards these temples were essentially sanatoriums, with strong religious overtones. In them patients were received and psychologically prepared, through prayer and sacrifice, to appreciate the past achievements of Aesculapius and his physician-priests. After the appropriate rituals were completed, the patient was allowed to enjoy "temple sleep." During the night patients would be visited by a "healer" who administered medical advice or interpreted dreams. In this way patients became convinced that they would be cured if the prescribed regimen (that is, diet, drugs, or blood-letting) was followed and that if they remained ill, it was due to their own lack of faith. Under this approach patients, not the treatment, were at fault if cure did not occur (Atkinson 1966).

One of the most celebrated of these "healing" temples was on the island of Cos—the reputed birthplace of Hippocrates, the acknowledged founder of Western medicine. Hippocrates was not so much an innovative physician as a systematic and comprehensive collector of existing remedies and techniques. Because he viewed the physician as a scientist rather than a priest, Hippocrates instilled an essential ingredient into the realm of medicine: the scientific spirit. Under his guidance diagnostic observation and clinical treatment began to replace superstition. Instead of blaming disease on the gods, Hippocrates taught that disease was a natural and rationally comprehensible process and that symptoms were reactions of the body to disease. The body itself, he emphasized, possessed its own means of recovery, and the function of the physician was to aid these natural processes. Hippocrates treated each patient as an original case to be studied and documented. His shrewd descriptions of diseases are models for physicians even today. Individuals trained by Hippocrates and the school of Cos migrated to the far corners of the Mediterranean to practice medicine and spread the philosophies of their preceptor (Sigerest 1951). With the work of Hippocrates, and the school and tradition that stems from him, came the first real break from magic and mysticism, and medicine as a rational art was founded.

As the Roman Empire reached its zenith, and as its influence expanded across half the world, it became heir to the great cultures it absorbed, including the medical advances made by them. Alhough the Romans themselves did little to further the advancement of clinical medicine (that is, the treatment of the individual patient), they did make outstanding contributions in the area of public health. They had a well-organized army medical service that not only accompanied the legions on their various campaigns to provide first aid on the battlefield but even established "base hospitals" for convalescents at strategic points throughout the empire. In addition the construction of sewer systems and aqueducts provided the empire with the medical and social advantages of sanitation. Insistence on clean drinking water and unadulterated foods allowed some control and prevention of epidemics and made urban existence, however primitive, possible. Unfortunately, because of inadequate scientific understanding of disease, the preoccupation of the Romans with public health could not completely avert the periodic medical disasters, particularly the plague, that befell its citizens.

Initially Greek physicians and their art were looked upon with disfavor by their Roman masters. As the years passed, however, the favorable impression that these disciples of Hippocrates made on the people became widespread. As a reward for their service to the peoples of the empire, Caesar (46 B.C.) granted Roman citizenship to all Greek practitioners of medicine in his domain. Their new status became so secure that when Rome suffered from famine that same year, these Greek practitioners were the only foreigners not expelled from the city. On the contrary they were even offered bonuses to stay.

Galen, considered the greatest physician in the history of Rome, was Greek. For Galen diagnosis became a fine art; in addition to taking care of his own patients, he responded to requests for medical advice from the far reaches of the empire. Galen wrote more than 300 books about his anatomical observations, his selective case histories, the drugs he prescribed, and his boasts. His account of human anatomy, however, was erroneous because he objected to human dissection and obtained his understanding solely from the study of animals. Moreover, because Galen so dominated the medical scene, and because Galenism was openly endorsed by the Roman Catholic Church as the basis for medical practice, medical inquiry was actually inhibited. Galen's views and writings became both the "bible" and "the law" as far as the pontiffs and pundits of the ensuing Middle Ages were concerned (Calder 1958).

With the collapse of the Roman Empire, the church became the main repository of knowledge. This knowledge, including medical knowledge, was literally dispersed among the many monasteries and orders of the church. The teachings of some of the leaders of the early Roman Catholic Church and the belief in divine mercy made inquiry into the causes of death unnecessary and even undesirable in the opinion of many. Curing patients by rational methods became viewed by some members of the church as sinful interference with the will of God. For them use of drugs signified a lack of faith on the part of both doctor and patient, and scientific medicine fell into disrepute. As a result, for almost a thousand years, medical research made little progress. Not until the sixteenth century, when the Renaissance was in full flower, did significant progress occur in the science of medicine.

Although deficient in medical knowledge societies in the Middle Ages were not entirely lacking in the Christian virtue of charity toward the sick poor, for Christian physicians were bound by their faith to treat the rich and poor alike. The church actually assumed considerable responsibility for the sick. Indeed the evolution of the modern hospital began with the advent of Christianity. With the rise of Constantine I (the first of the Roman emperors to embrace Christianity) in A.D. 335, pagan temples of healing were closed and hospitals were established in every cathedral city. The word *hospital* comes from the Latin *hospes*, meaning "host or guest"; the same root has provided the words *hotel*, *hostel*, and *hospice*. These first hospitals were simply places where weary travelers and the sick could find food, lodging, and nursing care provided by the attending monks and nuns.

As the Christian ethic of humanitarianism and charity spread throughout Europe so did its "hospital system." However, trained "physicians" still practiced their trade primarily in the homes of their patients, and usually only the weary travelers, the destitute, and those considered hopeless cases found their way to hospitals. Although conditions in these early hospitals varied widely—only a few were well financed, well managed, and treated their patients humanely—most were essentially custodial institutions keeping troublesome and infectious people isolated. In many such establishments crowding, filth, and high rates of mortality among both patients and attendants were commonplace. As a result the hospital became an institution to be feared and shunned.

Medical Care during the Renaissance

Events in the fifteenth and sixteenth centuries associated with the Renaissance and Reformation undermined the Roman Catholic Church's monopolistic hold on health care producing significant consequences for the science and practice of medicine. The desire to pursue the true secrets of nature was rekindled, and the advancement of medical science was once again stimulated. The study of human anatomy was advanced, and the seeds for further studies were planted by the artists Michelangelo, Raphael, Durer, and of course Leonardo da Vinci. These artists viewed the human body as it was and did not simply accept the beliefs of Galen. The Renaissance painters sketched actual people in both sickness and pain and, in the process, provided genuine insight into the workings of the heart, lungs, brain, and muscle structures. At the same time some physicians also began to approach their patients and the pursuit of medical knowledge in an empirical fashion. New medical schools emerged, the most famous of which were founded at Salerno, Bologna, Montpellier, Padua, and Oxford. These medical training centers embraced the Hippocratic doctrine that disease is a natural process amenable to therapeutic intervention based on empirical study and investigation.

During the Renaissance these fundamentals were examined more closely and the age of measurement began. In 1592 Galileo visited Padua, where he lectured on mathematics to large audiences of medical students and explained his theories and inventions, including the thermoscope, the pendulum, and the telescopic lens. Using these devices, one of his students (Sanctorius) made comparative studies of human temperature and pulse. William Harvey, who later graduated from Padua, applied Galileo's laws of motion and mechanics to the problem of blood circulation. His work, which permitted measurement of the amount of blood moving through the arteries, ultimately enabled the function of the heart to be determined (Calder 1958).

Galileo encouraged the use of experimentation and exact measurement as scientific tools that could provide physicians with an effective check against reckless speculation. Quantification meant that theories could be verified before becoming accepted. These new methods were quickly incorporated into the activities of medical researchers. Body temperature and pulse rate became measures that could be related to other symptoms to assist the physician in diagnosing specific illnesses or diseases. Concurrently the development of the microscope amplified human vision, and a previously unknown world came into focus.

Unfortunately these new scientific devices had little effect on the typical physician, who continued to let blood and disperse noxious ointments. Only in the universities did scientific groups band together to pool their instruments, talents, and knowledge.

Emergence of the Hospital
In England the medical profession found in Henry VIII a forceful and sympathetic patron who assisted doctors in their fight against malpractice and supported the establishment of the College of Physicians, the oldest purely medical institution in Europe. When he suppressed the monastery system in the early sixteenth century,

church hospitals were taken over by the cities in which they were located, bringing about a network of private, nonprofit, voluntary hospitals in which doctors and medical students replaced the nursing sisters and monk-physicians. As a result the professional nursing class became almost nonexistent in these public institutions. Only among the religious orders did nursing remain intact, further compounding the poor lot of patients confined within the walls of the public hospitals. These conditions were to continue until Florence Nightingale appeared on the scene three centuries later.

Yet another dramatic event was to occur. The demands made on England's hospitals, especially the urban hospitals, became excessive as the population of these urban centers continued to expand. These facilities would not accommodate the needs of so many. As a result, during the seventeenth century two of the major urban hospitals in London—St. Bartholomew's and St. Thomas—initiated a policy of admitting and treating only those patients who could be cured. Incurables were left to meet their destiny in other institutions such as asylums, prisons, or the almshouses.

Humanitarian and democratic movements occupied center stage primarily in France and the American colonies during the eighteenth century. The notion of equal rights had come of age, and as urbanization spread, American society concerned itself with the welfare of all its members. Medical practitioners again broadened the scope of their services to include the unfortunates of society and helped to ease their suffering by advocating the power of reason, spearheading prison reform, child care, and the hospital movement. Ironically, as the hospital began to take an active, curative role in medical care in the eighteenth century, the death rate among its patients continued to be excessive. In 1788, for example, the death rate among the patients at the Hotel Dru in Paris—thought to be founded in the seventh century, and the oldest hospital in existence today—was nearly 25 percent. These hospitals were lethal not only to patients, but also to the attendants working in them, whose own death rate hovered between 6 and 12 percent per year.

Clearly the hospital remained a place to avoid. Under these circumstances the fact that the first American colonists postponed or delayed building of hospitals is not surprising. For example, the first hospital in America, the Pennsylvania Hospital, was not built until 1751, and it was over 200 years after the founding of Boston before its first hospital, the Massachusetts General Hospital, opened its doors to the public in 1821 (Crichton 1970, Vogel 1980).

Not until the nineteenth century could hospitals claim to benefit a significant number of patients. This era of progress was due primarily to the improved nursing practices fostered by Florence Nightingale on her return to England from the Crimean War. She demonstrated that hospital deaths were caused more frequently by hospital conditions than by the disease. During the latter part of the nineteenth century, she was at the height of her popularity and influence, and few new hospitals were built anywhere in the world without her advice. During the middle of the nineteenth century, Nightingale forced medical attention to focus once more on the care of the patient. Enthusiastically and philosophically she expressed her views on nursing: "Nursing is putting us in the best possible condition for nature to restore and preserve health." And again: "The art is that of nursing the sick. Please mark, not nursing sickness" (Marks and Bealty 1972).

Although these efforts were significant, hospitals remained, until this century, institutions for the sick poor (Sigerest 1951). In the 1870s, for example, when the plans for the projected Johns Hopkins Hospital were reviewed, an allocation of 324 charity beds and 24 pay beds was considered appropriate. Not only did the hospital population before the turn of the century represent but a narrow portion of the socioeconomic spectrum, it also represented only a limited number of the type of diseases prevalent in the general population. In 1873, for example, roughly half of America's hospitals did not admit patients with contagious diseases, and many others would not admit incurables. Furthermore in this period surgical admissions in general hospitals were only 5 percent of total admissions, with trauma (that is, injuries incurred by traumatic experience) forming a major proportion of these cases.

American hospitals a century ago were rather simple institutions with no special research or technological facilities. In addition, because attending and consulting physicians were usually unsalaried and nursing costs were quite modest, the great bulk of the hospital's normal operating expenses were for food, drugs, and utilities. Not until the twentieth century did modern medicine in the United States come of age. And, as we shall see, technology played a significant role in its evolution.

The Modern Health Care System

Before 1900 medicine had little to offer the typical citizen because its resources were mainly the physician and his education and little black bag. In general, physicians seemed to be in short supply, but the shortage had rather different causes than more recent crises in the availability of health care professionals. Although the costs of obtaining medical training were relatively low, the demand for doctors' services was also very small because many of the services provided by the physician could also be obtained from experienced amateurs residing in the community. The home was typically the site for treatment and recuperation, and relatives and neighbors constituted an able and willing nursing staff. Babies were delivered by midwives, and illnesses not cured by home remedies were left to run their frequently fatal course. The contrast with contemporary health care practices, in which specialist physicians and nurses located within the hospital provide critical diagnostic and treatment services, is dramatic.

At the Turn of the Twentieth Century
The origins of the changes that occurred within medical science are found in the rapid developments that took place in the applied sciences (chemistry, physics, engineering, microbiology, physiology, pharmacology, and so on) at the turn of the century. This process of development was characterized by intense interdisciplinary cross fertilization and provided an environment in which medical research was able to take giant strides forward in the

development of techniques for the diagnosis and treatment of disease. For example, advances in organic chemistry made it possible to isolate (and in some important cases to synthesize) the active ingredients of drugs and anesthetics that had been previously available only in vegetable form. These developments not only alleviated suffering and permitted more controlled drug use (that is, put an end to the difficulties of prescribing correct dosages when the concentration and purity of most drugs (in herbs) were uncertain), they made the resulting products available in vast quantities. Sulfa drugs, antibiotics, cortisone, and many other milestone achievements of pharmacology became commonplace due to the expansion of industrialized techniques that made mass production possible. Similar achievements in radiology, cardiology, encephalography, and the like were made possible only because of corresponding feats in electrical and mechanical engineering that insured their widespread adoption. For example, in 1903 William Enthoven devised the first electrocardiograph to measure the electrical changes that occurred during the beating of the heart. In applying discoveries in the physical sciences to the analysis of a biological process, he gave birth to a new age for both cardiovascular medicine and electrical measurement techniques.

New discoveries in medical science followed one another like intermediates in a chain reaction. However, the most significant innovation for clinical medicine was the development of X-ray technology. When W. K. Roentgen described in 1895 these "new kinds of rays," the inner human was opened to medical inspection. Initially X-rays were used in the diagnosis of bone fractures and dislocations. In the process this modern technology became commonplace in most urban hospitals. Separate departments of radiology were established, and the influence of their activities spread to almost ever other department of medicine (surgery, gynecology, and so forth). By the 1930s X-ray visualization of practically all the organ systems of the body had been made possible through the use of barium salts and a wide variety of radioopaque materials.

X-ray technology gave physicians an enormously powerful tool that for the first time permitted the accurate diagnosis of a wide variety of diseases and injuries. Moreover, too cumbersome and expensive for the local doctor's clinic, X-ray machines had to be placed in the hospitals. Once there, *X-ray technology essentially triggered the transformation of the hospital from a passive receptacle for the sick poor to an active curative institution for all members of society.*

When reviewing some of the most significant developments in health care practices, one is astounded to find that they have occurred fairly recently—that is, within the last fifty to sixty years. Consider, for example, that electroencephalography (EEG)—the recording of the electrical activity of the brain—was not available until 1929 when it was developed by Hans Berger. The information provided by this instrumentation technique has proved as important in the diagnoses of cerebral disease as the electrocardiograph (ECG) has been in heart disease.

Further it was not until the introduction of sulfanilamide in the mid-1930s and penicillin in the early 1940s that the main danger of hospitalization—that is, cross-infection among patients—was significantly reduced. With these new drugs in their arsenals, surgeons were able to perform their operations without prohibitive morbidity and mortality due to infection. Also consider that, even though the different blood groups and their incompatibility were discovered in 1900 and sodium citrate was used in 1913 to prevent clotting, the full development of blood banks was not practical until the 1930s when technology provided adequate refrigeration. Until that time "fresh" donors were bled, and the blood was transfused while still warm (Knowles 1973).

The employment of the available technology assisted in advancing the development of complex surgical procedures. The Drinker respirator was introduced in 1927 and the first heart-lung bypass machine in 1939. In the 1940s cardiac catheterization and angiography (the use of a cannula threaded through an arm vein and into the heart with the injection of radioopaque dye for the X-ray visualization of lung and heart vessels and valves) were developed. Accurate diagnoses of congenital and acquired heart disease (mainly valve disorders due to rheumatic fever) also became possible, and a new era of cardiac and vascular surgery was established.

The second industrial revolution was associated with the astounding developments in electronics that began to come to fruition in the mid-twentieth century (Susskind 1973). Science and technology had leapfrogged past one another throughout recorded history, so that anyone seeking a causal relation was just as likely to find technology the cause and science the effect as the other way around; gunnery lead to ballistics, the steam engine to thermodynamics, powered flight to aerodynamics. The second industrial revolution changed the causal relationship between technology and science to one of systematic exploitation of scientific research. Whereas, during the first industrial revolution, machines were put in place of animals and human muscles, the second industrial revolution initiated the use of machinery for functions ordinarily performed by the senses and human mind.

The second industrial revolution uncovered numerous ways in which engineers interested in the solution of biomedical problems could help their colleagues in the life sciences that a new profession was created—biomedical engineering (Bronzino 1977, 1986). The practitioners of this new discipline receive a combined technical and scientific education and work on a variety of projects, ranging from purely scientific research (for example, learning the basic mechanisms for nerve cells) to designing entirely new kinds of medical instruments. The development of a new round of medical instruments such as electrocardiographs, spectrophotometers, electron microscopes, and radioisotope equipment all became possible due to the emergence and activity of these new professionals (Bronzino 1971).

The impact of these discoveries and many others was profound. The health care system that existed at the turn of the century, consisting primarily of the "horse and buggy" physician, was gone forever, and his replacement—the doctor backed by and centered in the hospital—began to change to accommodate the new technology. Thus it can be seen that the modern hospital in its contemporary, familiar form is essentially 50 years old.

After World War II, the evolution of comprehensive medical care accelerated rapidly, partly as a result of the technological insights obtained in the pursuit of military objectives that then

Figure 1.1

The significant changes that have occurred in just the operating room during the past 50 years are dramatically portrayed in this sequence of photographs. They characterize the surgical scene at (A) the turn of the century, (B) in the mid-1930s, and (C) today. The effect of technology is apparent; in the last photograph the foreground is filled with a heart-lung machine and devices required to administer pain-killing anesthetic gases to the patient. The surgical team is intentionally moved off center stage in this photograph to emphasize the increasing utilization and incorporation of technological tools in the medical arena.

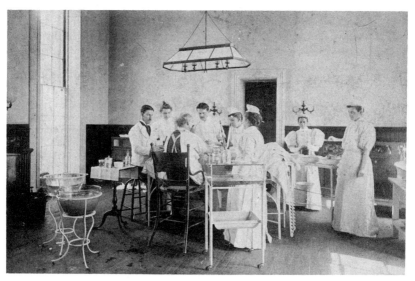

A

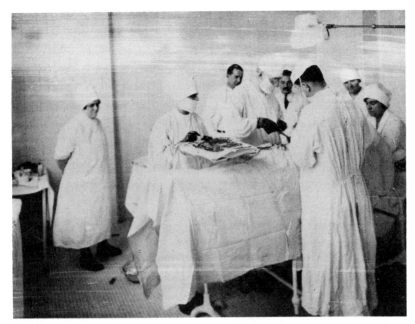

B

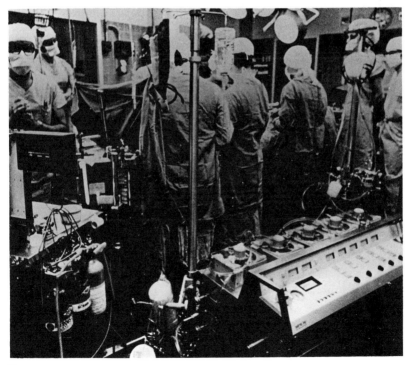

C

Figure 1.2
Technological innovations in the neonatal intensive care unit. (Top) Continuous personal attention was necessary in the early 1920s. (Bottom) Today computerized systems continuously monitor vital signs. (Courtesy of the Saint Francis Hospital and Medical Center, Hartford, Connecticut)

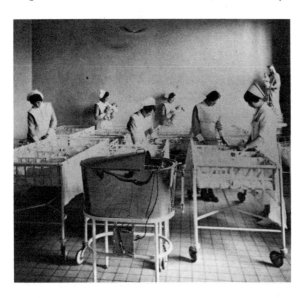

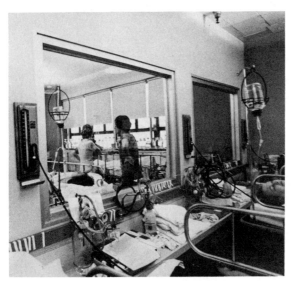

became available for peaceful applications. The medical profession benefited in many important ways from the resulting influx of technological discoveries. Consider the following examples:

1. The realm of electronics became prominent, making possible the mapping of the subtle electrical behavior of the fundamental unit of the central nervous system—the neuron—and the monitoring of beating hearts of patients in *intensive care units* (figure 1.3).

2. *Nuclear medicine,* an outgrowth of the atomic age, emerged as a powerful and effective approach in detecting specific physiological abnormalities. Nuclear medicine involves the use of various radioactive materials, which, when swallowed or injected, tend to gravitate toward specific body organs such as the thyroid gland, lungs, brain, or kidneys. Thus by holding a special radioactivity detector or probe over the particular organ area, it is possible to pick up the distribution of the radiation emitted from the radioactive materials present. With the aid of a computer, this information is assembled and displayed to provide a detailed image of the organ under study or to indicate the presence of abnormal masses under the "all-seeing eye" of the probe. This technique is also capable of providing information regarding the function of certain physiological systems such as the thyroid, liver, or lungs. Scanning systems presently available enable this information to be obtained quickly and with minimal hazard to the patient.

3. *Diagnostic ultrasound,* based on sonar techniques, has been so widely accepted that diagnostic ultrasonic studies are now part of the routine diagnostic evaluation required in many medical specialties. Ultrasound is based on the fundamental concepts of acoustics and involves no exposure to radiation, no injections, no swallowing of radioactive materials, in essence, no discomfort of any kind. A transducer is simply held against the skin, emits high-frequency sound waves, picks up the echoes, and then, once again with the aid of a computer, the gathered information is converted into a meaningful visual image. Diagnostic ultrasonic systems are presently being used by an increasing number of medical specialists to detect and localize the presence of abscesses, blood clots, and tumors (figure 1.4).

Figure 1.3
Central station computer system to aid medical staff in continuous electrocardi-ographic monitoring of up to eight patients.   This system consists of four principal components: digital computer, video display terminal, hard-copy device, and strip chart writers.   Typically these components are integrated into the central station of a coronary care unit. (Courtesy of Electronics for Medicine, Inc., White Plains, New York)

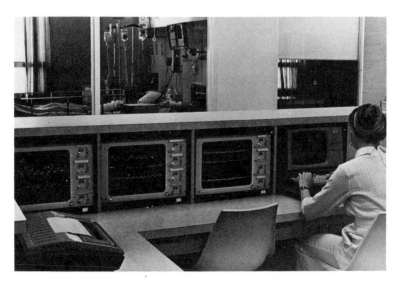

Figure 1.4
Ultrasonoscope used during pregnancy enables the clinician to view placenta, pockets of amniotic fluid, and size of the fetus's head. This device can also be used to study abdominal tumors and provide scans of the kidneys, pelvis, and pancreas. (Courtesy of The Hartford Courant, Hartford, Connecticut)

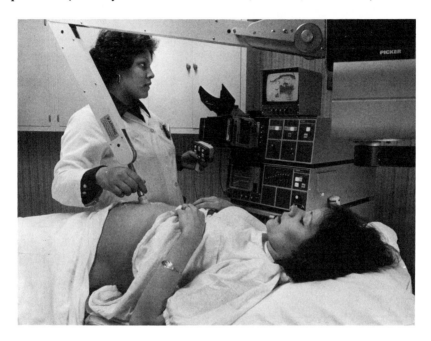

The significance to the medical profession of the technological developments that originated in other fields of research becomes even more impressive when the spin-offs from NASA's space ventures are considered. In many respects what was science fiction in the 1930s and 1940s is now commonplace. Of the technological innovations of the twentieth century, the electronic computer looms as a true giant. During the 1970s and 1980s the expansion of computer technology has been explosive, involving almost every facet of human activity. Today computers are nearly everywhere. Yet despite this tremendous growth spurt, computers are considered to be still in their infancy.

Impact of the Computer Revolution

The impact of computer technology has been extensive. The computer has become an important and powerful tool for collecting, recording, analyzing, and distributing tremendous amounts of information. It has in many ways irreversibly altered our way of life. The computer has freed us from having to perform difficult and time-consuming tasks. For example, tedious calculations, the monitoring and control of repetitive processes, and the maintenance of complex records have all been made easier through the use of the computer. Furthermore its successful application ranges from the dramatic to what is already considered routine. Recall the emotional joy of human accomplishment when Neil Armstrong issued his now famous phrase, ". . . one giant leap for mankind," from the moon. This adventure in space was possible only because the electronic computer was available to assist the astronauts in the guidance and control of the Apollo spacecraft during its 200,000-mile journey. Consider on the other hand the casualness involved as we pick up the telephone to make airline reservations for a flight anywhere in the world. The entire reservation process also has been made possible only through the use of a computer that keeps track of the flights involved, previous reservations, and so forth. This is done in such a way as to be continuously updated, so that when the reservationist verifies the availability of a particular flight, it is routinely accepted as "gospel truth."

Medicine and health care have also been affected by the availability and successful application of these valuable tools. Because of the intrinsic usefulness of the computer as a research aid, there has been a significant effort since the 1960s to apply quantitative analytical tools in medical and biological studies (Bronzino 1982). Impression has given way to quantification, which in turn has begun to free the investigator from clerical drudgery by eliminating the need for laborious hand calculations. Computers have helped life scientists perform research by permitting computations not otherwise humanly possible, either because it would take too long or because it could not be done at all. The fact that computers require explicit instructions and can do only what they are instructed to do has also stimulated the researcher to increase the precision of the experimental data and to organize the data better. The computer has enabled researchers to analyze their data faster and present them in a form that makes it easier for them to interpret the results of various experimental manipulations. Thus, convinced of both the merits of using computers in data analysis and the value of quantification of observable behavior, medical scientists have enthusiastically welcomed the computer into their world. Computer technologists have therefore become valuable allies in programming this machine to look at the acquired data with even more eloquent mathematical "spyglasses."

However, although the computer has long occupied a place of prominence in scientific investigation, it has also been effectively used to improve health care delivery in general. The business functions of the hospital, for example, were the first to be automated, and almost any hospital of any size has some type of computer-based billing, mostly employing simple modifications of approaches and equipment used in commercial settings. Other accounting functions such as inventory control, budgets, and planning have also been affected by computer application. However, because medical institutions are fundamentally different from commercial enterprises, computers affect them somewhat differently. In evaluating the effectiveness and relative cost of using computers to help health professionals provide better services for patients, it is essential to recognize the unique character of the health care industry.

Whereas the goal of a commercial enterprise is economic viability, the goals of the health care industry are much more complex, varied, and unquantifiable (Rothman and Mosman 1972). A hospital has goals in addition to economic viability—it must also provide services that society cannot easily put a price on. It must provide the facilities and services to save lives, even if such facilities are not economical to operate. If a hospital has few "customers" for its heart-lung machines and intensive care wares, it cannot simply eliminate the machines from the list of services provided to the public. This is not to say that such an institution should not be held economically accountable and establish arrangements (such as regional health care consortiums) to achieve cost effectiveness. It does mean, however, that the solutions required to achieve these varied goals are complex.

In many instances computers have proved cost effective. Research laboratories have found that they need computer technology to compete in terms of work accomplished per unit cost. Hospital administrators have found that applications of computers actually reduce the cost of services. This has certainly been true in the automation of instruments and procedures in clinical laboratories—which usually includes the disciplines of hematology, clinical chemistry, and clinical microbiology—with the breakdown of activities by discipline as follows (Williams 1969, Martinek 1972):

1. *Hematology* is concerned with monitoring the activity of a single physiological process: the formation and development of erythrocytes (red blood cells).
2. *Clinical chemistry* is based on the technical science of chemistry and is oriented solely to the application of clinical methods without regard for any specific physiological system or anatomical group of organs or diseases.
3. *Clinical microbiology*, which includes bacteriology, serology, immunology, virology, and other related disciplines, is concerned primarily with the detection of a large number of extraneous

biological agents (microbes) and blood bank operations limited to the collection and preservation of blood and its characterization for compatibility and therapeutic use.

Over the years the practice of medicine has become increasingly dependent on the proliferating number of diagnostic tests and procedures performed in clinical laboratories (Dickson 1969). These tests, ranging from a rather simple physical measurement of specific gravity to such sophisticated techniques as atomic absorption spectrophotometry, form the basis for most of the diagnoses and therapeutic procedures initiated by today's health care team.

As the demand for laboratory tests has continuously increased (doubling approximately every five years), it has become necessary for clinical laboratories to automate the large volume of analyses and incorporate computer processing of raw data for the generation of daily reports, periodic summaries, and permanent data records. In addition to their routine workload, these laboratories, often with limited staff, have been bombarded with requests to perform various combinations of tests and evaluations on emergency patients. Laboratories have also been required to keep up-to-date with the latest methodology. In viewing demands and requirements, these laboratories have turned to automated equipment in performing high-quality specimen analysis.

With the availability of automated systems such as the SMAC (sequential multiple analyzer plus computer) system, it is now possible to process hundreds of samples every hour (figure 1.5). This data-processing approach enables the laboratory scientist to perform up to twenty different tests on each blood specimen, offers improved chemistries, and requires a smaller sample size (about one-tenth the amount otherwise needed to perform these tests). This makes it easier on infants, children, and elderly patients, from whom it is sometimes difficult to obtain blood samples.

Such an automated system can offer savings also in reagent and staffing costs. Because the computer part of this system performs process control, it can calibrate each channel at preselected intervals, troubleshoot all channels and modules, identify samples,

Figure 1.5
The sequential multiple analyzer plus computer (SMAC) system. (From Bronzino 1982)

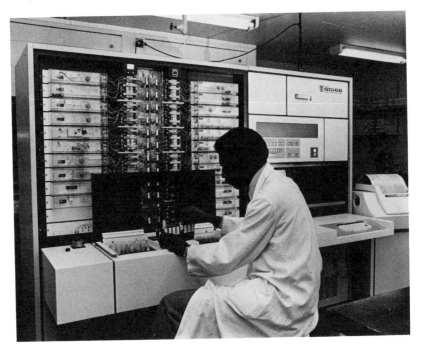

calculate data, and report results, enabling one technician to perform the tasks of many.

These systems represent only the beginning. Throughout the United States there are approximately 7500 clinical laboratories (including those within hospitals) supervised by clinical chemists or clinical pathologists. It is surprising, however, that these laboratories account for only 25 percent of the total clinical laboratory work performed in the United States each year. The other 75 percent is performed by the small laboratories in general practitioners' and internists' offices and by the small commercial laboratories scattered throughout the country. Thus the magnitude of the economic impact of clinical laboratory services is substantial and will only increase as the activities of even the smaller laboratories are tied into regional systems.

In recent years the automation of many clinical laboratories, to a large extent, not only has been successful in technical terms but also has become widely accepted within the medical community. It is generally agreed that these automated hospital facilities can provide a worthwhile contribution to the physician in managing overall patient care. As the operations within the clinical laboratory have become (and continue to be) automated, with the resulting data appropriately organized within the computer, it becomes apparent that this information and methodology represents the cornerstone of an overall patient record management system (Collen et al. 1974).

The availability of automated clinical laboratories and the services they provide to the physician have also stimulated the development of computer systems designed to create permanent records of various laboratory work. These records must be physically incorporated into the patient's charts. In the past the need for such complete records has been met either by inserting laboratory slips into the chart or by transcribing the values to a cumulative record page. In one variation of the many transcription procedures, the laboratory transcribes the values to a patient master record and forwards a duplicate of the updated record to the chart for insertion.

All of these methods suffer from delays and expense, however, and the records themselves are often unreliable and unreadable.

In effort to replace this approach with a printed record that is clear, complete, accurate, and inexpensive, computer-generated permanent record systems have been designed and incorporated within the clinical environment. In these systems, once the data are correctly entered into the computer system, the computer does not lose, misfile, or incorrectly transcribe data. The record is produced with data conveniently arranged for rapid review and can be read by anyone.

Through the use of these new types of patient management systems, medical and nursing staff have instant access to vast bodies of information and to all other hospital departments (laboratory, radiology, admissions, pharmacy, food service, and so forth) (figure 1.6). Professionals no longer have to rely totally on their own memory and can have all the patient data presented at once in a form that facilitates the decision-making process. Such a system permits medical staff to be in continuous control of the patient's status from preadmission to discharge. Thus computer-stored, integrated medical records are essential to the functioning of a computerized medical data system for any large medical program.

Intensive care units are designed for the management of patients experiencing trauma, respiratory insufficiency, coronary episodes, and renal failure. These units provide specialized medical, nursing, and technical personnel, as well as appropriate monitoring devices, to increase both the efficiency and competence of medical care during periods of life-threatening illness. To achieve their objective, ICUs must provide the physician or nurse with relevant information regarding the patient's condition so that appropriate decisions can be made as quickly as possible. Although the use of electronic monitoring equipment at the bedside has permitted the recording of many physiological events, it has remained difficult for the physician or nurse to keep track of all of the information being monitored and at the same time carry out administrative

Figure 1.6
Computerized patient data management system allows unprecedented ease and flexibility in entering and displaying patient data. Data logging and trend displays can be provided at the bedside while the nursing staff at a central station keeps track of each patient's status. Data-taking extends from the time the patient is admitted to the intensive care area until discharged and virtually eliminates the need for handwritten logging of much of the day-to-day patient data.

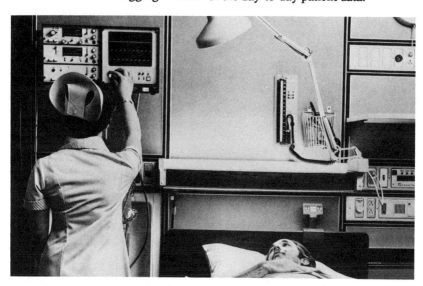

functions and implement treatment procedures. The manual operation of this seemingly ever-increasing number of monitoring devices and the expanded record-keeping requirements placed on the nurses, coupled with a shortage of nursing personnel, have made it necessary for hospitals to look toward automated patient monitoring systems as a possible "aid" in enabling these units to function effectively. As a result a number of automated patient monitoring systems have come on the medical scene with the objective of taking over the monotonous, incorrectly repetitive tasks that make it difficult for the nurse or physician to maintain a constant state of vigilance. Because a computerized monitoring unit requires no sleep or coffee breaks, its use can actually increase the attention paid to the individual patient. This does not imply that the available automated systems can match the skills of the physician, but they can remove a significant part of the data collection and monitoring burden from her shoulders.

In spite of the apparent advantages of such an approach, the thought of adding computers to monitoring systems is not always greeted with enthusiasm. Nevertheless the experience acquired in recent years with the use of monitoring systems and computers in hospitals indicates clearly that automated patient monitoring systems can assist health professionals in the management of critical illness. For example, in the critical care environment the amount of data presented to the nurse by various physiological monitors is often so great that in some cases, by sheer volume, it obscures the essential features required to arrive at proper diagnoses. Automated patient-monitoring systems, however, have been helpful not only in providing for the recording and display of a vast array of physiological phenomena but also by establishing their significance.

Adequate monitoring of physiological events is essential in acute life-threatening situations because changes in vital functions may occur rapidly (such as a sudden drop in blood pressure) and unexpectedly (as in the case of arrhythmias). It is imperative that members of the intensive care team become aware of the occurrence of these compromising events if corrective action is to be taken in time. In essence the concept behind utilizing automated equipment in critical care situations is very similar to that behind enabling humans to fly jet aircraft. In both cases the human is part of a "feedback" system (that is, he has the ability to correct his own performance, direction, and objective) in which he is supplied with information regarding the relative status of the object he must maintain at a prescribed level. In both cases automated systems pick up some of the workload by monitoring many parameters, performing the necessary calculations, and providing information that will enable specific corrections or modifications to be made. The health professional's ability to integrate the presented information and select appropriate courses of action makes him the key factor in assuring that the patient will receive immediate and proper treatment if necessary.

Some decisions can be made automatically. For example, alarm facilities to alert the medical staff can be automatically called into action if a particular parameter exceeds the limits that have been set for it by the attending physician. In this way necessary care can be given to patients before it is too late. Arrhythmia monitoring systems are commercially available for the early detection and warning of ventricular premature beats and most other premonitoring or life-threatening cardiac arrhythmias. It is now possible to provide a continuous display of the current status of a large number of patients, enabling the attending nursing staff to immediately verify the condition of each. Each patient's status can be continually updated so that whenever an alarm sounds, the nurse need only look quickly at the screen to determine the patient and condition causing the alarm. Although this type of system does not recognize all possible cardiac arrhythmias and is not intended to totally replace human observation, it does provide highly efficient, round-the-clock surveillance of coronary care patients. Vital information, if not noted by the staff, is not lost. Indeed the system becomes the "eyes" of the staff when their own eyes are not fixed on displays looking for and warning of life-threatening arrhythmias. As technology continually changes, so does the power of these monitoring marvels.

Throughout history a persistent goal of medicine has been the development of procedures for clearly defining the basic cause of a patient's distress. The search for tools capable of "looking into" the human organism with minimal harm to the patient has been considered extremely important as well. Yet until quite recently the availability of devices capable of providing "images" of normal and diseased tissue within a patient's body has been quite limited. Today new noninvasive imaging modalities make internal structures of the body far more accessible than ever before. These modern imaging devices, based on fundamental concepts in physicsal science (for example, X-ray physics, nuclear physics, and acoustics) and incorporating the latest innovations in computer technology and data-processing techniques, have proved extremely useful in patient care. With the power provided by these new imaging tools, modern medical practice has entered an entirely new era.

One of the most significant advances in medical imaging has been the development of *computerized axial tomography* (CAT), or simply *computed tomography* (CT), and the first computer-based medical instrument—the computerized brain and body scanner. This diagnostic equipment, introduced to the medical world in 1972, allows the visualization of structures and masses in inaccessible regions of the body, an amazing feat accomplished through the most fundamental use of the computer. In addition to enhancing the value of a diagnostic measurement, performing a job faster, and interpreting a physiologic function more accurately, the computer is even more vital in computed tomography. The picture or image provided by this approach is made possible only by the power of the computer. The computer itself generates the image and produces the diagnostic features to be interpreted by clinicians. Consequently the resident computer is the most important component— the very heart—of this new medical wonder.

For almost three-quarters of a century, conventional medical X-ray images were obtained in the same fashion, that is, using a broad X-ray beam and photographic film. Using this standard technique, X rays pass through the body, projecting an image of bones, organs, air spaces, and any obstructions, tumors, or foreign bodies onto a sheet of film. The shadowgraph images obtained result from the variation in the intensity of the transmitted X-ray beam after it has passed through tissues and body fluids of different densities. When this procedure is used to project three-dimensional objects onto a two-dimensional screen, however, difficulties are usually encountered. Because there is often an overlapping of structures represented on the film, it becomes difficult to distinguish between tissues that are similar in density. For this reason conventional X-ray techniques have been unsuccessful in obtaining images of the various sections of the brain, which consist primarily of soft tissue. To overcome this deficiency, attempts have been made to obtain shadowgraphs from a number of different angles in which the internal organs appear in different relationships to one another and to introduce a medium (such as air or iodine solutions) that is either translucent or opaque to X rays (Fischer and Thomson 1978). However, these efforts are usually time consuming, sometimes difficult, and often just not accurate enough.

Computed tomography represents a completely different approach. Consisting primarily of a scanning and detection system, a computer, and a display medium, it combines image-reconstruction techniques with X-ray absorption measurements in such a way as to facilitate the display of any internal organ in three dimensions. The starting point is quite similar to that used in conventional approaches. A collimated beam of X rays is directed through the section of body being scanned to a detector located on the other side of the patient. With a narrowly collimated source and detector system, it is possible to send a narrow beam of photons to a specific detection site. Some of the energy of the X rays is absorbed, and the remainder continues to the detector and is measured.

Surprisingly CT scanners are relatively simple to use (figure 1.7). The patient simply lies down, placing the portion of the body to be scanned and imaged in a section of the device that passes an X-ray beam through it. The radiation emerging from the tissue is absorbed by a detector that in turn supplies the resident computer with information regarding the intensity of the radiation striking it.

Figure 1.7
A modern CT scanner

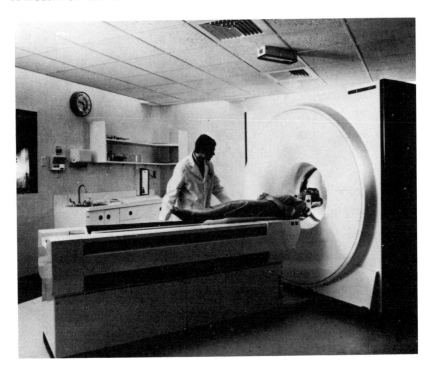

The computer performs the necessary calculations, compiles the results, uses image-reconstruction algorithms, and provides an image or visual display for review by the physician. The detail in the resultant image provided by CT scanning is so remarkable that it has completely revolutionized the field of radiology and gained wide acclaim throughout the medical world.

In a relatively short time CT scanners have become a required diagnostic tool in many medical institutions (Bronzino 1977). The "scanning fever" is evident everywhere. And yet we have witnessed only the beginning of an era in medical imaging. Researchers are actively using the techniques of computed tomography to fashion equally powerful, but quite different, diagnostic equipment. Magnetic resonance imaging systems—that is, machines with strong magnetic fields and radio-frequency scanning probes—have been built that are capable of interrogating even the atoms in living tissues to determine the physical, biochemical, and mechanical status of an organ, presumably without the risk of long-term danger associated wih radiologic technology (Marx 1980). Other large computer-controlled tomographic units that map the distribution of *positron-emitting* radiopharmaceutics have also been developed and used to construct detailed images of organ metabolism, physiology, and function. At the same time the images obtained by CT scanning have continued to improve, indicating that CT is capable of providing far more detail and insight than was originally thought possible. Such activity will certainly continue to enhance the medical imaging capabilities available to clinicians throughout the remainder of the twentieth century.

Throughout this process of continual technological innovation, the hospital has remained the central institution for the provision of diagnostic and treatment services. The role of the hospital as the provider of technologically sophisticated services is likely to continue in the future. At the same time, however, technological developments also offer opportunities for changing health care practices. Technology can certainly be used to provide health care for individuals living in remote rural areas by means of closed-circuit television communication links to regional health care centers. Multiphasic screening systems can be used more extensively as vehicles for providing preventive medicine to a vast majority of the population, thereby limiting admission to the hospital to only those in need of the specialized diagnostic and treatment facilities housed there (figure 1.8). Automation of patient and nursing records is being carried out and can be extended to enable the physician to be aware of the status of patients not only during the hospital stay but also while they are at home. The creation of central medical records systems will permit individuals moving or becoming ill away from home to have their records made available to attending physicians in other areas cheaply and quickly.

Many medical technological innovations have done much to improve the quality of health care in the United States and at the same time have reduced the costs of specific treatments to patients. Although other medical technologies create important opportunities for extending life or improving the quality of life for patients, they are extremely expensive and have changed our understanding of the human organism. Thus the medical technological innovations of the twentieth century have confronted American society with difficult economic and ethical issues. Not only must the economic consequences of the vast outpouring of technological innovations be fully understood, the moral and ethical questions they have raised must be addressed if this technology is to be effectively and efficiently utilized in the future.

Figure 1.8
Cutaway diagram of the actual floor plan of an automated health testing
laboratory (AHTL). Various stations are arranged logically to facilitate orderly
movement of patients through the system. (From Bronzino, J. D. 1974 (Nov).
Industrial Research 17:60-67.)

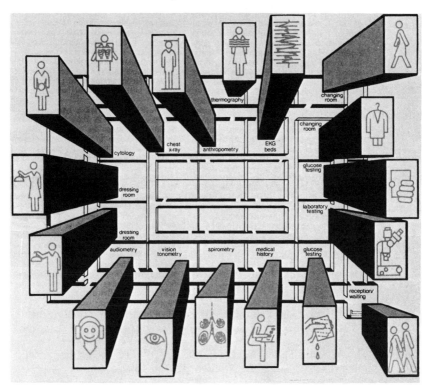

References

Atkinson, D. T. 1966. *Magic, Myth and Medicine.* Fawcett Publications Inc.

Bronzino, J. D. 1971. The biomedical engineer—the roles he can play. *Science* 174:1001-1003.

Bronzino, J. D. 1977. *Technology for Patient Care.* C. V. Mosby.

Bronzino, J. D. 1982. *Computer Applications for Patient Care.* Addison-Wesley.

Bronzino, J. D. 1986. *Biomedical Engineering: Basic Concepts and Instrumentation.* PWS Publishers.

Calder, R. 1958. *Medicine and Man.* George Allen and Unwin Ltd.

Collen, M. F., Davis, L. S., Van Brunt, E. E., and Terdiman, J. F. 1974. Functional goals and problems in large-scale patient management and automated screening. *Federation Proceedings* 33: 2376-2379.

Crichton, M. 1970. *Five Patients—The Hospital Explained.* Alfred A. Knopf, Inc.

Damadian, R. 1977. Fonar image of the live human body. *Physiol. Chem. Phys.* 9:97.

Dickson, J. F. 1969. Automation of clinical laboratories. *Proceedings of the IEEE* 57:1974-1985.

Fischer, H. W., and Thomson, K. R. 1978. Contrast media in coronary arteriography: A review. *Investig. Radiol.* 13:450-459.

Hounsfield, G. N. 1973. Computerized transverse axial scanning (tomography); part 1, description of system. *Br. J. Radiol.* 46:1016-1022.

Knowles, J. 1973. The hospital. *Scientific American* 229:128-137.

Lerch, I. A. 1980. Beyond the CT scanner: In search of a point of light. *The Sciences* 20:6-9.

Marks, G., and Bealty, W. K. 1972. *Women in White.* Charles Scribner's Sons.

Martinek, R. 1972. Automated analytical systems. *Medical Electronics and Data* 3:33-39.

Marx, J. L. 1980. NMR opens a new window into the body. *Science* 210:302-305.

Rothman, S., and Mosman, C. 1972. *Computers and Society.* Science Research Associates, Inc.

Sigerest, H. E. 1951. *A History of Medicine*. Oxford University Press.

Susskind, C. 1973. *Understanding Technology*. The Johns Hopkins University Press.

Vogel, M. J. 1980. *The Invention of the Modern Hospital: Boston 1870-1930*. University of Chicago Press.

Williams, G. Z. 1969. Automation in clinical laboratories. In Dickson, J. F., and Brown, J. H. V. (eds): *Future Goals of Engineering in Biology and Medicine*. Academic Press, Inc.

2

Economics of Medical Technology

Although the development and adoption of new medical technologies by the American health care industry has produced great benefits for American society, it has also produced large increases in health care costs. Expenditures on health care have become so large that frequently they have been perceived as too burdensome for the economy as a whole. In an effort to respond to this perception, President Carter in 1977 initiated a call for retrenchments in federal and state spending on Medicare and Medicaid programs. This initiative and those of the subsequent Reagan administrations have resulted in some modifications to the federal health care programs that are intended to control the rate of growth of health care costs. At the same time insurance companies, under pressure from large corporate and trade union clients to cut premiums for group health care insurance plans, have also begun to adopt cost containment measures. As a result of these policy adjustments, the rate of growth of health care sector costs significantly slowed between 1982 and 1986, and the rate at which new medical technologies are being introduced has also been reduced.

Cost containment policies are directed toward the clearly defined objectives of reducing the levels of government and private sector expenditures on health care, but these are not the only objectives of policymakers. The maintenance and improvement of the health status of the population at large is still a major concern. If access to health care is restricted to contain costs, ideally the restrictions should be designed to minimize any reduction in the quality of health care delivered to patients. Policymakers therefore have become increasingly concerned about the costs and benefits associated with any given medical technology. This concern has led to the use of *benefit/cost analysis* and the related technique of *technology assessment* to evaluate the likely benefits and costs to society of new or emerging medical technologies. The findings of such studies in turn provide guidance to the federal government and insurance companies about which medical therapies should be paid for under Medicare and private insurance plans. (See, for example, a 1978 report by the Office of Technology Assessment (OTA 1978).)

The objectives of this chapter are therefore twofold. The first is to present a brief history of the evolution of the economic policies that have affected the American health care system since 1950. In the process the question of whether medical technology has been responsible for the subsequent explosive growth of health care costs is examined. Other possible causes of increased health care expenditures also are identified, including (1) direct and indirect government subsidies for purchases of health care, (2) rising per capita incomes, (3) population growth (each of which has increased private demand for health care services), and (4) government subsidies for research through federal agencies such as the National Institutes for Health and the National Science Foundation. Second, this chapter deals with the basic methodologies of benefit-cost analysis and technology assessment and identifies some of the major difficulties that arise in their use as guidelines for allocating scarce resources between competing medical technologies. Benefit-cost analysis provides a useful framework for assessing the economic ramifications of medical technologies and consequently is often used to evaluate the economic viability of the medical technologies discussed in subsequent chapters.

The Economics of Health Care Expenditures since 1950

The present state of the American health care system is the result of significant developments within the health care industry and in federal health care policy that have taken place since 1950. An examination of the relation between rising health care costs, the creation of new medical technologies, and other possible causes of rising health costs strongly indicates that federal policy has been particularly important. Direct subsidies to users of health care services have been a prominent feature of the federal government's health care policy since 1966 when the Medicare and Medicaid programs were implemented as a result of provisions included in the "Great Society" legislation embodied in the 1965 Social Security

Act. Indirect subsidies in the form of income tax exemptions for employer and employee contributions to health insurance plans provided incentives for expansions in coverage under third-party comprehensive private health insurance plans throughout the entire period. Both types of subsidy have stimulated increased use of health care services by individuals. Additionally increases have occurred in the prices of "inputs" or "resources" used in the production of health care, such as the wages paid to physicians, nurses, and other health care professionals; the prices of drugs and basic equipment (for example, beds and linen); the costs of constructing new health care facilities and maintaining existing ones (for example, clinics and hospitals); and the prices of services used to transport sick people (for example, basic ambulance services). Aggregate or nationwide expenditures on health care have also risen to some extent because of increases in the demand for health care services that would have taken place without government intervention. The most important nongovernment causes of increased demand for health care have been growth in population and the purchasing power of family incomes. All of these issues are examined in some detail.

Health Status

The general health of the population of the United States has improved significantly since 1950. A crude but useful indicator of health status is life expectancy (table 2.1). In 1950 the life expectancy for the average American citizen was 68.2 years; by 1983 that figure had risen to 74.7 years (U.S. National Center for Health Statistics). These aggregate data disguise significant differences in the behavior of life expectancies for men as opposed to women, and blacks and other minorities as opposed to whites. In 1950 life expectancy was 65.6 years for men and 71 years for women. By 1983 those figures had risen to 71 years and 78.3 years respectively. Thus during this period the absolute gap between life expectancies for men and women rose from 5.5 years to 7.3 years because life expectancy for men rose by 5.6 years, or 8.2 percent, and life expectancy for women rose by 7.3 years, or 10.3 percent.

Table 2.1
Life Expectancies in the United States by Sex and Race:
1950 and 1983

	Average Life Expectancies (yrs)	
Group	1950	1983
Total population	68.2	74.7
Men	65.2	71.0
Women	71.0	78.3
Blacks	60.8	71.3
Men	59.1	67.1
Women	62.9	75.3
Whites	69.1	75.2
Men	66.5	71.6
Women	72.2	78.8

From U.S. National Center for Health Statistics: *Vital Statistics of the United States*; various issues

A breakdown of life expectancies by race is also revealing. In 1950 the life expectancy for a black person was 60.8 years; for a white person it was 69.1 years. By 1983 life expectancy had become 71.3 years for blacks and 75.2 years for whites. These changes translate into percentage increases of 17.3 percent for blacks and 8.8 percent for whites. Among blacks, between 1950 and 1983, black women enjoyed the largest increase in life expectancy, from 62.9 years to 75.3 years, a percentage increase of 19.7 percent. Life expectancy for black men also increased, but not quite as rapidly, rising by 13.5 percent from 59.1 years in 1950 to 67.1 years in 1983. Among whites life expectancy for women rose from 72.2 years to 78.8 years, or by 9.1 percent, whereas life expectancy for men rose from 66.5 years to 71.6 years, or by 7.7 percent.

Clearly life expectancies for all groups in American society increased substantially between 1950 and 1983. Among broad subcategories, women benefited more than men, and blacks benefited more than whites. Life expectancy, however, is only one indicator of the health status of a population. Since 1950 other aspects of the nation's health status have improved. The quality of life experienced by patients suffering from particular diseases has been enhanced by new medical therapies, including the use of new drugs, prosthetic devices (such as the artificial hip), and new diagnostic and surgical techniques that permit early recognition and less costly treatment of specific illnesses.

Morbidity involves not only pain and suffering but also costs to the individual and society as a whole in the form of lost work time and reduced productivity. Those costs are often substantial. Reductions in both the incidence of specific diseases and the length of time over which individuals are incapacitated by them have provided substantial benefits to American society in the past thirty years. For example, cardiovascular disease killed 5.1 people per thousand of the population in 1950. That figure rose to 5.16 in 1965, but then declined dramatically to 4.24 by 1981. The average length of hospital stay for cardiac patients also fell from 10.6 days in 1975* to 8.6 days in 1983. Length of stay for hospitalizations, irrespective of cause, declined from 7.8 days in 1965 to 6.9 days in 1983, a substantial decline of 11.5 percent.

These reductions in mortality and morbidity, however, cannot be attributed solely to innovations in medical technologies. Such innovations have made important contributions in the areas of prevention (through the development of new vaccines, new delivery methods for vaccines, and diagnostic techniques that lead to prevention through life adjustments) and the amelioration of the effects of disease. The health status of the population has also been affected by better working conditions, reductions in pollution levels, and, perhaps most significantly, improved nutrition associated with rising per capita incomes, more extensive education, and government programs to subsidize food for the poor (for example, the food stamp program and the school lunch program).

---

*Data on hospital stay length were not collected on a systematic basis until 1975.

The effect of federal food subsidy programs on the health status of the poor, and especially poor children, has been substantial. Between 1955 and 1983 the number of participants in the food stamp program rose from zero to 21.6 million people (reaching a peak of 22.6 million in 1981). The number of children participating in the school lunch program increased from 10.13 million in 1950 to 23.2 million in 1983 (although the latter represents a decline from the 1979 peak participation level of 27.0 million children (U.S.D.A., Agricultural Statistics)). Food programs directed to the needs of the elderly poor also have been expanded. Since the introduction of nutrition programs for the elderly, the total number of meals provided under the program has risen from 11 million in 1935 to 200 million in 1983.

These programs have been directed toward the nutritional status of low-income households. Poor black and Hispanic families have had higher participation rates than poor white families. In 1982, 61 percent of poor black families participated in the food stamp program and 78.9 percent in the school lunch program. Corresponding rates for poor Hispanic families were 51.4 percent and 78.1 percent respectively. Much of the improvement in the health status of blacks and other minorities relative to whites can be attributed to better nutrition. But largely because of the policy initiatives that are discussed in the following, it is also associated with improved access to health care services.

The Concepts of Supply and Demand
Before we examine in detail the behavior of health care expenditures in the United States, it is useful to examine the components of those expenditures from a theoretical perspective. Therefore brief descriptions of several basic economic concepts are provided.

Total expenditures on any commodity over a given time period are equal to the price P at which each unit of the commodity is purchased multiplied by the quantity Q of the commodity purchased during that time period; that is,

total expenditures $= P \cdot Q.$

Events that cause either or both variables to increase will cause total expenditures to rise. Events that cause the value of one of the two variables to increase and the value of the other to fall may either increase or decrease expenditures on the product, depending on the relative sizes of the adjustments in opposite directions.

The types of events that tend to raise expenditures can be better understood if a little time is spent grappling with some elementary concepts used in *supply and demand analysis* of market behavior. A market brings the buyers and sellers of a specific commodity together for purposes of trade. In the case of health care the buyer is either the patient, the patient's family, or a health care professional acting on behalf of the patient. The seller is a hospital, physician, dentist, nursing home, pharmacy, or some other type of health care provider.

In supply-demand analysis it is often useful to characterize the *economic* behavior of buyers through the use of demand curves and the economic behavior of suppliers through the use of supply curves. A *demand curve* is a very simple construct that shows the quantity of a product that buyers will purchase during a given period of time (if the product is available in appropriate quantities) at each given price in the range of prices that are likely to occur, when other factors that might affect consumer buying decisions (such as income levels or the prices of competing products) remain fixed. An example is presented in figure 2.1. The vertical axis of the diagram in figure 2.1 is used to measure the market price of the commodity (say, cosmetic surgery). The horizontal axis is used to measure the quantity of the commodity demanded by consumers at a given market price. The demand curve for this commodity is represented by the line DD^1. Point A on the demand curve indicates that at a market price of $4, consumers want to buy 50 units of the product (cosmetic surgery) per time period. Point B indicates that at a market price of $3 consumers want to buy 80 units of the product. Point C indicates that at a market price of $2, consumers want to buy 150 units of the product. Other points on DD^1 correspond to similar price–quantity demanded combinations. The

Figure 2.1
The demand curve. DD¹ represents the demand curve for a hypothetical commodity. Each point on the demand curve shows the quantity that consumers of the product will purchase at a given market price over a given time period.

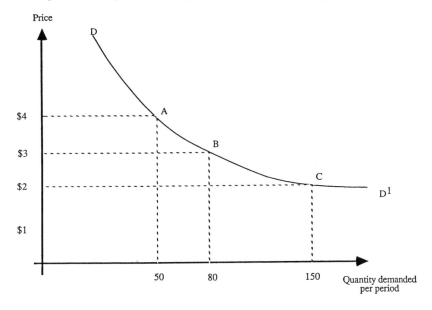

example in figure 2.1 is very simple, but it reflects one important belief that economists hold about the demand for a commodity: if no other events take place, a fall (rise) in the market price of the commodity itself will cause the quantity purchased to increase (decrease). This belief is generally described as the *law of demand*.

The supply of a product in a given market place can be examined using a similar framework. A *supply curve,* also a very simple construct, shows the quantity of a product firms will offer for sale during a given period of time at each price in the range of prices that is likely to occur, as long as factors other than product price that might affect sellers' production and marketing decisions do not change. The supply curve for the product (cosmetic surgery) is represented by the line SS^1 in figure 2.2. Point J on the supply curve indicates that at a market price of $2, suppliers will offer 30 units of the product for sale. Point K indicates that at a market price of $3, suppliers will offer 80 units, and point L indicates that at a price of $4, they will offer 120 units. Other points on the supply curve represent other price–quantity supplied combinations that reflect the behavior of suppliers. The supply curve in figure 2.2 also reflects an important belief held by economists about the behavior of firms: when other factors that may influence the decisions of producers remain unchanged, a rise in the price of a product will increase the quantity of that product offered for sale. (This assumption has been widely confirmed by observations on the behavior of firms.)

Sellers and buyers of goods and services do not operate in isolation. Exchanges take place between them in market places. Other things being equal, if sellers offer more of a product than buyers are willing to purchase at the current market price, that price will not persist for very long. There would in fact be a *surplus* of the product. If, in the case of our imaginary product, the market price happened to be $4 per unit, the buyers would only purchase 50 units per time period, but sellers would offer 120 units for sale. The sellers would be left with a surplus of 70 units, which would either accumulate on their shelves or be disposed of through a sale.

Figure 2.2
Supply curve. SS¹ represents the supply curve for a hypothetical commodity. Each point on the supply curve shows the quantity that sellers are willing to offer for sale at a given market price.

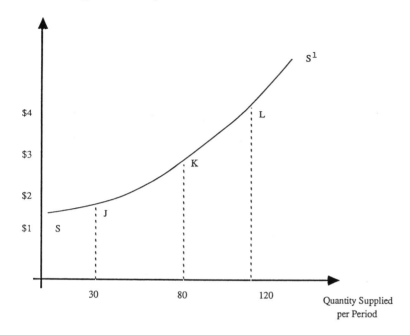

The surplus situation is represented in figure 2.3, in which the supply curve has been superimposed on the demand curve. At market price P, ($4) sellers offer OQ_2 units (120 units) of product for sale, whereas buyers purchase only OQ_1 units (50 units), leaving an unsold surplus of Q_1Q_2 units (70 units). The solution to a surplus situation is for sellers to lower the price of their product. The result of the price cut will be increases in purchases by buyers and reductions in the amounts offered for sale by producers.

The opposite situation occurs if the market price for a product is relatively low. Suppose, in the example in figure 2.3, that the market price for the product is $2. Producers offer 30 units for sale at that price, whereas buyers wish to purchase 150 units, leaving a *shortage* of 120 units between the wants of buyers and the offerings of sellers. The likely outcome is that some buyers, frustrated by their failure to acquire the goods they want, will offer payments in excess of $2 for units of the good, which sellers will be only too happy to accept. In other words the shortage will lead to increases in the market price of the product. The price increases will encourage additional output, discourage purchases, and reduce or eliminate the shortage.

A market clearing, or *equilibrium,* situation occurs when the market price falls or rises to a level at which producers offer amounts of the commodity identical to the amounts buyers are willing to purchase. In the simple example presented in figure 2.3, this outcome occurs at a market price of $3. At this price producers offer 80 units of the product and buyers want to purchase 80 units of the product over the same time period; there is a surplus neither of product nor of customers. The equilibrium situation is represented by point E in figure 2.3, the point of intersection between the supply curve and the demand curve for the product. Notice that total expenditures on all units of the product, the total *costs* of enjoying the commodity, amount to $240 (the market price of $3 multiplied by the 80 units sold and purchased).

At this point it is informative to extend our theoretical discussion to the case of health care expenditures. Total expenditures on a commodity rise when prices or quantities, or both, rise or when

Figure 2.3
Surpluses, shortages, and market equilibrium. At market price \$4 (P_1) buyers wish to purchase only 50 units of product OQ_1, whereas sellers offer 120 units of product OQ_2, leaving a surplus of 70 units of product $Q_1 Q_2$. At a market price of \$2, buyers wish to purchase 150 units of product OQ_4, whereas sellers offer only 30 units (OQ_3), leaving a shortage of 120 units (Q_3, Q_4). At a market price of \$3, because buyers and sellers both wish to trade 80 units of product OQ_0, a market equilibrium is established.

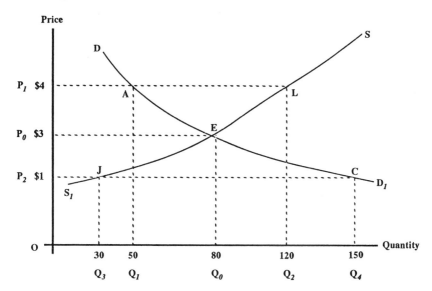

quantities increase by more (proportionately) than prices decline or prices increase by more (proportionately) than quantities decline. But, for expenditures to change, prices or quantities must change! Now consider figure 2.4. Imagine that the curves labeled $S_0S_0^1$ and $D_0D_0^1$ represent the market conditions for medical services in 1950. The market price is P_0, and total purchase of "units" of health care services is Q_0. Expenditures on health care services in that year are therefore represented by the rectangular area $OP_0E_0Q_0$.* The question at issue is, What causes changes in total health care expenditures? Suppose forces are at work that cause an increase in demand, that is, increases in the amounts of the product consumers want to buy at any given product price. The implications can be identified using figure 2.4. At the market price P_0 and at all other prices, buyers will attempt to purchase more than they did before; at P_0 they wish to purchase, say, Q_3 units instead of Q_0 units, and at P_2 they wish to purchase Q_4 units instead of Q_2 units. In terms of the diagram, for some (as yet unspecified) reason the entire demand curve for health services has shifted to the right, from D_0D_0 to D_1D_1. Because the supply curve has not adjusted, at the initial (1950) market price of P_1, sellers still offer OQ_0 units of health care services to buyers, but buyers now wish to purchase OQ_3 units. A shortage has come into existence because of the increase in demand. As a result of the shortage, market prices rise, encouraging higher output levels of health care services than OQ_0 and cutbacks in desired purchases below OQ_3 until a new equilibrium or balance between output and desired purchases is achieved. This situation occurs at market price P_1, at which point producers want to sell Q_1 units of product and buyers want to purchase Q_1 units of product.

The upshot of the increase in demand for health care expenditures can now be clearly understood. Both price and output have increased (from P_0 to P_1 and Q_0 to Q_1 respectively), and levels of health care expenditures, represented in the diagram by rectangle

*The price of the product, OP_0, represents the height of the rectangle, and the point quantity sold, OQ_0, represents its base. The area of a rectangle is its base multiplied by its height, or P_0 multiplied by Q_1, and hence the area of rectangle $OP_0E_0Q_0$ represents total expenditures.

Figure 2.4
The effects of increase in demands in health care expenditures. The shift in the demand curve from $D_0D_0^1$ to $D_1D_1^1$ leads to an increase in market price from P_0 to P_1, an increase in the market quantity from Q to Q_1, and an increase in total expenditures from $OP_0E_0Q_0$ to $OP_0E_1Q_1$.

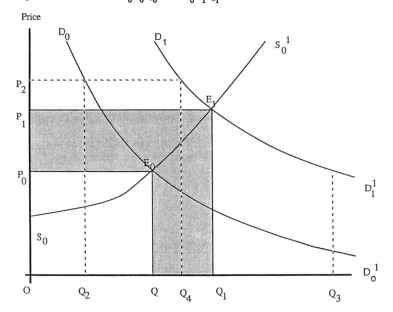

$OP_1E_1Q_1$, have increased.* The above insight is fundamental to any understanding of what has happened to health care expenditures since 1950. Numerous forces have worked to increase the demand for health care expenditures. They include population growth, federal government insurance programs, tax subsidies that lower the costs of private health insurance and increases in the wealth and real purchasing power of household incomes.†

Health Care Expenditures
The data on health care expenditures in the United States presented in table 2.2 are astonishing. In 1950 total health care expenditures in the United States amounted to $12.7 billion. By 1986 they had risen to $458.2 billion. Even when these figures are adjusted to take account of general inflation over the period, the increase is still substantial. In 1950 total health care expenditures, if measured in terms of 1984 prices, would have amounted to $55.2 billion. By 1986 they had risen to $432.3 billion (again in terms of 1984 prices). Thus over the 36-year period total health care expenditures (in constant dollar terms) increased by more than 783 percent.

Over the same time period, as the data presented in table 2.3 indicate, the general index of prices for medical services increased

---

* The new expenditure rectangle $OP_1E_1Q_1$ exceeds the original expenditures rectangle OP_0E_0Q by the shaded area $P_0P_1E_1E_0 + QQ_1E_1E_0$.

†These effects are examined in some detail in this chapter, but at this stage it is important to address one other major issue. The effects of economywide inflation are included in most data on health care costs. These effects should not be confused with additional increases in total expenditures associated with increases in the price of medical services in excess of the economywide rate of inflation. As illustrated in figure 2.4, one of the effects of a rise in demand for a commodity is a rise in its price.

That price increase is the direct result of the increasing scarcity of the resources needed to produce the commodity. The prices of such resources always tend to increase when the demand for the product itself increases because the need and demand for those resources on the part of suppliers has increased. In fact the price indexes for medical services have risen more sharply than prices in general, a result which now should not come as a surprise.

Table 2.2
U.S. Health Care Expenditures, 1950-1984*

	Total Health Care Expenditures	
Year	Current Dollars	1984 Dollars†
1950	12.7	55.2
1955	17.7	67.9
1960	26.9	91.3
1965	41.9	131.3
1970	75.0	191.3
1975	132.7	245.9
1980	248.0	323.7
1984	391.1	391.1
1986	458.2	432.3

From U.S. Health Care Financing Adminstration:
Health Care Financing Review; various issues.

*Selected years; expenditures in billions of dollars.

†These figures were obtained by adjusting the figures in the
Current Dollars column using GNP "price deflator" price index
series presented in table 2.3. The adjustment formula is

$$\text{health care expenditures} = \frac{\text{health care expenditures in current dollars for year T} \cdot \text{GNP price deflator price index for 1984}}{\text{GNP price deflator price index for year T}}$$

from 37.7 in 1950 to 275.2 in 1986, indicating an absolute increase
of 730 percent over that period. When the medical price index is
compared with a general index of all prices in the economy such as
the GNP (gross national product) deflator (presented in table 2.3 for
selected years), it is apparent that medical services have become
more expensive relative to other types of goods and services. The
GNP deflator index increased from 53.6 in 1950 to 238.8 in 1984,
indicating an absolute rise of 445 percent in the general price level.

Table 2.3
Indexes for the General Level of Prices and Prices of
Medical Goods and Services, 1950-1984*

Year	GNP Deflator Price Index† (1972 = 100)	Medical Services and Goods Price Index (1972 = 100)
1950	53.6	37.7
1955	60.8	45.5
1960	68.7	55.1
1965	74.4	62.8
1970	91.4	84.6
1975	125.8	118.3
1980	178.6	183.3
1984	224.9	255.0
1986	238.8	275.2

*Selected years

†U.S. Department of Commerce: *Survey of Current Business*; various issues.

Over that 36-year period the prices of medical commodities and the services of health care professionals increased by approximately 63 percent relative to the prices of other commodities and the incomes of workers in other occupations.

The rise in expenditures on health care services has by no means been simply the result of higher prices for medical commodities. One way of examining the change in the quantity of health care services is to look at annual total expenditures on health care when these expenditures are adjusted for changes in health care prices that have occurred from year to year.   These data are presented for selected years in table 2.4.   Between 1950, when health care services (measured in 1984 prices for health care services) stood at $89.4 billion, and 1985, when they stood at $387.4 billion, purchases of health care services increased in "real terms" at least by

Table 2.4
Expenditures in Health Care Services in the United States
Measured in Terms of Health Care Resources, 1950-1984*

Years	Health Care Resources
1950	89.4
1960	128.5
1970	235.1
1980	352.1
1984	391.1

*These figures, for selected years in billions of dollars, were obtained by adjusting the 1989 expenditures in health care presented in column 2 of table 2.2 using the price index for medical goods and services presented in column 3 of table 2.3.

a factor of nearly four.† Other indicators of the rate of utilization of health care services strongly confirm this finding. In 1950, the average number of days spent in hospital per 1000 people was 1165 days. In 1982 that figure had risen to 1186 days. As the population had increased from 152 million people to 236 million people, hospital utilization had effectively increased by a factor of 1.58. Visits to physicians also rose substantially from 396 million visits in 1970 to 429 million visits in 1981.‡

Another frequently used method of placing health care expenditures in a broader context is to examine the share of the country's GNP for which they account. These data are presented for the United States for selected years in table 2.5. A country's GNP is

†The simple procedure used here does not account for improvements in the quality of health services that undoubtedly took place over the time period. For example, better training and new discoveries almost certainly improved the skills of physicians, whereas discoveries in the fields of biochemistry, pharmacology, and other health-related sciences improved the efficiency of many treatments.

‡Data on physicians' visits are not available for previous years.

defined as the market value of all final goods and services produced in a given year. The word "final" needs further explanation: final products are products purchased by households for their own use, by firms as investment goods, or by the government. They do not include intermediate products, such as flour bought by a baker to make bread to be sold to a consumer the next day. The bread the consumer buys is the final good—the value of the flour used by the baker is incorporated in the price charged to the consumer for the bread. Health care services are final commodities because they are "enjoyed" directly by consumers.

Expenditures on health care services in the United States accounted for only 4.4 percent of GNP in 1950. By 1966, the year in which federal Medicare and Medicaid programs were introduced, that figure had risen to 6.0 percent, in 1984 it stood at 10.6 percent, and by 1986 it had risen to 10.8 percent. For purposes of comparison it is worth noting that over the same period, 1954 to 1984, the share of GNP allocated to defense and to primary and secondary education, respectively, changed from 13.8 percent* to 6.1 percent and from 2.04 percent to 6.5 percent. The data lead to the inescapable conclusion that health care, which used to be a significant component of the U.S. economy, now may be *the* most important industry in the national economy. Small wonder then that in recent years serious attention has been paid to the size of health care expenditures and the nature of the industries that provide health care services.

The issue has become even more pointed because other countries spend considerably smaller portions of their GNPs on health care expenditures. The most dramatic contrast is provided by the United Kingdom, which spends approximately 6.5 percent of its GNP on health care, but many other developed countries spend much less of their incomes on medical care than does the United States. For example, although health care expenditures in Canada have increased substantially since 1950, for the most part they have not increased as rapidly as in the United States.

*Of course, in 1950 the United States was involved in the Korean War.

Health care expenditures in the United States have increased so substantially in both absolute and relative terms that the question, "Who pays for those expenditures," has become very important. The answer to this question begins to shed some light on the reasons why health care expenditures have grown so rapidly. In the mid- and late twentieth century, four types of payers have underwritten the health care bills incurred by the average citizen: (1) the citizen himself, (2) private insurance companies, (3) federal or state government through tax-subsidized public insurance schemes, and (4) private philanthropy (or charity). The data presented in table 2.6 indicate that, in 1950, 56.3 percent of all expenditures in health care were met out of the consumer's own pocket, 7.8 percent by private insurance companies, 27 percent by public funds, and 8.8 percent by philanthropy. In 1965, the year before the implementation of the Medicare and Medicaid programs, payment shares were 44.6 percent from patients' out-of-pocket payments, 21.6 percent from private insurance, 26.7 percent from public funds, and 7.8 percent from philanthropy and other sources. In 1984 payment shares had changed substantially. Consumers' out-of-pocket payments represented only 26.6 percent of all health care expenditures, private insurance payments accounted for 31.1 percent, public sources for 41.4 percent, and philanthropy for only 2.9 percent.

In absolute terms consumer out-of-pocket expenditures rose from $7.13 billion in 1950 to $95.4 billion in 1984. Although this represents a substantial rise in both nominal and real terms, it is worth noting that the relative burden of out-of-pocket health care expenditures on household incomes has changed very little. In 1950 out-of-pocket health care expenses amounted to 4.4 percent of personal disposable income;* in 1965 that figure stood at 3.7 percent and in 1985 at 3.6 percent. Thus the direct costs of health care to the average person have become less burdensome relative to his or her income. Over the same period the health care industry's share of GNP rose from 4.6 percent in 1950 to 5.9 percent in 1965 and 10.6 percent in 1984. Philanthropy has not taken up the

---

*Personal disposable income is the after-tax income that a household is free to spend or save in any way that its members choose (without breaking the law).

Table 2.5
Health Care Expenditures as a Percentage of Gross
National Product, 1950-1984 (selected years)

Expenditures Year	Health Care Share of GNP
1950	4.4
1960	5.3
1965	5.9
1966	6.0
1970	7.6
1975	8.6
1980	9.4
1983	10.8
1984	10.6
1986	10.8

From National Health Care Financing Administration:
Health Care Financing Review; various issues.

Table 2.6
Percentage Share of U.S. Health Care Expenditures by
Source of Payment, 1950, 1965, and 1984

Payment Source	1950	1965	1984
Out-of-pocket	56.3	44.6	26.6
Private insurance	7.8	21.6	31.0
Federal and state	27.1	26.7	41.4
Private philanthropy	8.8	7.9	2.9

From U.S. Health Care Financing Administration: *Health Care
Financing Review*; various issues

slack left by the relative decline in out-of-pocket expenses, but
obviously private insurance schemes and public, tax-funded payments have. In both cases we can attribute the reasons, at least in
part, to federal policy decisions.

Table 2.7
Direct Payments for Health Care by Patients as a Fraction
of Personal Disposable Income, 1950-1984 (selected years)

Year	Share of Personal Disposable Income
1950	4.6
1965	3.7
1984	3.6

The Growth of Private Health Insurance Schemes

In the private sector federal policy has been particularly important in stimulating the growth of private group health insurance schemes. Such schemes, which predominate in the insurance industry, are almost always associated with employers or trade unions. For our purposes it is sufficient to consider only the nature of typical employer-based private health care plans to understand why federal policy has encouraged the growth of group insurance schemes. At the outset it is important to understand that these programs are intended to generate profits. Consequently insurance companies design programs that are expected to yield (in the form of participant payments—premiums—to the insurance company) more revenues than the company will have to make on expenditures for the health care services used by participants. The surplus of revenues over costs associated with a plan are required to cover administrative costs (including the wages of insurance salesmen, processing costs for claims, and so forth) and to provide a return on invested capital. Thus, in any given year, monies paid into the plan must be more than sufficient to cover health care expenditures. The average participant therefore makes payments (or has payments made on his or her behalf) into the scheme that exceed average expenditures (per participant) on health care services covered by the scheme.

If such is the case, why do people participate in private group schemes? Would they not be better off simply paying premiums into their own private "health care fund" and using that fund to buy health care? For several reasons the answer is, in many cases, No. First, an insurance company may be able to negotiate lower prices for health care services on behalf of participants. Second, the insurance company may be able to invest the "health care fund" premiums more efficiently than an individual could and, by causing invested funds to yield higher returns, reduce the amount the individual needs to put away for health care protection. Third, the individual cannot reliably predict what his or her health care expenditures will be because of the unpredictable incidence of many serious diseases (with very costly treatments). He or she therefore uses health care insurance as the traditional hedge against unexpected and catastrophically large financial outlays.

These reasons are the ones a layperson typically gives for buying insurance. A fourth is that the individual participating in a group insurance scheme does *not* have to pay all of the premiums—his or her employer subsidizes the cost through employer contributions that, in many cases, amount to 100 percent of the premiums. The employee seems to get something for nothing or virtually nothing. In fact that perspective is wrong. Economists are perhaps too fond of claiming that "there is no such thing as a free lunch," but they are perfectly correct in claiming that there is no such thing as free health insurance. The health insurance premium payments made by an *employer* are part of the employee's total compensation. They are part of the cost of hiring the employee just as are wages and employer's contributions to social security. If a firm did not give a worker health insurance, it would have to pay her a higher salary. From the employee's viewpoint, however, there is a crucial difference between receiving an extra dollar of wages and an extra dollar of health insurance. The wage dollar will be taxed, but the health insurance dollar will not. Thus in effect health services, if acquired through private employer-based group insurance schemes, are heavily subsidized and therefore fundamentally cheaper to the worker as a payer of income tax.

Between 1950 and 1960 two factors came into play that have made the private group insurance subsidy much more important. First, a large percentage of the population became subject to income tax, because both real incomes (incomes measured in terms of their purchasing power) and nominal incomes (incomes measured simply in dollars) rose much faster than the minimum income threshold for tax liability. Second, for the same reasons, more people moved into higher income tax brackets at the margin (that is, they ended up paying a higher rate of tax on an additional dollar received in the form of income than used to be the case), and therefore employer health insurance contributions became more valuable to them. Thus, for example, in negotiations with employers, employee groups such as traditional trade unions and professional associations actively pursued employer-subsidized health insurance plans as a complement to, or in lieu of, wage increases.*

The income tax structure has also allowed people who itemize deductions to claim a certain fraction of their health care expenditures as a tax deduction. Thus to some degree out-of-pocket payments for health care services have also been subsidized as a result of federal (and state) tax policy. However, the importance of the tax subsidy for out-of-pocket expenses has been much less significant than the employer's health insurance contribution tax subsidy because most people do not itemize deductions and only a fraction of out-of-pocket health care expenses can be deducted. In contrast a tax subsidy exists for virtually all recipients of employer contributions to health insurance for all of those contributions and hence for all services purchased under such group insurance plans.

The effects of the tax subsidy on private group insurance have been substantial, as is clear from the rapid growth in the importance of private health insurance as a source of payments for health care. The effects of private insurance on the demand for health care services have also been dramatic. If a person participates in a health insurance program, in most cases that person is required to pay only a fraction of the total cost of the health care services that are used. Suppose this fraction, usually called the coinsurance rate, is 20 percent (in which case the insurance company will be

---

*Pauley (1986) has pointed out that since 1982, when income tax rates were cut, such pressures have weakened because the value of employer-based health care contributions to employees has declined.

willing to pick up 80 percent of the bill). Suppose too that without the insurance coverage, a person would purchase a particular medical treatment for a cold if it cost $20 or less, but not if it cost more than $20. Once that person joined the employer-subsidized insurance scheme, he or she would still be willing to pay only $20 for the treatment out of his or her own pocket. But that is all that would have to be paid because, if the treatment costs a total of $100, the insurance scheme would pay for $80 (80 percent) of the cost of the treatment. Effectively the person has massively increased his or her demand for the health care service because of participation in the health insurance scheme, which itself was (probably) encouraged by a huge tax subsidy on employers' contributions to such schemes. Such tax subsidies have substantially increased the demand for health care services and as a result encouraged the development of medical technologies that facilitate provision of those services.

The Effect of Medicare and Medicaid

Two other federal programs have had especially dramatic effects on the demand for health care services: Medicare and Medicaid. Between 1950 and 1965 public expenditures on health care increased from $3.4 billion to $10.1 billion. Most of this increase represented an expansion in the purchase of health care services for various categories of U.S. citizens, including aid to veterans, the elderly, and the poor under a variety of state and modestly funded federal programs. The relative importance of government intervention, at least in terms of its share of health care expenditures, which amounted to 27.2 percent of those expenditures in 1950 and 26.9 percent in 1965, did not change significantly over that time period. The size of federal expenditures on health care relative to the federal budget also changed little, rising slightly from 5.0 percent in 1950 to 5.6 percent in 1965.

The situation changed rapidly subsequent to the implementation of the Medicare and Medicaid programs on July 1, 1966. These programs were introduced as a major component of President Lyndon B. Johnson's "Great Society" legislative initiatives that were incorporated in the 1965 Social Security Act. After the introduction of the Medicare and Medicaid programs, federal health care expenditures increased dramatically. In 1966 they stood at $10.1 billion. Within two years (by 1968) they had increased by over 80 percent to $18.1 billion, accounting for most of the increase in total health care expenditures during that time. By 1970 that figure had further increased to $26.9 billion, representing 37.5 percent of total health care expenditures, a dramatic increase over the 1965 figure. Government expenditures in health care also increased relative to total government expenditures in all activities, amounting to 8.8 percent of all government outlays in 1970.

During the 1970s health care expenditures funded under federal and state programs continued to rise sharply as amendments to the provisions of the 1965 Social Security Act in 1972 expanded the number of people eligible for Medicaid. By 1975 public health care expenditures amounted to $58.3 billion or 42.4 percent of total health care expenditures. In 1980 they had reached $105.5 billion, approximately 42.5 percent of total health care expenditures. The latter part of the 1970s thus reflected some degree of stabilization in the U.S. government's share of total health care expenditures, but health care expenditures themselves were growing rapidly during that period, rising from 8.8 percent to 9.4 percent of GNP and also representing an increasing proportion of total government expenditures—from 11.7 percent in 1975 to 12.5 percent in 1980.

In a period of fifteen years, almost exclusively as a result of the Medicare and Medicaid programs, public health care expenditures rose from modest to massive proportions, from small items in state and federal budgets to amounts generating serious debate about needs for large increments in federal budgets or cutbacks in other major public programs such as education and defense. Serious questions were raised also about the viability of continued government support for health care. Arguments were made that the Medicare and Medicaid programs were structured to encourage health care expenditures for the most marginally beneficial therapies, including high-tech procedures that were also extremely costly.

There is some justification for those criticisms, though probably not as much as some polemicists might want us to believe. In 1965 the Medicare program was designed to guarantee health services for the elderly. Similarly the Medicaid program was intended to provide basic health services to many economically disadvantaged citizens. Both programs, prior to and at the time of their inception, were regarded as extremely suspect by the medical profession itself, which feared that the thin end of a nationalized health service wedge might be disguised within the initial drafts of the Medicare/Medicaid legislative proposal. President Johnson, Congress, and the upper administration of what is now the Department of Health and Human Services therefore designed the provisions of the legislation to ensure sufficient voluntary participation by health care professionals to guarantee that the new programs would achieve their goals.

The critical provision of the 1965 legislation, at least from that perspective, concerned reimbursement for in-patient hospitalization services rendered. Doctors and others were to be paid on a "cost-plus" or "reasonable cost" basis; that is, they would be reimbursed for all costs in treating Medicare patients. This provision reduced the price of health care to zero for qualifying program participants. Moreover health care providers were placed in the position of being able to submit and receive payment for any medical procedures deemed likely to provide even marginal benefit to the patient. Treatment procedures were evaluated in this context for the federal government by medical peer review groups using criteria that focused on potential therapeutic benefits but not on costs. The environment created by the cost-plus provision was therefore extremely favorable for the adaption of new therapies based on new technologies.

The Medicare legislation put into effect in 1966 under the terms of the 1965 Social Security Act was intended to protect the elderly from the economic burdens of large medical bills through public health insurance. The bill had two parts: A and B. Part A provided hospital insurance for up to 90 days during each benefit period. For a given patient a benefit period begins on the first day of hospitalization and ends when the patient has not been in a hospital or "skilled nursing facility"* for 60 days. Under Part A the patient,

*A skilled nursing facility might be a nursing home or other type of in-patient, posthospitalization care facility staffed by trained nurses.

after being discharged from the hospital, is also covered for stays of up to 100 days in a skilled nursing facility. There is no limit to the number of benefit periods one patient may take. Furthermore each individual has a special one-time allotment of 60 additional hospital days that can be used if the 90-day hospitalization allocation is exhausted in a benefit period.

Part A clearly does not provide a comprehensive health insurance plan for eligible participants. Part B covers many of the gaps left by Part A through supplementary medical insurance (SMI), which has to be purchased by the participant,* though at greatly subsidized rates. In 1985, 97 percent of all persons covered under Part A for health insurance also purchased supplementary health insurance (SHI). In so doing, they obtain coverage for physicians services and supplies provided in relation to physician-based care, outpatient care and some clinic services. Since 1983, on a provisional basis, hospice benefits also have become available for medicare participants. Even with SHI Medicare does not provide 100 percent coverage, but comes close to that goal. There are various deductibles (for example, in 1985 the hospitalization deductible was $400). Many Medicare participants purchase private insurance to cover the Medicare gap; thus they end up facing virtually no costs when they decide to use medical services.

The population eligible for Medicare includes the elderly (over 65) and, since 1972, the disabled and people suffering from end-stage renal disease who typically require dialysis. The total eligible population numbered 19.1 million people in 1966, but had risen to 27.1 million people by 1984, of which only 2.88 million were included under the provision for the disabled and kidney dialysis patients introduced in 1972. The above data underline an important factor that has led to and will continue to generate increases in the demand for health care services. The number of elderly people in the United States has grown very rapidly (by over 26 percent between 1966 and 1986) and will continue to increase in the future in both absolute and relative terms. The older people get, the greater becomes their demand for health care services.

*In 1985 the monthly premium for SMI was $15 to $50 per person.

Thus an aging population coupled with ready access to virtually free health care is a prescription for an explosion in the use of health care services and total health care expenditures. Such proved to be the case between 1966 and 1984. In 1967 Medicare expenditures amounted to $4.7 billion (9.2 percent of total U.S. health care expenditures). By 1984 that figure had risen to $64.6 billion (16.7 percent of total health care expenditures).

Medicaid, introduced at the same time as Medicare, was designed for a different population—the "deserving poor"—which the government regards as including the blind, disabled, and elderly who receive supplemental social security payments and children and mothers receiving support under the Aid to Families with Dependent Children (AFDC) programs. The "undeserving poor" (single persons, married couples, and others) are not necessarily protected by any government program.

Under Title XIX of the 1965 Social Security Act, Medicaid programs provide important basic services such as hospital inpatient and outpatient care, laboratory and X-ray services, skilled nursing facility care (for eligible people over 21 years of age), physician services, family planning services, and rural clinic services to eligible participants. There is no way of simply describing "the Medicaid program" as in fact each state is allowed to develop its own Medicaid eligibility standards because each state sets its own standards for defining eligibility for the AFDC and Supplementary Security Insurance programs. These standards vary considerably among states with respect to income and other criteria. States, at their option, also may define people not automatically covered under Medicaid (because of participation in other welfare programs) as medically needy, using state-specific criteria with respect to an individual's incomes and assets.

The total number of Medicaid eligible recipients has varied over time. In 1973, for example, 19.6 million people (85 percent of the population with incomes below the poverty line) received help under the program. In 1976, 22.8 million people (91 percent of the below-poverty population) received Medicaid. By 1984, however, participation had fallen to 21.4 million people (64 percent of the below-poverty population). These variations have occurred because of changes in eligibility criteria and changes in the policies of the states and the federal government with respect to the provision of welfare payments to the poor.

Under Medicaid, as under Medicare, payments for health care services were originally made on a "vendor" basis. However, partly because of variations in eligibility requirements, Medicaid expenditures grew less rapidly than Medicare expenditures, from $3 billion in 1967 to $14.1 billion in 1975 and $26.8 billion in 1980. Between 1980 and 1984, however, Medicaid expenditures jumped quite sharply to $38.7 billion (an average annual rate of increase of 9.6 percent). Part of that growth is accounted for by increases in payments for intermediate care facilities for the mentally retarded, which in itself is not directly related to developments in the provision of high-tech medical services.

In 1982 a major change was introduced in the payments system for Medicare. No longer could health care providers of hospital services be paid on a cost-plus basis. Instead a *prospective payments mechanism* was established. Under the new system each patient is assigned to one of 486 diagnosis-related groups (DRGs), and a payment schedule determines what fixed amount the hospital (and physician) will receive for treating the patient. If the costs of treating a patient are less than the predetermined payment, the hospital may pocket the difference; if treatment costs exceed the prospective payment, the hospital must bear the loss. To protect the Medicare program against abuses such as incorrect diagnoses of patients that place them in unnecessarily expensive DRG categories, a peer review organization (PRO) system was developed. Under the PRO system review boards carry out systematic examinations of hospital policies to ensure appropriate compliance with the DRG prospective payments system.

The new payments mechanism, legislated under the Tax Equity and Fiscal Responsibility Act of 1982, became effective in October 1983. Its immediate consequences were significant: the admissions rate to hospitals and the average length of stay for Medicare patients both fell. Nevertheless overall Medicare expenditures rose substantially in 1984, partly because of an expanding elderly population whose oldest members were increasingly infirm.

The changes in Medicare implemented in 1982 also made it easier for Medicare participants to qualify for health maintenance organization (HMO) services. Thus an important fillip was provided to a relatively new form of health service organizations. HMOs charge participants a fixed monthly membership fee, guaranteeing them comprehensive health services in return. The fixed per capita fee approach is regarded as providing incentives for cost control. Although evidence on this point is ambiguous, the extension of the Medicare system to include such programs reflects a serious concern at the federal level about cost control, rather than just the provision of state-of-the-art health care.

The implications of prospective payments schemes for medical technology are significant. Such schemes encourage hospitals to identify procedures that reduce treatment costs. Innovative therapies that may improve the quality of care but are also more costly become unattractive. The Medicare payments mechanism is therefore likely to discourage many innovative approaches to health care. It will tend to channel the effects of commercial research and development programs toward the development of lower cost techniques of providing acceptable standards of care. The effects of the prospective payment system on publicly funded research programs are harder to predict.

The adoption of prospective payments mechanisms and other innovations designed to control costs has not been limited to federal programs. Since the 1970s some insurance companies have taken similar actions to control costs, and most recently many private insurers have adopted the Medicare DRG scheme as the basis for their own payment schedules. Insurance companies have also developed relationships with HMOs through competitive bidding processes. Under such arrangements HMOs compete for the right to serve individuals participating in the insurance company's health care plan. The contract is then given to the low-bidding HMO, which becomes the source of health care for people participating in the scheme.

These private-sector innovations imply radical adjustments for the provision of health care under private insurance programs and reflect the cost-control concerns of insurance companies' clients as much as the desires of insurance companies to reduce or at least contain the sizes of claims. Many employers offering health care benefit packages have become increasingly concerned about the burden of the premiums they must pay to maintain a given degree of coverage, while at the same time federal tax cuts implemented since 1982 have reduced the value of health benefits to employees. The consequences for medical technology of the increased emphasis on cost containment by the private insurance sector are likely to be similar to those resulting from the shift in the Medicare program. The focus of technological change will shift toward cost-reducing innovations and away from developments that enhance quality of care but at increased costs (Sloan et al. 1986).

Federal policy initiatives carried out over the past forty years have had massive effects on the demand for health care. Tax exemptions for employers' (and, to some extent, employees') private insurance contributions and Medicare and Medicaid have catalyzed increased use of health care resources and stimulated the adoption of new medical therapies based on new technologies. Throughout the period from 1950 to 1985, federal programs have also been directed toward encouraging the production of new technologies through the provision of grants and other types of subsidies for medical research programs.

The most important vehicles for such subsidy programs have been the National Institutes of Health (NIH) and The National Science Foundation (NSF). Of the two NIH has provided more support. Such support increased from $79 million in 1950 to $4.801 billion in 1984 (table 2.8). This represents a huge increase over the entire period, but does not reflect more recent trends in funding levels. In terms of real purchasing power, the expansion has been less dramatic. If measured in 1984 dollars, federal medical research expenditures in 1950 would have been $331 million. In fact the rate of growth of federal medical research funding requires more detailed examination. Between 1950 and 1960 research funding measured in real terms grew rapidly at an annual rate of 18.3 percent, from $331 million to $1.761 billion. Federal funding continued to grow significantly during the 1960s,

Table 2.8
Federal Government Research Expenditures
in the U.S., 1950–1984*

Year	Research Expenditures
1950	331
1960	1761
1970	3865
1980	4875
1986	4801

From U.S. Health Care Financing Administration:
Health Care Financing Review; various issues.

*Selected years; expenditures in millions of dollars.

reaching $3.865 billion in 1970, and again in the 1970s, reaching $4.875 billion in 1980. Since 1980, however, federal policy has dramatically changed. There has been almost no alteration in the level of federal funding for medical research, which in 1986 amounted to only $4.801 billion. Because the prices for resources used in medical research have increased more rapidly than the prices of commodities in general, less federally funded medical research was carried out in 1986 than in 1980.

The decision to reduce federal support for medical research was deliberate. It reflected the views of the Reagan administration that government expenditures should be cut and that the private sector should and consequently would be more active in medical research and development. The latter prediction has not been confirmed by industry's response in relation to funding for medical research. Between 1980 and 1983 private funding for medical research (which in both years was less than one-tenth of federal funding) increased more slowly than did the general price level, implying an overall decline in private sector support for medical research.

The implications of the recent cutbacks in the funding of medical research for medical technologies seem fairly clear. Just as the expansion of research funding between 1950 and 1980 led to the development of a huge array of potential new medical technologies, so the decline in research funding is likely to reduce the rate at which medical technologies are identified and integrated into medical treatment therapies and procedures. In such an environment it is particularly important for a society to support the research and development of medical technologies that seem most promising in terms of the benefits they yield in relation to their costs.

Benefit-Cost Analysis

If medical technologies are no longer to be developed and adopted simply because they hold out some promise for improvements in the quality of medical care, then presumably only technologies that yield the largest amounts of benefits from the available health care budget should be adopted. In fact economists have long argued on grounds of social welfare considerations that only commodities whose benefits exceed their costs should be provided and that this principle applies no more or less to health care technologies than to any other commodity (see, for example, Mishan 1976 and Gramlich 1981). The real force of that argument becomes apparent when it is realized that by "cost" the economist means the benefits that would have been enjoyed if resources used in the production of the service of interest were directed to the manufacture of other products.

The calculus of benefit-cost analysis involves the evaluation of benefits gained as opposed to benefits given up as a result of a given action. The benefit-cost appraisal of any medical technology therefore requires an accounting of the social benefits it yields and the social costs it imposes. Benefits of medical technologies usually include reduction in physical pain and suffering from disease, improvements in quality of life associated with improved motor skills, and increases in length of life. These benefits are not always easy to measure, but certainly they are difficult to estimate in the common measure of value used by society (dollars, in the case of the United States). Equally (if not more) difficult to appraise are the benefits that take the form of reductions in the mental anguish and the financial burdens that an illness places on family and friends. More readily measurable are reductions in cost of care that result from the use of new technologies.

These benefit categories deserve more than casual comment. First, consider the last category of benefits. Suppose a certain type of pneumonia requires an average of six days of hospital care at a daily cost of $1,000 per patient under current state-of-the-art procedures. Now suppose a new technology is developed that reduces the required length of hospital stay to three days. The benefit per patient in reduced hospitalization costs would be $3,000. This type of technology would be a classic example of a cost-reducing technology.

Now consider the second category of benefits. Suppose that before the use of a new technology, one in five patients died of the disease. Subsequently none died. In addition suppose that the typical patient, if he or she survives, can expect an additional ten years of life. Further suppose that, through the patient's actions in avoiding risks through choice of occupation and the like, the average person considers a year of additional life to be worth $20,000. Clearly the benefits associated with additional life to *that* person would be substantial.

The fact that this type of benefit is spread over the future leads to what is known as the discount problem. The problem arises because a dollar received now is not the same as a dollar to be received one, two, or ten years from now. One could invest a dollar received today and have that dollar and the accumulated interest in the future. To receive one dollar in some future point of time, one needs *now* only to invest less than that dollar. The discount problem is the problem of determining the amounts that would have to be invested today to yield a fixed sum at a given future. That amount is called the *present value* (PV) of the future fixed sum. In the problem at hand one would need to know the PV of the $20,000 of value received by the patient in his first year of extra life, and to it add the PV of the $20,000 he receives in his second year of extra life, and so on to estimate the PV of all the benefits he obtains through a longer life (see appendix).

Now consider the next type of benefit or cost avoided as a result of the new technology—physical, emotional, and other types of wear and tear on friends and relations. Reduction in length of illness reduces the amount and cost of travel needed for hospital visitation, time lost from work and other activities, and worry about the loved one's illness. Some of these costs are hard to measure in terms of money, but in some instances a monetary assessment is feasible (for example, travel costs, value of lost work time, and stress-related medical costs for third parties involved in the crisis).

The final set of benefits enjoyed directly by the patient are associated with reduced morbidity. In the example, because the wonder technology has reduced the duration and (perhaps) the intensity of pneumonia, the patient is better off; he avoids three days of pain and misery. Not only are his hospitalization costs reduced but his quality of life improves, and he returns to work sooner. These gains provide both the patient and society clear and measurable monetary benefits.

All of these phenomena are clear and distinct benefits for both individuals and society at large. What are the costs of the new technology? The answer to that question is perhaps more obvious. The development of new technology requires investments in research and development. Monies must be expended on basic scientific research (for example, biology, neurology, biochemistry, biomedical engineering, clinical trials, pilot modeling trials, and the establishment of production facilities). In addition the costs of providing the therapy must be included. These costs may be quite subtle. They should include outlays associated with the production of the therapy and also any costs that may be associated with any adverse effects of the therapy experienced by the typical patient.

It is important to note, before leaving the discussion of types of costs and benefits, that actions have both benefit and cost effects that are not associated with the effects of those actions on the individual directly associated with them. In the case of medical technology, questions concerning the distribution of benefits come into play. Distributional questions arise with respect both to patient groups *and* income groups. A particular medical technology (say, dialysis) may yield benefits far in excess of its costs to a (relatively) small number of patients. However, the health care resource requirements of the therapy may draw substantial amounts of resources away from another type of health care treatment (say a certain type of prenatal care) that might yield fewer aggregate benefits (for the same costs) but to many more patients. On the grounds of social justice, policymakers might then choose not to provide the more exclusive treatment (dialysis) but the less benefit-intensive therapy (prenatal care). Alternatively the provision of certain types of health technology might be judged to benefit high-income groups rather than low-income groups (for example, heart transplant operations), whereas other types of health care technologies (prenatal care) might benefit low-income groups rather than high-income groups. When society is obviously concerned about such questions, benefit-cost analysis should include the society-wide implications of the development of a medical technology.

From the point of view of social optimality (doing what is best for society), new medical technologies should be adopted when the benefits associated with them exceed their costs, even if the difference between the two is already small. Remember that costs, in this context, are the benefits that other services would have provided if the new technology were not developed; thus new medical technologies should be developed when their benefits exceed the benefits that would otherwise have been enjoyed. The nature of the social decision thus seems very straightforward; one simply adds all the benefits and all the costs associated with a new therapy and chooses according to a mechanistic rule.

Unfortunately the world is not so simple. Before new technologies are developed and tested, it is extremely difficult to know what all the consequences of their use will be. For example, a new, untested therapy for pneumonia (or some other disease) might harm or help the patient. Even when a technology has been subjected to clinical testing, its actual benefits may be very difficult to ascertain, especially in light of possible unforeseen adverse effects. Thus at the most fundamental level of concern (benefits to patients from technologies), decision makers may lack any reliable basis for a benefit-cost assessment. The cost side of the equation is equally tricky because technological research and development costs are notoriously difficult to forecast, and sometimes seem akin to the cost of nuclear power plants with respect to their propensity to expand over time. Implementation costs for therapies may rise or fall over time. The original cardiac pacemakers, for example, were much more expensive than pacemakers used today. A multitude of other problems compound the difficulties associated with benefit-cost analysis, including whether and what monetary values should be attached to human lives and how future costs and benefits should be discounted.

Benefit-cost analysis is therefore no panacea for social decision making. But does it have any value? Most economists would answer that question affirmatively. If one fails to assess, even at the most general level, the relative benefits of actions, one is truly acting in ignorance. Benefit-cost analysis provides a framework for systematically comparing alternative courses of action on the basis of available information. Other approaches amount to a vote for ignorance, and society, in making its decisions, becomes analogous to the blindfolded child trying to pin a tail onto a picture of a donkey. Often the child does not even find the picture, and if he does, the place he chooses for the tail is often most unfortunate.

Summary

This chapter has provided an overview of the economic growth of the U.S. health care delivery system and the effects of government policy and economywide expansion on the rate at which new medical technologies have been identified, developed, and diffused throughout the health care system since 1950. We have shown that the period from 1950 to 1988 can be roughly divided into two separate phases: 1950 to 1978 and 1978 to 1988. Between

1950 and 1978 innovations in government programs directly targeted at the health care delivery system encouraged the development and adoption of new technology. Federal funding for health research massively increased over the period from 1950 to 1980, generating a plethora of discoveries involving technology-based therapies for the treatment of disease. A gradual expansion of federal funding for health care provision programs took place between 1950 and 1965. Then it turned into a flood of additional support under the aegis of the "Great Society" legislation, which created the Medicare and Medicaid programs. These programs, which permitted "cost-plus" reimbursement to health care providers, effectively guaranteed payments for any new or existing therapy that provided even marginal benefits to the patient. A climate particularly conducive to the adoption of new therapies based on innovations in medical technology was therefore created by the Medicare and Medicaid legislation. At the same time, indirectly and perhaps accidentally, changes in the tax code that raised marginal income tax rates increased the demand for employer-based private group health insurance. The effect of the upward drift in marginal tax rates throughout the fifties, sixties, and seventies was therefore to increase private demand for health care, further enhancing incentives for technical innovation and adoption in the industry.

Between 1978 and 1982, however, all of the policy incentives were reversed. Beginning in 1978, reimbursement procedures for Medicare and Medicaid were altered and subsequently based on prospective payments schedules, thus increasing incentives to health care providers for cost control and reducing incentives for innovation in the quality of care. With the advent of the Reagan administrations, federal funding for health research was cut in terms of the real resources made available to the academic community. In addition, as a result of various tax reform initiatives, marginal tax rates were reduced with consequent adverse effects on the demand for health insurance and health care in the private sector. Thus during the 1980s incentives for innovation and the development of new medical technologies have diminished. Federal policy may change again under the Bush administration. If more emphasis is placed on research, federally funded catastrophic health insurance becomes a reality, and taxes do actually have to be raised to cope with the federal budget deficit, then better times could lie ahead for innovators in health care.

Also in this chapter, some fundamental economic concepts, principles, and models have been introduced that are used to examine the economic issues associated with specific medical technologies in chapters 4, 5, 6, and 7. These include basic concepts of supply and demand, models of market behavior, and the principles of cost-benefit analysis. Other economic concepts, principles, and models are introduced in the following chapters in relation to economic issues associated with specific technologies. They include principles associated with risk (chapters 4 and 7), market structure (chapters 4, 5, and 7), criteria for determining optimal investment strategies (chapter 6), different concepts of cost, including concerns about private versus social cost (chapters 6 and 7), and the role of rationing in the provision of high-cost medical technology (chapter 8).

Appendix
The Mechanics of Benefit-Cost Analysis

Benefit-cost analysis is complex and difficult to carry out largely because the benefits and costs associated with a project involving the development of new technologies are difficult to measure. Once the benefit and cost streams of a project have been estimated (or "guestimated"), the mechanics of cost-benefit analysis are relatively straightforward. The purpose of this appendix is to illustrate those mechanics through a simple example.*

Before presenting the example, however, some preliminary concepts have to be introduced. The concept of a *present value* (PV) is critical to the mechanics involved in benefit-cost analysis. In and of itself a PV is a simple idea. Suppose a hospital knows it will receive different amounts of dollars from an endowment fund (amounts x_1, x_2, x_3, x_4, and x_5) at the end of each of the next five years. However, the owners of the hospital would like to have a large sum of cash today to finance the construction of an intensive care unit (ICU). They decide to borrow funds to help finance the ICU but only to the extent they can use the hospital's guaranteed annual income streams from its endowment to pay off the debt and all of the interest associated with the debt over the next five years. The amount the hospital can borrow against that income stream is the income stream's PV.

Now what is the income stream's PV? Suppose the hospital can borrow funds at an annual simple interest rate of r. Consider the income of x_1 dollars to be received by the hospital at the end of the first year. The hospital could borrow an amount v_1 today and pay back that amount and the interest charge rv_1 associated with it if the sum of the principal and interest charge was equal to x_1; that is,

$$v_1 + rv_1 = x_1 \quad \text{or} \quad (1 + r)v_1 = x_1, \tag{1}$$

*This appendix abstracts from many complex issues, including the possibility that interest rates will vary over the life of the project, the measurement of benefits and costs, the treatment of uncertainty and risk, accounting properly for inflation effects, and the choice of the discount rate. For thorough treatments of these issues, see Mishan 1976 and Gramlich 1981.

or

$$v_1 = \frac{1}{1+r} x_1. \tag{2}$$

Thus v_1 is the PV of x_1 dollars to be received one year later (that is, v_1 is the amount that x_1 dollars received one year in the future is worth today). For example, suppose that x_1 is $110 and r, the annual interest rate, is 0.1 (one-tenth or 10 percent of the principal). Then the PV of the $110 to be received one year from now is

$$v_1 = \frac{1}{1.1} 110 = \$100.$$

If the hospital borrows $100 today at 10 percent, it will have to pay its creditor $110 one year from now.

Now consider the x_2 dollars to be received two years from now. The hospital can borrow v_2 dollars against this future revenue, but must pay simple interest in the amount of rv_2 at the end of the first year. Assuming interest is compounded annually, it must also pay interest on the principal and the first year's interest at the end of the second year. Thus at the end of the first year, the hospital would owe $v_2 + rv_2$ or $(1 + r)v_2$. At the end of the second year, it would owe this amount, $(1 + r)v_2$, plus the second year's interest charge on this amount, $r(1 + r)v_2$. If the hospital is to repay its debt of v_2 plus the interest charges accrued over the *two-year* period with x_2, then

$$(1 + r)v_2 + r(1 + r)v_2 = x_2, \quad \text{or}$$

$$(1 + r)(1 + r)v_2 = x_2, \quad \text{or}$$

$$(1 + r)^2 v_2 = x_2,$$

and therefore

$$v_2 = \frac{1}{(1 + r)^2} x_2. \tag{3}$$

Thus v_2 is the PV of x_2 dollars received two years from now.

A similar argument can be used to show that the PV of x_3 dollars to be received three years from now, v_3, is

$$v_3 = \frac{1}{(1+r)_3} x_3. \tag{4}$$

In fact the approach is perfect in general. Suppose x_n dollars is to be received n years from now. Its PV, v_n, is

$$v_n = \frac{1}{(1+r)^n} x_n. \tag{5}$$

The factor $1/(1+r)^n$, by which the future value x_n is converted into its equivalent PV, v_n, is called the *discount* factor. Notice that equation 4 can be applied to all of the cases considered previously. If, for example, $n = 3$, and this value for n is substituted into equation 5, we have the formula for the PV of an amount x to be received three years from now contained in equation 3. Similarly, if $n = 1$, then equation 5 provides the formula for the PV of an amount x to be received one year from now contained in equation 1.

What about the PV of the hospital's stream of incomes to be received over the five-year period? It is simply the sum of the PVs of each of the amounts received at the end of each of the five years; that is,

$$v = v_1 + v_2 + v_3 + v_4 + v_5$$

or, substituting for the v_i's,

$$v = \frac{1}{(1+r)} x_1 + \frac{1}{(1+r)_2} x_2 + \frac{1}{(1+r)_3} x_3 = \frac{1}{(1+r)_4} x_4 + \frac{1}{(1+r)_5} x_5,$$

or

$$v = \sum_{i=1}^{5} \frac{1}{(1+r)^i} x_i,$$

where v is the present value of the entire income stream. For example, suppose $x_1 = x_2 = x_3 = x_4 = x_5 = 1000$, and $r = 0.1$. Then

$$v = \frac{1}{(1 \cdot 1)} 1000 + \frac{1}{(1 \cdot 1)^2} 1000 + \frac{1}{(1 \cdot 1)^3} 1000$$

$$+ \frac{1}{(1 \cdot 1)^4} 1000 + \frac{1}{(1 \cdot 1)^5} 1000$$

$$= 909.1 + 826.5 + 748.8 + 680 + 617$$

$$= 3781.9.$$

If the hospital knew that it had a guaranteed income stream of $1000 per year for five years, the PV of that income stream would be $3781.90. This is the amount it could borrow at an interest rate of 10 percent and pay back in full (including interest charges) at the rate of $1000 per year over five years.

Now suppose the (potentially variable) income stream is received over a period of n years. The approach we have outlined is directly applicable; the PV formula becomes

$$v = \frac{1}{(1 + r)} x_1 + \frac{1}{(1 + r)^2} x_2 + \ldots + \frac{1}{(1 + r)^n} x_n$$

or

$$v = \sum_{i=1}^{n} \frac{1}{(1+r)^i} x_i. \tag{6}$$

We are now in a position to return to the question of cost-benefit analysis. PVs can be used to compare two different income streams; for example, the total benefits and total costs associated with a new health care technology. Typically a new technology's benefits occur in different amounts and in different years than its costs. At first sight comparing the benefits stream with the cost stream appears to be an intractable "apples and oranges" problem. But there is a way out. The PV of each stream can be computed. If the PV of the benefits shown exceeds the PV of the cost stream, then, in a very real sense, the technology will pay for itself. The outlays, including the interest charges associated with them, are more than paid for by the benefits stream. On the other hand, if the PV of the benefits stream is less than the PV of the cost stream, then the reverse holds true. The benefits stream is insufficient to pay for the cost stream.

Consider the following example. Suppose a new type of pacemaker is to be developed. Three years of research and development will be required before it can be used. The R & D program will require outlays of $3 million per year incurred at the outset of the project (immediately), at the end of year 1 (the beginning of year 2), and the end of year 2 (the beginning of year 3). These outlays will provide research materials, labor, and cover the costs of various laboratory experiments and initial clinical trials using volunteer patients. Thereafter, in each of the next four years, benefits will accrue in the form of extended patient lives and reduced morbidity. Simultaneously costs associated with the production of the new pacemakers and their implantation also will be incurred. These benefits and costs in fact will accrue continuously, but for the purpose of the example and to avoid some difficult mathematical problems, we will assume that they occur at the end of years 4, 5, 6, and 7 in the respective amounts of $5 million and $2 million per year. At the end of the seventh year, the pacemaker will become obsolete because of the introduction of newer, more sophisticated and more efficient therapies.

Table 2.9 provides details of the benefits and cost streams. Should the project be adopted? If the benefits and costs are simply added up, the answer appears to be Yes.* Over the entire course of the project, total benefits amount to $20 million, whereas total costs

---

*When this is done, the analyst is simply assuming that the discount rate r is zero; in other words a dollar is equally valuable whether it is received today or 100 years from now.

Table 2.9
The Benefit and Cost Streams for the New
Pacemaker Technology (in millions of dollars)

Year	Costs	Benefits
0	3	0
1	3	0
2	3	0
3	0	0
4	2	5
5	2	5
6	2	5
7	2	5
Undiscounted total costs and benefits	17	20

only equal \$17 million. But the benefits will occur later in the project (years 4–7) than many of the costs (years 0–3). Suppose that the appropriate interest charge, or discount rate, is 10 percent. The PV of the cost stream can be computed using equation 6 with one extension. In the example an outlay of \$3 million is incurred at the outset of the project (year 0). Its PV is \$3 million. The \$3 million cost is incurred today; paying it back today requires exactly the same amount of money; that is $v_0 = x_0$. Note that equation 5 can be applied even in this case. Let $n = 0$. This implies that

$$v_0 = \frac{1}{(1 + r)^0} x_0.$$

But $(1+r)^0 = 1$; therefore $v_0 = x_0$. The PV of x_0 dollars to be received today is simply x_0 dollars. Noting this result, we can modify equation 6:

$$v = x_0 + \frac{1}{(1+r)^1} x_1 + \frac{1}{(1+r)^2} x_2 + \ldots + \frac{1}{(1+r)^n} x_n, \text{ or}$$

$$v = \sum_{i=0}^{n} \frac{1}{(1 + r)^i} x_i.$$

(7)

Equation 8 can be used to compute the PV of the cost stream v_{cost} in the above example:

$$v_{cost} = 3 + \frac{3}{1.1} + \frac{3}{1.1^2} + \frac{0}{1.1^3} + \frac{2}{1.1^4} + \frac{2}{1.1^5} + \frac{2}{1.1^6} + \frac{2}{1.1^7}$$

$$= \$11.95 \text{ million.}$$

Similarly it can be used to compute the PV of the benefit stream:

$$v_{benefit} = 0 + \frac{0}{1.1} + \frac{0}{1.1^2} + \frac{0}{1.1^3} + \frac{5}{1.1^4} + \frac{5}{1.1^5} + \frac{5}{1.1^6} + \frac{5}{1.1^7}$$

$$= \$11.91 \text{ million.}$$

In this example, *given* a discount rate of 10 percent, the PV of the project's benefits are exceeded by the PV of its costs. The project should not be carried out.

But note the critical role of the discount rate. Suppose it were to fall to 5 percent. The PV calculations would change. The PV of the cost stream would become

$$v_{cost} = 3 + \frac{3}{1.05} + \frac{3}{1.05^2} + \frac{0}{1.05^3} + \frac{2}{1.05^4} + \frac{2}{1.05^5} + \frac{2}{1.05^6} + \frac{2}{1.05^7}$$

$$= \$14.96 \text{ million,}$$

and the PV of the benefit stream would become

$$v_{benefit} = 0 + \frac{0}{1.1} + \frac{0}{1.1^2} + \frac{0}{1.1^3} + \frac{5}{1.1^4} + \frac{5}{1.1^5} + \frac{5}{1.1^6} + \frac{5}{1.1^7}$$

$$= \$15.26 \text{ million.}$$

Given the lower discount rate, now the PV of the benefits exceeds the PV of the costs, and the project should be undertaken. The reason is quite straightforward. The lower discount rate increases the value of income streams and costs that occur farther in the future relative to those that occur closer to the present. Thus it raises the value of the benefits stream relative to the cost stream because the benefits occur later in the income stream than do most of the costs.

This example has been constructed to illustrate the mechanics of benefit-cost analysis. It also demonstrates the importance of PVs in the analysis and the significance of the discount rate used to compute those PVs. It is not surprising that groups who favor the development of new medical technologies, whose benefits almost always occur (often much) later than their costs, argue for the use of the low discount rates. Those opposed to the new technologies tend to argue for high discount rates.

All of this discussion raises the question of what constitutes the correct discount rate. The economist's response is that it is the rate of return that could be obtained from the investment of the re-sources required for the project being examined in the best avail-able alternative projects. That rate has been the subject of great debate. Estimates range from a low of zero (after all many investments carried out by firms and governments yield not a dime) to the rate of interest charged on riskless loans in the bond market, which has been as high as 24 percent over the past ten years. Most economists adopt the view that a positive discount rate should be used and that the rate should be the expected return on marginal investment projects that are of identical duration to the project being considered. If inflation's effects on benefit and cost streams are ignored, this criterion suggests that a discount rate of between 4 percent and 9 percent should be used to evaluate future projects where benefit and cost streams are relatively certain.* Discount rates should be higher for projects whose outcomes are difficult to forecast because those projects are riskier.

---

*Interest rates for long-run (almost) riskless debt have ranged between 9 percent and 14 percent over the past five years, and anticipated inflation rates appear to be about 5 percent. The "real" interest rate, discounting for inflation, is (approxi-mately) the market interest rate less the anticipated rate of inflation. Thus, if the market interest rate is 14 percent, and the anticipated inflation rate is 5 percent, then the real interest rate is 9 percent. Similarly, if the market interest rate is 9 percent, and the anticipated inflation rate is 5 percent, then the real interest rate is 4 percent.

References

Allman, S. H., and Blenson, R. 1979. *Medical Technology—The Culprit Behind Health Care Jobs?* Department of Health, Education and Welfare.

Gramlich, E. M. 1981. *Benefit Cost Analysis of Government Programs.* Prentice Hall.

Mishan, E. J. 1976. *Cost-Benefit Analysis.* Praeger.

Pauley, M. V. 1986 (June). Taxation, health insurance and market failure in the medical economy. *Journal of Economic Literature* 25(2):629–675.

Sloan, F. H., Valranoni, J., Perrin, J. M., and Adamache, K. W. 1986 (March). Diffusion of surgical technology: An exploratory study. *Journal of Health Economics* 5(1):31–61.

U.S. Bureau of Labor Statistics. 1970–1988. *Consumer Price Indexes for Selected Items and Groups.* Various issues. Washington, DC.

U.S. Congress Office of Technology Assessment. 1978. *Assessing the Efficiency and Safety of Medical Technologies.* Washington, DC.

U.S. Department of Agriculture. 1967–1986. *Agricultural Statistics.* Various issues. Washington, DC.

U.S. Department of Commerce. 1970–1986. *Survey of Current Business.* Various issues. Washington, DC.

U.S. Health Care Financing Administration. 1974–1987. *Health Care Financing Review.* Various issues. Washington, DC.

U.S. National Center for Health Statistics. 1950–1986. *Vital Statistics of the United States.* Various issues. Washington, DC.

U.S. Congress Office of Technology Assessment: *Assessing the Efficiency and Safety of Medical Technologies,* Washington, DC, Congress of the United States, 1978.

3

Ethics of Medical Technology

The innovations in medical technology described in chapter 1 have
endowed modern medical care providers with therapeutic capabilities far beyond those of their predecessors of even a few decades
ago. Today's clinicians can prolong life, prevent and cure many
diseases and disabilities, and otherwise ameliorate a wide range of
undesirable physical and mental conditions for which little could
have been done in the recent past. Contemporary Western medicine has much to offer, and the impressive innovations in medical
technology that have occurred in the last fifty to sixty years are
assuredly among the main sources of its various benefits. Benefits
even more impressive can be expected in the near future as technology continues to expand the therapeutic preventive and rehabilitative powers of medicine. To see only the benefits of modern
medical technology, however, is to fail to see the whole picture.
Along with these genuinely impressive benefits have come a host
of difficult and perplexing ethical issues. Precisely because of its
newly achieved potency, medical care is now, more than ever, a
subject of intense ethical concern.

This chapter has two aims. The first is to inform readers that
ethics and morality are not simple matters of opinion or religious
faith but are subjects that can be approached rigorously on a secular
basis. This aim is accomplished by introducing readers to the
philosophical study of morality. The second aim is to acquaint
readers with traditional medical morality and some of the challenges to it posed by modern medical technology. Discussion of
these challenges will raise some of the issues treated in greater
detail in chapters 4 through 7.

Ethics and Morality

Although the terms *ethics* and *morality* are often used interchangeably in ordinary discourse, philosophers tend to assign distinctly
different meanings to them. In the lexicon of philosophers, *ethics*

refers to a certain sort of study, whereas *morality* refers to the distinctive object of that study. Accordingly a discussion of ethics best begins by focusing on its subject matter, morality. The morality of a person, a nation, a culture, or any other entity that can sensibly be said to have a morality consists of a body of moral judgments. So understanding what morality is requires understanding what distinguishes moral judgments from other sorts of judgments. All moral judgments are normative judgments of a particular kind. Normative judgments can often be recognized simply by the terminology used to express them. Such judgments are typically expressed by means of evaluative terms such as *good, bad, right, wrong, ought, ought not, obligatory, and nonobligatory,* to note a few. The following assertions are examples of *normative judgments:*

Stealing is wrong.

Gold is an excellent investment.

Toyota is the right choice for smart car buyers.

Voluntary euthanasia should not be legalized.

Everyone ought to have access to an education.

Each of these judgments expresses an evaluation. Each conveys a positive or negative attitude toward some state of affairs. Accordingly each carries some implication for human conduct. Each is intended to play an action-guiding function. When one says, "Gold is an excellent investment," one implies that investors should purchase gold. When one says, "Everyone should have access to an education," one implies that arrangements should be made to ensure that the opportunity to have an education is available to all who wish to have it. This prescriptive or action-guiding function distinguishes normative from nonnormative judgments.

Nonnormative judgments explain, predict, describe, and categorize states of affairs. They do not express evaluations of those states of affairs and are not usually meant to convey favorable or unfavorable attitudes toward anything. Hence by themselves they imply nothing concerning how persons should act. Nonnormative judgments usually can be recognized simply by the absence of evaluative terms. Examples of *nonnormative* judgments are

Stealing exposes one to the likelihood of arrest.

Gold is a very malleable metal.

Toyota sells more cars than any other car manufacturer.

Voluntary euthanasia is likely to remain illegal.

Access to an education cannot be provided to everyone.

None of these judgments expresses an evaluation. Each makes some claim concerning what is, was, or will be the case. Each is descriptive rather than prescriptive.

Although normative judgments usually can be distinguished from nonnormative judgments by the presence or absence of evaluative terms such as *good, bad, right, wrong,* and the like, this is not always the case. For example, if Jack knows that Jill is in the market for a new car and says to her, "Toyota sells more cars than any other manufacturer," he probably means this statement to be taken as equivalent to the statement "Toyota is the right choice for smart car buyers." In many instances determining whether a statement expresses a normative or nonnormative judgment requires knowing the intended meaning of the statement. That is, one will need to know if the speaker is making a judgment intended to influence conduct.

Judgments that comprise a morality then are those that express values and are intended to guide human conduct. However, not all normative judgments are moral judgments. *Some normative judgments are legal judgments.* When I say that dog owners should not let their pets run free, I am not necessarily saying that it is morally wrong to do so or that persons who do so should be subjected to moral condemnation. Rather I am saying that this form of conduct is proscribed by law. Indeed I may believe that letting one's pet dog run free has no moral significance whatsoever. *Some normative judgments are judgments of etiquette.* When I say, for example, "One should never take food from another's plate," I am merely saying that such conduct is rude. Holding this view is perfectly consistent with believing that there is nothing at all morally praiseworthy about politeness or morally blameworthy about rudeness. *Some normative judgments are aesthetic judgments.* When I utter the words "A man should never wear pink socks with blue trousers," I merely give expression to my views concerning what is aesthetically pleasing in male dress, and again these views need not imply that I take such conduct to have any moral significance at all. Although this brief list is not exhaustive, it shows that even when a judgment is clearly normative, it is not always clear whether the judgment is moral or nonmoral.

How then is the particular kind of normative judgment distinctive to morality distinguished from other kinds of normative judgments? The most important way is by determining the kind of standard on which the judgment is based. *Moral judgments are based on moral standards.* Consequently the problem of distinguishing those normative judgments that are moral from those that are nonmoral resolves into distinguishing moral from nonmoral standards. Though philosophers are not in complete agreement on this issue, ethicist Manuel Valesquez (1988) has isolated the four characteristics most often cited in the literature of ethics as the distinguishing marks of moral standards:

1. "[M]oral standards deal with matters that are (or are thought to be) of serious consequence to our human well-being. That is, they are concerned with behavior that can seriously injure or seriously benefit human beings (or that is believed to be capable of having these consequences)."

2. "Moral standards are not made up by particular bodies, nor does their validity rest on the particular decisions of particular persons.

The validity of moral standards rests instead on the adequacy of the reasons that are taken to support and justify them, and so long as these reasons are adequate, the standards remain valid."

3. "[M]oral standards are supposed to override self-interest. That is, if a person has a moral obligation do something, then the person is supposed to do it even if it is not in the person's own interests to do so."

4. "[M]oral standards are based on 'the moral point of view,' that is, a point of view that does not evaluate standards according to whether or not they advance the interests of a particular individual or group, but that goes beyond personal interests to a 'universal' standpoint in which everyone's interests are impartially counted as equal."

The first characteristic noted here, the seriousness of the matters to which moral standards pertain, is sufficient to distinguish moral standards from standards of etiquette, grammar, and aesthetics, for instance. This characteristic also reveals why health care is a particularly apt subject for moral concern. Little, if anything, has greater significance for human well-being than life and health, and the effect that medical technology has on health care makes it an appropriate subject for moral scrutiny as well. The second characteristic distinguishes moral standards from legal standards. For unlike moral standards legal standards depend on particular authoritative bodies, particularly those bodies that constitute a society's legislature. Taken together, all four characteristics should suffice in most cases to distinguish moral standards from other standards.

How does ethics differ from other sorts of studies of moral judgments? After all moral judgments are studied by sociologists, anthropologists, historians, and psychologists as well as by ethicists. Perhaps the most important difference between ethics and these other types of studies of moral judgments is that the latter are empirical studies. That is, they primarily seek to discover, report, and categorize the moral judgments that people actually have made, do make, or will make and to explain why those judgments have been made, are made, or will be made. Ethics, although also concerned with the moral judgments that are in fact made, is concerned chiefly with determining which moral judgments are valid in various circumstances and thus with determining the moral judgments that should be made and used as the bases for human conduct. The philosophical study of morality is usually divided into two distinct but interdependent branches: *metaethics,* the analysis of moral concepts, and *normative ethics,* the attempt to arrive at and derive judgments from valid fundamental standards of morality—moral principles. Most of the ethical concerns of this text fall within the category of normative ethics.

Metaethics

Metaethics is concerned primarily with analysis of moral concepts, and because concepts are usually expressed by means of words, metaethics largely focuses on clarifying moral terminology. This activity is crucial to any attempt to determine which moral judgments are valid. If terminology used to express moral judgments is not clearly understood, determining when those judgments are valid or invalid is likely to be extremely difficult. If language used to state a moral problem or dilemma is not clearly understood, little progress can be made in arriving at an acceptable solution. For example, some commentators on the justifiability of euthanasia argue that continuing life-support efforts against a patient's wishes is a violation of that person's moral right to a natural death. Is there such a right? This question cannot be fully answered until the meanings of *moral right* and *natural death* are clear. Only then can we even partially determine whether there is such a right and whether continuation of life-support efforts against a patient's will violates that right. Because metaethics is a form of conceptual analysis, an understanding of how metaethical inquiry proceeds must begin with an explanation of conceptual analysis. In conceptual analysis the main aim is to establish the conditions necessary and sufficient for the proper use of particular concepts (or, more strictly speaking, for the correct use of the words chosen to express them).

What then is meant by the assertion that something is a necessary condition of something else? To hold that X is a necessary condition of Y is to claim that Y can be the case only if X is also the case. If X is not the case, then neither is Y. For example, no object that is not also a plane closed figure can be a rectangle. Being a plane closed figure is a necessary condition of being a rectangle. Given this, use of the term *rectangle* to refer to an object that is not a plane closed figure is misuse. To hold that X is a sufficient condition of Y is to assert that whenever X is the case, Y must also be the case. If X is the case, then so is Y. For example, any object that is a plane closed figure with four sides or four interior angles must also be a rectangle. Accordingly being a plane closed figure with four sides or four internal angles is a sufficient condition of being a rectangle. So any thing that is a plane closed figure with four sides or four internal angles is properly referred to by the term *rectangle.*

A condition can be necessary without being sufficient. For instance, a circle is also a plane closed figure, but a circle is not a rectangle. Therefore although being a plane closed figure is a necessary condition of being a rectangle, it is not a sufficient condition. Similarly a condition can be sufficient without being necessary. For example, anything that is a bicycle is also a vehicle. So being a bicycle is a sufficient condition of being a vehicle. But there are many vehicles—cars, trucks, and space shuttles, for instance—that are not bicycles. So being a bicycle is not a necessary condition of being a vehicle.

To completely analyze a concept and thereby completely illuminate the meaning of the term by which it is expressed, a complete set of necessary and sufficient conditions for the use of that concept must be given. Unfortunately this can not always be accomplished. In many instances a complete set of necessary and sufficient conditions for the use of a concept cannot be provided, and the conditions that can be given are sometimes unavoidably imprecise. Although we are quite sure that a rectangle must have no fewer and no more than four sides, we are unable to be similarly certain in the analysis of other concepts. The concept of baldness is an often cited example of a nonmoral concept that seems to defy precise analysis. Few people are unable to distinguish a bald person from one who is not bald, yet none are able to specify exactly the amount of hair a person must lose to be correctly regarded as bald. Nonetheless the aim in conceptual analysis is to be as complete and precise as possible in specifying the necessary and sufficient conditions for the use of any given concept. The nearer to total completeness and precision the analyses of concepts approach, the greater the likelihood that we correctly understand the questions and judgments in which those concepts function.

A lesson that can safely be drawn from metaethical investigations that have been conducted is that many of the most important moral concepts are not as precise as a mathematical concept such as the rectangle. Indeed much ethical debate centers on the proper analysis of moral concepts. Consider the concept of informed consent, a concept of great importance in medical ethics. Most commentators on medical practice, as well as most clinicians, agree that no medical procedure should be performed on a patient (except perhaps in cases of dire emergency where immediate action is necessary to save a person's life) unless informed consent has been obtained. But along with this agreement goes a great deal of disagreement about what actually constitutes informed consent. Is consent genuine when given under the stress of the fear and suffering that often accompanies disease and disability? How much of the information given about a medical procedure must the patient actually understand before consent to its use can be considered informed? (This last question is especially important now that medical care often involves highly specialized and extremely complex technology that many patients may not comprehend.) Even a cursory perusal of the literature of biomedical ethics concerned with informed consent shows that there are few answers to these and related questions that garner unanimous agreement. Though some progress has been made in understanding important ethical concepts such as informed consent, many such concepts are like the concept of baldness and will always defy complete analysis to some degree. In any case analyzing moral concepts to the fullest degree possible, and thus coming to understand as fully as possible the terminology of moral questions and judgments, are crucial to efforts to determine which moral judgments can reasonably be accepted as valid.

Normative Ethics

Moral judgments are normative judgments derived from moral standards, but how are such standards generated? This question leads to the other main branch of ethics, normative ethics. The overriding goal of normative ethics is to determine the most fundamental moral standards—that is, moral principles—on which all less basic moral standards as well as all valid moral judgments are ultimately based. In normative ethics, as in most philosophical endeavors, fundamental disagreement exists. On the one side there are those who argue that all questions concerning the moral rightness or wrongness of conduct should be answered in the terms of the consequences that ensue from the various courses of action open to persons. According to this school of thought, known as consequentialism (or teleological ethics), the morally right action is always the one among the open alternatives that has the best consequences. An important implication of consequentialism is that no actions or courses of conduct are automatically ruled out as immoral or ruled in as morally obligatory. That is, no action or form of conduct is intrinsically wrong or right. The wrongness or rightness of an action or course of conduct is completely contingent on its consequences. On the other side is the school of normative ethical thought known as nonconsequentialism (or deontological ethics). Proponents of this point of view deny exactly what is asserted by consequentialists, namely, that the moral evaluation of human conduct should always and only be based on the consequences of that conduct. Nonconsequentialists are united in the conviction that factors other than consequences are relevant to the moral assessment of our actions and thus reject the claim that the course of conduct with the best result is always the morally right one.

Utilitarianism

The most widely accepted form of consequentialism is utilitarianism. The first detailed and systematic formulation of this ethical theory is usually credited to two English ethicists, Jeremy Bentham (1748–1832) and his most famous student, John Stuart Mill (1806–1873). Remember that according to consequentialism, actions are to be morally evaluated solely in terms of their consequences. The action, among the available alternatives, that has the best consequences is always the right action—the one that is morally obligatory. Any other action would be the morally wrong course to take. To actually utilize a consequentialist approach to determine how one ought to act, some standard by which to evaluate the consequences of various courses of conduct is needed. The standard offered by Bentham and Mill is hedonism, the view that the only thing that is intrinsically good is pleasure or the absence of pain and the only thing that is intrinsically bad or evil is pain or the absence of pleasure. As Mill put it, "[A]ctions are right in proportion as they promote happiness; wrong as they tend to produce the reverse of happiness" (Mill 1957). On this view the course of conduct among the available options that has the best consequences is the one that brings about the greatest balance of pleasure over pain, happiness over unhappiness. The right choice, when deciding what to do, is always the choice that maximizes pleasure and/or minimizes pain. Whose pleasure and whose pain? All who are affected by the actions in question.

Accordingly, for classical utilitarianism, there are two steps to determine what ought to be done in any situation: First, determine what courses of conduct are open. Second, determine the consequences of each alternative. This requires determining who will be affected by each alternative and how they will be affected. When this has been accomplished, the course that is the morally right one is known—the one that maximizes pleasure or minimizes pain or both. Indeed classical utilitarianism can be reduced to one basic moral principle, which hereafter is referred to as the principle of utility: Maximize pleasure and minimize pain.

Few, if any, modern utilitarians accept hedonism as the standard by which to evaluate the consequences of our actions. The apparent impossibility of actually measuring pleasure and pain along with the apparent impossibility of comparing the pleasures and pains of different persons has prompted many modern utilitarians to adopt preference-satisfaction as the standard by which to evaluate the consequences of actions. On this view the best course of conduct among the alternatives is always the one that results in the most preference-satisfaction or the least preference-frustration. Thus most modern utilitarians recommend determining what ought to be done in any set of circumstances by determining what options are open and by determining the effect, satisfaction or frustration, that each option would have on those whose preferences would be affected by it. As with classical utilitarianism the ultimate focus is on maximization of the good and minimization of the bad. Because the good is satisfaction of preferences, and the bad is frustration of preferences, the right course to adopt is always the one that maximizes satisfaction of preferences or minimizes frustration of preferences. Here the principle of utility would be stated: Maximize preference-satisfaction and minimize preference-frustration.

Before further discussion of utilitarianism it should be noted that both classical and modern utilitarianism have some inherent appeal as approaches to the moral problems posed by modern medical technology. Each seems to capture something important about health care. Disease and ill health typically cause people pain or unhappiness and frustrate their ability to satisfy at least some of their preferences. Good health, though, seems to be a source of pleasure as well as a prerequisite to the ability to obtain various pleasures and satisfy various preferences. Because the overriding motivation behind innovation in and use of medical technology is to increase the ability of health care practitioners to maintain health and to cure and prevent disease, an obvious virtue of both forms of utilitarianism is that each would assess the development and use of medical technology in terms of what many believe makes health valuable, either the attainment of pleasure and avoidance of pain or the attainment of preference-satisfaction and avoidance of preference-frustration.

An important distinction, which applies to both forms of utilitaranism, is the distinction between act-utilitarianism and rule-utilitarianism. This important distinction concerns how the principle of utility is applied to generate specific moral judgments. The act utilitarian applies the principle of utility directly to individual actions, whereas the rule utilitarian applies it to moral rules to determine which rules are justified. An act utilitarian, in determining what ought to be done in a given situation, should first determine all of the available options in that situation and then determine which particular option will produce the best consequences. That option is morally obligatory. A rule-utilitarian, on the other hand, should determine what is required by consulting justified moral rules—that is, rules that would produce the best consequences if followed by everyone. Again the central difference between these forms of utilitarianism is that on the former one applies the principle of utility directly to individual actions and is obligated to do whatever maximizes the good, whereas on the latter one's obligation is to do whatever is required by justified moral rules—those selected by the principle of utility, those which would produce the best consequences if adhered to by all.

Although both forms of utilitarianism regard the principle of utility as *the* fundamental principle of morality, a course of action that is morally justified under one form need not be justified under the other. This can be illustrated by returning to the issue of informed consent. Suppose the question at issue is, Must the informed consent of a patient be obtained before performing a medical procedure and using the relevant medical technology? From an act-utilitarian perspective the correct answer depends on the particular circumstances of individual cases. In some cases the good consequences of securing informed consent may be outweighed by the undesirable consequences. Providing information necessary to ensure that a patient's consent is informed may seriously impair the treatment's efficacy. For instance, in a case where the only treatments with at least some chance of success have nonetheless little prospect of success, an optimistic attitude on the part of the patient may be crucial. Yet informing the patient of the small likelihood of success could engender exactly the opposite of optimism. Here the course that maximizes utility may be the one that withholds relevant information from the patient and thus omits consent that is informed. Or consider a patient who, athough well informed, refuses to consent to a medical procedure necessary for her continued survival. Here the course that maximizes utility may require that the physician proceed against the patient's wishes. Neither blanket endorsement nor blanket rejection of informed consent will ensue from taking an act-utilitarian position. Although the decision procedure is the same for each case, the correct answer need not be the same, for the consequences of obtaining informed consent will not necessarily be the same for every case.

From a rule-utilitarian perspective the correct answer depends on what is required by the relevant valid moral rule or rules. Suppose only two such rules are to be evaluated: Always secure the patient's informed consent before beginning treatment, and Never secure the patient's informed consent before beginning treatment. The rule that would have the best consequences, if adhered to by everyone, would be the rule to apply. Even if not securing the patient's informed consent is clearly the choice that would produce the best consequences in a particular case, if the rule requiring informed consent is the valid one, the patient's informed consent must be secured anyway. What may be morally obligatory under act utilitarianism may not be morally obligatory under rule utilitarianism.

Before discussion of nonconsequentialist normative ethics, some sense of the reason many ethicists reject utilitarianism should be provided. Ethicists Jeffrie Murphy and Jules Coleman (1984) provide an excellent account, compellingly illustrating why utilitarianism is not universally accepted:

Suppose that cancer and all other major diseases could be cured and the vast majority of human beings thus rendered much more happy and secure by rounding up a few persons against their will and subjecting them to painful and ultimately fatal medical experiments. (To avoid any complexities that might be raised by the suggestion that we search for volunteers, let us suppose that the experiments require a certain enzyme secreted by the brain only when persons are aware that they are being coerced, hurt, and threatened with death.) Intuitively, most of us would be inclined to say that—even given the general good consequences that would flow from this experimental process—adopting the process would still be immoral. We might say that it is *unjust* to treat people in this way or that they have a *right* not to be treated in this way and say that if utilitarianism cannot accommodate such rights (and thus agree with us in condemning the practice) then so much the worse for utilitarianism. But how can utilitarianism accommodate a respect for rights and justice and thus condemn the practice? Does not its commitment to "the greatest good for the greatest number" require that it approve of treating people in this awful way if the general welfare will be promoted, if happiness will be maximized? The moral then is clear: it *is* so much the worse for utilitarianism, and all morally sensitive people will thus reject it as an ethical theory.

The point made by Murphy and Coleman is that utilitarianism can countenance any sort of treatment of persons as long as the result is the maximization of the good. But, or so it seems to many, people have rights not to be subjected to certain forms of treatment, and to violate those rights is treat them unjustly no matter how good the consequences of such treatment may be. On this view the fatal flaw of utilitarianism is that it can justify injustice. This criticism is certainly to the point if the form of utilitarianism under consideration is act utilitarianism. Does this criticism succeed against rule utilitarianism?

Whereas act utilitarianism applies the principle of utility directly to individual acts, rule utilitarianism applies it to moral rules to determine which are valid and then judges particular acts by means of the resultant rules. Given this difference, rule utilitarianism might seem immune to the criticism. The rule utilitarian can argue that even if the hideous practice of victimizing some persons for the general welfare does have the best consequences in the short run, in the long run prohibiting such practices will produce much more happiness. Again Murphy and Coleman (1984) provide an apt account of the matter:

We might, as utilitarians, approve of certain general rules or practices that assign certain rights to persons, e.g., the right not to be experimented on without one's consent. We might even build this rule into law, i.e., make a legal rule (a statute) that forbids such experimentation. The reason we adopt this rule or practice, however, is because of our belief that the majority of people will be happier in the long run living in a society having a rule (conferring a right) of this nature. . . . .Whatever the short-run gains of performing a particular act of victimization, the long-run disutility of allowing the state to act this way would be enormous. Thus there sometimes are good grounds for refusing to perform such acts because such acts are condemned by social rules or practices that are themselves good to adopt because having them (and the rights they assign) promote the general welfare in the long run, i.e., a society that accords its citizens such rights as the right to refuse being experimented on without consent is likely to be a much happier society than a society that does not accord such rights.

So rule utilitarianism seems to escape the objection that utilitarianism justifies injustice. For the rule utilitarian can argue that a rule conferring a right not to be experimented on without having given informed consent will have the best results if followed by everyone. But appearances are deceiving. Consider the following rules: (1) Never experiment on persons without obtaining their prior consent, and (2) Only experiment on persons without obtaining their prior consent when doing so maximizes happiness. Rule 1 confers a right on persons never to be experimented on without prior consent. Rule 2 confers the same right, but allows it to be set aside whenever doing so will maximize utility. Consider again the experiment described by Coleman and Murphy. A few people are coercively subjected to painful and ultimately fatal experiments to cure cancer and all other major diseases and thereby provide the vast majority with great happiness. Rule 1 would prohibit this, but rule 2 would not if the experiment could be conducted so as not to undermine the long-term happiness of the majority. And it is surely imaginable that this could be done. For instance, the experiment could be conducted in secret so that the vast majority is left ignorant of how their happy circumstance came about. If this could be done, a great deal of benefit could be engendered without making the majority fearful that any one at any time could be sacrificed to maximize aggreate pleasure or preference-satisfaction. Rule 2 is therefore clearly superior on utilitarian grounds to rule 1. For rule 2 allows a good-maximizing course of action that is prohibited by rule 1. So if act utilitarianism is deficient because it can justify injustice, rule utilitarianism must be regarded as deficient in exactly the same respect. Admittedly the hypothetical example used here to criticize utilitarianism is rather farfetched. Nonetheless it effectively illustrates the reason for much scepticism toward utilitarianism, namely, that if the fundamental principle of morality determines the rightness or wrongness of treatment of persons solely in terms of maximization of aggregate good, no form of treatment of persons can be ruled out. This means that any right that persons might have not to be treated in certain ways can be set aside whenever doing so serves the maximization of aggregate good whether that good is conceived as pleasure or the satisfaction of preferences. For this sort of reason many ethicists reject utilitarianism, and indeed all forms of consequentialism, in favor of a normative ethical perspective that does not make moral evaluation of human conduct depend entirely on consequences.

Kantianism

Most contemporary versions of nonconsequentialism are forms of Kantianism. That is, they are in some measure based on the views of the German philosopher Immanuel Kant (1724–1804). Kant's ethical theory is a radical rejection of the notion fundamental to utilitarianism, that is, morality is primarily a matter of promoting human welfare whether this is defined as maximizing pleasure or maximizing the satisfaction of preferences. As many commentators have noted, Kant's position is an attempt to preserve in secular form a fundamental claim of Christian ethics, which Murphy and Coleman (1984) state as follows:

[T]here is something uniquely precious about human beings from the moral point of view. There are, for example, certain special moral requirements (rights) that attach to human beings that do not attach to any other animal, e.g., the requirement that we not kill and eat them for food, or hunt them for sport, or experiment on them for medical science. They are, in short, owed a special kind of respect simply because they are people.

Because no forms of treatment of persons can be ruled out a priori by a consequentialist approach to ethics—after all any sort of treatment of persons is allowable on this approach as long as it generates the best consequences—Kant rejects all forms of consequentialism. Because consequentialism cannot set any absolute limits on permissible treatment of persons, "it cannot capture the special respect that, according to Christianity and Kant, is owed to human beings because of their special moral status" (Murphy and Coleman 1984).

To understand Kant's theory, it is necessary to understand just what in his view makes human beings morally special entities deserving a unique type of respect to which no other entities are entitled. Christianity explains the moral uniqueness of human beings on the basis of a doctrine known as ensoulment. This doctrine holds that God has given eternal souls to human beings, thereby differentiating them from all other beings. This unique property of having a soul, according to Christian ethics, makes human beings the only entities with inherent value and is precisely why they deserve respect and have rights not to be subjected to certain forms of treatment. Kant provides a secular version of the doctrine of ensoulment by claiming that human beings possess autonomy and are thereby made morally unique and deserving of respect. Autonomy is understood by Kant to be the capacity to make choices on the basis of rational deliberation. The central task of ethics then is to specify what is required by respect for the moral specialness, the unique dignity, of human beings.

In common with the utilitarians Kant argues that there is a single fundamental principle of morality from which all valid moral standards and specific moral judgments can be derived. Whereas *the* fundamental principle of morality in utilitarianism is the principle of utility, Kant's fundamental principle is the categorical imperative. To say that a moral principle is an imperative is simply to say that it commands human actions. To say that it is categorical is to say that its validity is independent of specific aims and ends. An example of a noncategorical or hypothetical imperative would be, "Brush three times a day." This imperative is binding only on those with some aim that it promotes such as healthy teeth or unoffensive breath. For those unconcerned about the condition of their teeth or breath, this imperative has no relevance. To say that the fundamental principle of morality is categorical is to say that its validity is completely independent of all such considerations.

Kant offered four versions of the categorical imperative, each of which, he alleged, is equivalent in meaning to the others. The version that best fits this account of Kant is, "Always act so that you treat persons as ends in themselves and never as means only." First, what does it mean to treat persons as a means only? A person is treated as a means only when treatment of her is based solely on the benefits it will produce. To do this is, according to Kant, to treat a person as a mere instrument, a thing valued only for its usefulness. Return to Coleman and Murphy's hypothetical example—a few persons are subjected against their will to painful and ultimately fatal experiments to bring about massive benefits for the majority of people. This example illustrates precisely what Kant means by treating persons as means only. Those few who are sacrificed for the greater good of the majority are being treated as mere things that have value only to the extent that they can be used for this purpose. Anyone who has experienced the indignity of discovering that others value him merely for his usefulness has experienced being denied the respect that Kant, in agreement with Christianity, urges is due all human beings.

What then does it mean to treat persons as ends in themselves? It means treating them as having intrinsic value quite apart from their usefulness. Only beings with autonomy—human beings— have this sort of value, on Kant's scheme. Persons are treated as having this sort of value when their rights are respected. In the case of the experiment that causes pain and death for a few to greatly benefit many, such treatment is morally allowable only if it can be achieved without coercion, only if those who are subjected to it have given their consent. So a right, such as the right not to be experimented on without giving prior consent, functions to ensure respect for the special status that autonomy confers on human beings. Most ethicists who reject consequentialism do so because they accept, in some form, the Kantian notion that individuals have rights not to be treated in certain ways and therefore that treatment of them should not be based solely on the results produced.

Although Kant himself regarded consequences as irrelevant to morality, few other nonconsequentialists are quite this extreme in their rejection of consequentialism. Most do take consequences to be morally relevant. Like Kant they do not believe that consequences constitute the entire moral picture. Unlike Kant they do believe that consequences should be brought to bear on moral assessment, but with the requirement that doing so does not compromise respect for human dignity. Some nonconsequentialists are even more moderate and argue that when, but only when, an extremely large evil can be averted thereby or an extremely important good can be attained thereby, respect for human dignity can be compromised. Indeed the notion that human dignity is a value of extremely high importance, but can be sacrificed on some rare and exceptional circumstances, probably represents the position of most people. In any case conflicts between an essentially utilitarian concern for the common good and an essentially Kantian concern for the individual are endemic to public policy decisions, including those that affect the development and use of medical technology.

Beneficence, Nonmaleficence, and Technological Progress
Though various moral codes and oaths have been formulated for and by medical practitioners since the beginnings of Western medicine in classical Greek civilization, two moral norms have remained constant throughout all historical periods as the fundamental norms of Western health care, namely, *beneficence*—the

provision of benefits—and *nonmaleficence*—the avoidance of doing harm. These norms can be traced back to a body of writings from classical antiquity known as the Hippocratic Corpus. Although these writings are associated with the name of Hippocrates, the acknowledged father of Western medicine, medical historians remain unsure whether any, including the Hippocratic Oath, were actually the product of his authorship. Some portions of the Corpus are believed to have been written as early as the sixth century B.C., whereas others are believed to have been composed as late as the beginning of the Christian era. Scholars of medical history do agree that many, and perhaps even most, of the specific moral injunctions contained in the Corpus represent neither the actual practices nor the moral ideals of the majority of physicians of ancient Greece and Rome.

Nonetheless the general injunction to physicians found throughout the Hippocratic corpus in various forms—"As to disease, make a habit of two things—to help or, at least, to do no harm"—was an ideal accepted by at least some of the physicians of classical antiquity. With the rise of Christianity at the decline of Hellenistic civilization, beneficence and nonmaleficence became even more firmly established as the fundamental ethical norms of Western medicine. Whereas in both ancient Greece and ancient Rome, beneficence and nonmaleficence were accepted as merely concomitant to the craft of medicine, the object of which was to maintain bodily health, in Christendom, with its emphasis on compassion and the fellowship of humankind, these norms increasingly came to be regarded as the only acceptable motivations for the practice of medicine. Even today, in the vastly more secular cultures of modern Western societies, no ethical norms are more fundamental to health care than these. Stress on the provision of benefits and the avoidance of doing harm can be found in virtually all contemporary Western codes of conduct for health professionals with no less prominence than in the codes and oaths accepted by medical practitioners in past centuries.

Traditionally nonmaleficence has been given greater priority than beneficence in the ethics of medical care. This was not due to a belief that the duty to avoid harming patients was intrinsically more important than the duty to improve their well-being. Rather the priority of nonmaleficence reflected the fact that for the greater portion of its history, medicine's capacity to harm patients far exceeded its therapeutic capabilities. Before the advent of scientific medicine and its associated technologies, health care providers possessed many treatments that posed clear and often deadly risks to their patients, but had little if any genuine promise of benefit. Genuine remedies were the exception rather than the rule. In this context the traditional stress on the avoidance of harm over the provision of benefits appears quite reasonable. Where iatrogenic (physician-induced) injury is very probable, the patient's well-being is surely best served by those health care providers who are careful at least to avoid adding medical insult to the patient's already injured state.

With the advent of modern science matters changed radically. Medical diagnosis and treatment took on forms dictated by knowledge generated in laboratories, tested in the clinics, and verified by statistical methods. This new but ongoing alliance between medicine and science became the primary source of the sophisticated technologies that now pervade medical care, providing health care professionals with levels of therapeutic efficacy exceeding anything even imagined by their predecessors. With this dramatic increase in therapeutic, preventive, and rehabilitative power, beneficence has come to the forefront of medicine's ethical concerns. Some have been sufficiently impressed with modern medicine's capabilities to argue that the old medical ethic, "Above all, do no harm," should be replaced by "The patient deserves the best," a new medical ethic regarded as more appropriate to modern medicine's technologically based efficacy. But the technologies that have so rapidly increased the therapeutic capabilities of medicine have also produced genuine quandaries about what is most beneficial or least harmful for the patient.

Utilitarianism, Kantianism, and
Traditional Medical Morality

To say that beneficence and nonmaleficence have traditionally been regarded as the *fundamental* moral norms of Western health care is to say that, for most of its history, Western medicine has taken provision of benefits and avoidance of harm to patients as its highest duties. It is to say that all the practices and aims of health care providers are supposed to be subordinated to and compatible with these norms. From a utilitarian point of view, though, no duty, medical or otherwise, can be more basic than the duty to conform to the principle of utility, the duty do whatever maximizes pleasure (or preference-satisfaction) and minimizes pain (or preference-frustration). Beneficence and nonmaleficence should be strived for only to the degree that doing so is required by the principle of utility.

From an act-utilitarian perspective whether treatment of a patient should seek beneficent or nonmaleficent results depends on the particular circumstances of the case. Whereas beneficence or nonmaleficence may maximize utility in some cases, it may not in others. An act-utilitarian health care provider would often find herself at odds with the fundamental norms of traditional medical morality. From a rule-utilitarian perspective beneficence and nonmaleficence should be adopted as moral rules for the provision of health care only if adherence to them by all health care providers would have better consequences than similar adherence to any alternative rule. So although rule utilitarianism might be able to endorse beneficence and nonmaleficence as moral rules for medical care, it cannot treat them, no less than act-utilitarianism, as fundamental norms of medicine. On either version of utilitarianism no norm is more fundamental than the principle of utility.

Indeed a health care provider who adopted either form of utilitarianism as her medical morality would thereby adopt an orientation entirely different from that of traditional medical morality. For although utilitarianism in all forms focuses on aggregate utility, the well-being of all affected by a given course of conduct, traditional medical ethics has limited the medical professional's focus to the therapeutic relationship—the relationship between the patient and the health care provider. Whereas traditional medical morality makes the well-being of the individual patient the highest end, utilitarianism can regard it only as a means to aggregate well-being.

From a Kantian perspective no moral duty ranks higher than the duty to act always so as to respect the inherent dignity unique to human beings. This means that human beings are never to treat one another as mere instruments of each other's will. Each person is to refrain from violating the rights of others regardless of the utility such violation might generate. A Kantian health professional would seek the beneficent or nonmaleficient treatment of a patient only when and insofar as doing so could be accomplished without violating the patient's rights. One such right is clearly the patient's right to liberty, the right to do as one pleases as long as one does not impose undeserved harm on others. This right has been given recognition in modern medicine in the requirement of informed consent, for example. That this requirement has garnered widespread endorsement may indicate that Kantian moral thought has gained considerable credibility in modern medical morality. As with utilitarianism Kantianism cannot endorse beneficence and nonmaleficence as the highest duties of medical morality. For even if a course of treatment is clearly beneficent or nonmaleficent, it may violate some of the patient's rights. Traditional medical morality has always been somewhat paternalistic. The medical professional is traditionally charged with promoting the patient's good even if this means, for example, deceiving or coercing him. The patient's rights were not regarded as moral boundaries beyond which the health care provider should not step. From a Kantian perspective the duty not to step beyond these boundaries is the highest of moral duties for health care providers and patients alike. Medicine is practiced morally, from this point of view, only if it respects the unique inherent dignity of the patient above all else.

Death

An apt place to start illustrating the ways the moral aspects of medical care have been made more complex and problematic is with the modern life support systems found in the intensive care unit, including devices such as the respirator. These devices enable clinicians to sustain respiration and circulation in patients who have suffered brain damage severe enough to cause total and irreparable loss of brain function. The use of this technology raises the question, When is a human being dead? Though this question might seem to be a straightforward factual matter, it is not. For it cannot be answered simply by attending to all the relevant facts. It may be known, with no reasonable doubt, that a patient's brain has suffered trauma sufficient to deprive it of all function irreparably. It may also be known that such an individual's heart and lungs would cease to function if all artificial support were withdrawn. Yet this knowledge does not determine whether it is morally permissible to treat such an individual as a corpse. Answering this question requires determining or, perhaps more accurately, deciding which features of living persons are essential to their status as living persons. In other words it is necessary to specify which qualities, if irreparably lost, render a person identical to a corpse in all morally relevant respects and thus, for all intents and purposes, dead. Once the relevant qualities have been specified, and not until then, deciding whether an person with total and irreparable loss of brain function is dead would be a straightforward empirical matter. For then one would simply have to determine whether such an individual in fact lacked those features that make a living person morally nonidentical to a corpse.

Traditionally the criterion for determining death has been the irreparable termination of heartbeat, respiration, and blood pressure. Before the development of modern supportive technology, such termination would have occurred relatively quickly in anyone who had suffered massive or total destruction of the brain. With the use of modern supportive technology, though, heartbeat, respiration, and blood pressure can be artificially maintained indefinitely. Because such technology exists and has been implemented, the traditional criterion for death may no longer be adequate. To decide that it is not adequate is to decide that continued respiration and circulation are not in themselves sufficient to distinguish a person from a corpse. In many states total and irreparable loss of brain function (variously referred to as brainstem death, whole brain death, and, simply, brain death) has been adopted as the legal criterion for death. In those jurisdictions an individual in a state of brainstem death is legally indistinguishable from a corpse and so may legally be treated as one. But this does not settle the moral question because valid legal practices are not automatically morally correct. The moral issue remains. Is a brainstem-dead person morally indistinguishable from a corpse? Many would answer this question affirmatively, noting that once death of the brain stem has occurred, the brain cannot function at all, and the body's regulatory mechanisms cease to function unless artificially maintained. Thus mechanical sustenance of a person in a state of brainstem death is merely postponement of the inevitable. And indeed it is merely the mechanical intervention that differentiates such an individual from a corpse. On this view a mechanically ventilated corpse is a corpse nonetheless.

Even if society reaches a consensus that brainstem death is death and thus that a brainstem-dead person is a corpse, hard cases would remain. Most notable is the sort of case in which a person is in a persistent vegetative state, also known as neocortical death. In this sort of case, although severe damage to the brain has occurred, there is sufficient brain function (particularly in the brainstem) to make mechanical sustenance of respiration and circulation unnecessary. Patients in a persistent vegetative state exhibit no evidence of self-awareness and no purposeful responses to external stimuli. Their eyes open periodically, however, and they may exhibit sleep-wake cycles. Some patients yawn or make chewing motions and may swallow spontaneously. Unlike the complete unresponsiveness of brainstem dead persons, a variety of simple and complex reflex responses can be elicited from a person in a persistent vegetative state. But there is very little chance that such a patient will ever regain consciousness. Modern supportive technology, including artificial feeding, kidney dialysis, and the like, makes it possible to sustain an individual in a state of neocortical death for decades. (The longest recorded survivor, Elaine Esposito, lapsed into a persistent vegetative state on August 6, 1941 and did not die until November 11, 1978.)

If brainstem death is death, is neocortical death also death? Again the issue is not a straightforward factual matter. For it too is a matter of specifying which properties of living human beings distinguish them from corpses and so make treatment of them as corpses morally impermissible. The classical criterion for death, irreparable cessation of respiration and circulation, would entail that a person in a persistent vegetative state is not a corpse and so, morally speaking, must not be treated as one. The brainstem-death criterion for death would also entail that a person in a state of neocortical death is not yet a corpse. On this criterion what is crucial is that brain damage be sufficient to cause the body's regulatory mechanisms to fail.

Does a person who has suffered neocortical death possess the characteristics that distinguish the living from cadavers any more than does a person whose respiratory and circulatory functioning continues only through mechanical means? This depends on what those characteristics are, and that in turn is a matter society must decide. It is not a matter that can be settled by increased medical knowledge or better medical technology. Until society decides, it will not be clear what would count as beneficent or nonmaleficent treatment of an individual in a state of neocortical death.

Euthanasia

If society decides that an individual in a state of neocortical death is not morally identical to a corpse, the justifiability of euthanasia,* of ending a person's life for her own good, arises. For even if a neocortically dead person is something more than a cadaver, we still must ask if is it impermissible to terminate life support of this person. Withdrawing life support would be, on the assumption that a persistent vegetative state is not death, deliberate termination of a human life. Here the central issue is whether the quality of life of such a person can be so low or even so much a liability to her that

---

*Although the discussion here raises some of the questions that are central to appreciating the moral dimensions of euthanasia, an argument justifying euthanasia is provided in chapter 5.

deliberately taking action to hasten death, or at least not to postpone it, is morally defensible. Can the quality of an individual's life be so low that the value of extending the quantity of that life is completely negated? If so, then Western medicine's longstanding commitment to beneficence and nonmaleficence would seem to render the termination of life support efforts obligatory in such a case. For in such a case to extend life would certainly not be to the patient's benefit, but would instead be a failure to avoid harming the patient.

Consider the case of Paul Brophy, a 48-year-old resident of Easton, Massachussetts, who in March 1983 suffered a brain hemorrhage and lost consciousness as the consequence of a ruptured aneurysm. While hospitalized Brophy regained consciousness, and after waiting two weeks for his cranial swelling to subside, he underwent a craniotomy. Never regaining consciousness from this operation, Brophy was hospitalized in a persistent vegetative state, a state of neocortical death, and was maintained by a surgically implanted gastronomy tube, which drips liquid nutrients from a plastic bag directly into the stomach. Mr. Brophy required seven and a half hours of nursing care daily. This care included shaving, mouth care, grooming, turning him and positioning him in bed, and attending to his bowels and bladder.

In a legal suit brought against New England Sinai Hospital, Mr. Brophy's wife, Patricia Brophy, sought the legal right to have all medical treatment of her husband, including nutrition and hydration, withheld or discontinued, recognizing that doing so would result in her husband's death. A preponderance of evidence was presented during court proceedings to show that if Brophy had been able, he would have expressed a desire to have artificial nutrition ended. The evidence was sufficient to move the presiding trial judge to conclude that if Mr. Brophy were "presently competent" his preference would be to forgo the provision of food and water by means of a gastronomy tube, and thereby terminate his life. Nevertheless a ruling against Mrs. Brophy was rendered.

Although the legal issues raised by this case are interesting in their own right, our concern is with the central moral issue. Is the quality of Brophy's life sufficiently low to make termination of his life morally permissible? While alive, Brophy made it clear on many occasions to both family and friends that if ever he was in a condition where consciousness was irretrievably lost and mainte-nance of his life was possible only by mechanical means, he would prefer to be allowed to die. What Brophy's own position on the issue would have been is clear. Deciding whether that position is a correct one is a matter of determining what capacities and qualities make life worth living and whether their absence justifies deliberate termination of a life even when such would be the wish of the individual in question. In the absence of this determination, the traditional norms of medical ethics—beneficence and non-maleficence—provide no guidance. For until this determination is made, whether termination of life support constitutes benefit or harm to the patient cannot be known.

An even more difficult kind of case concerning the justifiability of euthanasia is illustrated by the case of Elizabeth Bouvia. Bouvia, a lifelong quadriplegic victim of cerebral palsy, is often in pain, is completely dependent on others for all of her needs, and spends virtually all of her time bedridden. After deciding that she did not wish to continue a life of complete dependence, Bouvia entered Riverside General Hospital in California. Her intention was to have hospital personnel keep her comfortable while she starved to death. Although hospital officials, backed by the courts, consistently refused her requests to be allowed to starve while being provided medical support, Bouvia remained adamant.

Many who would regard the quality of Paul Brophy's life as sufficiently low to justify termination of life support, especially given the overwhelming evidence that this would be his wish, would not agree that this is true of Elizabeth Bouvia's case. After all whereas Brophy was completely devoid of consciousness and incapable of any purposive interaction with his environment, Bouvia is fully conscious and mentally alert. She has been married (though now divorced) and has completed a college education. Indeed televised interviews with Bouvia have shown that she is quite intelligent and very skilled and persuasive at presenting her reasons for wishing not to have her life continued by artificial means of nutrition. Nonetheless she believes that the quality of her life is sufficiently low that she should be allowed to deliberately choose to starve to death. Before the existence of life support technology, maintenance of Elizabeth Bouvia against her will may not have been possible and certainly would have been more difficult.

Are we prepared to accept Elizabeth Bouvia's judgment? In part her case is more difficult than Brophy's because unlike Brophy she is capable of interacting purposively with her environment. He could not speak or otherwise meaningfully interact with others, and so regarding him as nothing more than living matter, a "human vegetable," was not difficult. We can not easily see Bouvia this way. She is conscious, mentally alert, and intelligent. She can and does engage in meaningful interaction with others. Even though her life is one of discomfort, indignity, and complete dependence, she clearly is more than a "human vegetable."

Despite the differences between the two cases, both force the same issue. Can the quality of an individual's life be so low that deliberate termination of that life is morally justifiable, so that it is not a case of unjustifiable homicide at worst or unjustifiable and fatal neglect at best? Again how that question is answered is a matter of what level of quality of life, if any, is regarded as low enough to justify medical practitioners in not doing all they can to extend life. If we believe that there is such a level, then we will be forced to conclude that the use of life support technology to extend life is not always beneficent or nonmaleficent and so in some cases must be forgone.

Another issue of major importance here is respect for individual autonomy. For both cases concern the moral justifiability of voluntary euthanasia, that is, euthanasia voluntarily requested by the patient. A longstanding commitment, vigorously defended by various schools of thought in Western moral philosophy, is the belief that competent adults ought to be left free to do as they please, as long as they do not impose undeserved harm on others. Does this commitment entail a right to die? Some clearly believe that it does. After all, if anything is one's own, surely one's life is. Elizabeth

Bouvia and Paul Brophy did not seek to impose undeserved harm on anyone else, nor would allowing them to satisfy their wish to die cause undeserved harm to others. What justification can there be then for not giving in to their wishes?

One answer with some merit cites the very respect for individual autonomy that is at issue here. Respect for autonomy means, at least on some views, doing what is necessary to protect autonomy. And in some instances this could require that an individual's autonomy be restricted. For instance, an important consideration justifying laws that prevent even competent adults from selling themselves into lifelong slavery is that such an exercise of autonomy has the consequence of undermining autonomy altogether. Similarly when an individual acts to bring about his own death, his exercise of autonomy puts that very autonomy at risk of total and irrevocable destruction. Many would regard this as adequate justification for using the force of criminal law to prevent suicide, for example. Though this line of argument does not fit Brophy's case—his autonomy had already been destroyed by the very events that rendered him neocortically dead—it does seem to fit Bouvia's. There is much evidence that she is fully competent and that her choice is as complete an example of the capacity for autonomous action as any. Yet her choice, if allowed to come to fruition, would destroy her autonomy as it destroys her. On this line of thought, her case is a paradigmatic instance of respect for autonomy requiring limitation of autonomy, and thus one in which the lifesaving power of modern medical technology must be used, even against the patient's wishes.

Obtaining Human Organs for Transplantation

Further examples of the ways in which the moral situation of modern medical practitioners has been made more complex and perplexing can be obtained by focusing on how certain medical resources are obtained. Medical resources can be broadly distinguished into two kinds—human resources and nonhuman resources (McConnell 1982). Human resources are those obtained directly from the bodies of human beings. Examples are blood, corneas, skin, organs for transplantation, and bone marrow. Nonhuman resources are those obtained from nature or created by humans. Though this distinction is of little importance when considering how medical resources are to be allocated, once they have been obtained, it becomes very important when the issue is one of obtaining scarce medical resources in the first place. For when the resources being obtained are human resources, obtained directly from the bodies of human beings, how they are obtained is a matter of great moral significance.

Human medical resources can be distinguished according to the effect of their acquisition on the donor of the resource. In some cases the donor of the resource is a living human being, and the loss suffered is temporary. Resources that fall into this category include blood, bone marrow, sperm, and skin. These sorts of resources can be completely replaced by natural processes in the donor so that their removal does little or nothing to damage the donor's long-term well-being.

In other cases the donor of the resource is living and the loss suffered is permanent. This is sometimes the case for kidney donation. When a kidney is obtained from a living relative of the prospective recipient, the chances that the transplant will be rejected by the patient's body are lower than when the organ is obtained from a cadaver. Unlike blood and bone marrow, a kidney cannot be replaced by completely natural processes in the donor. Kidneys do not regenerate. A living kidney donor undergoes a major surgical procedure that has no therapeutic value for her and may do her serious harm. Because the removed kidney cannot be replaced, her future well-being depends on the remaining kidney and would be seriously jeopardized should it become impaired. The potential donor can be subjected to intense pressures from the prospective recipient and other family members. Because the transplantation physician is morally bound to seek the best for his patient, and this means using an organ obtained from a relative, the physician may also be a source of pressure on the potential donor. How are the physician's traditional moral duties of beneficence and nonmaleficence to be carried out in this sort of case? Obtaining the kidney for transplantation from a relative is obviously beneficial for the recipient, whereas giving the organ puts the donor's health at some risk in both the short and long term. What is best for the one is not necessarily best for the other. Were it not for relatively recent innovations in supportive technology, immunosuppressant drugs, surgical techniques, patient monitoring, and the like, which make organ transplantation possible, this sort of moral quandary could not have arisen.

Finally, in some cases the resources can only be obtained when the donor is dead. Hearts, livers, and often kidneys for transplantation are examples of this type of human medical resource. A number of ethical issues are raised by the use of cadavers as organ donors. According to McCormick (1978),

Although it is generally agreed that care of the severely injured patient should be the responsibility of a physician who is not a member of the transplant team in order to prevent a conflict of interest, how much should the transplant physician be involved in the terminal care of the prospective donor in order to assure proper fluid balance and good kidney function that make kidneys (or other organs) usable for transplantation? Although there have been some instances where a patient who was expected to die recovered because of such an intervention, some have questioned the ethics of treatment instituted primarily for the sake of the needed organ. The ethics of asking a family struck by the sudden death of a young loved one arises frequently: Not only must they be asked for permission for the transplant in what is a most tragic moment for them; they must be pressed to make a decision quickly in order to facilitate the required medical procedures.

Although the lives of many people have been saved because technology exists that makes transplantation of cadaver organs feasible, a persistent difficulty is that the number of potential recipients far exceeds the number of actual donors. The modern technologies that make cadaver organ transplantation possible cannot solve this difficulty. It is true that if such technology did not exist, there would be no shortage of cadaver organs for transplantation, but only because organ transplantation would be impossible. Though perfection of artificial organs or interspecies organ transplantation may some day resolve this problem, that day seems rather far off. In the meantime technology has made it possible to save lives through organ transplantation, the demand for organ transplants exceeds the supply of cadaver organs available for transplantation, and certainty about what is beneficent or nonmaleficent will not bridge the gap. Until a practical alternative to human organ transplantation is developed, some way to obtain more human cadaver organs for transplantation must be found if medicine's generally beneficent objective of saving lives is to be met as fully as possible.

There are three approaches to obtaining human cadaver organs: voluntary donation, exchange, and salvaging (McConnell 1982). The first approach relies entirely on voluntary donations. While alive, a person gives permission to the appropriate authorities to remove and use any of his organs upon his death. Present U.S. policy for obtaining organs for transplantation, under the Uniform Anatomical Gift Act, drafted in 1968, is an example of the voluntary donation approach to obtaining cadaver organs (Dukeminier 1978):

This act [the Uniform Anatomical Gift Act] provides that any individual of sound mind and over eighteen years of age may donate all or part of his cadaver to any hospital, medical school, organ bank, surgeon, or physician in medical research or transplantation, or to any specified individual for therapy or transplantation needed by him. The gift may be made by a duly executed will or by a written document signed by the donor in the presence of two witnesses who must sign the document in his presence. No delivery or filing of the document is necessary. The document can be a card carried on the person, which can be revoked by destroying, cancelling, or mutilating the card. The Uniform Act also provides that, in the absence of actual notice of contrary wishes of the deceased person, the next of kin can donate all or any part of the deceased person's body for the purposes of the act.

The central advantage of the voluntary approach is its respect for individual autonomy. It is consistent with the widely recognized moral right of every competent adult to do with his own body as he wishes so long as he does not thereby do undeserved harm to others. A major disadvantage of this approach, as confirmed by experience in the United States, is that it does not work. Although polls continue to indicate that most people in the United States favor the use of cadaver organs for transplantation, very few have donated their organs or the organs of their relatives. (In 1983, 5,000 people needed livers, yet only 145 liver transplantations were performed. Because doctors are unable to keep a patient alive once failure of the liver occurs, the majority of persons needing liver transplants dies. Three hundred to 400 of those who die annually because of scarcity of livers for transplantation are children. Each year around 14,000 people need heart transplants. Yet in 1983 only 172 heart transplantations occurred.) Although all fifty states have enacted the Uniform Anatomical Gift Act or some equivalent, the effect on the gap between organs needed and organs donated has been negligible. "For various reasons, including the difficulty of imagining one's own death and the fearful anxiety caused thereby, most people, and particularly young people with organs most suitable for transplant, do not go out of their way to sign instruments of gift" (Dukeminier 1978).

A second disadvantage of the voluntary donation approach is that to successfully reduce the disparity between the supply of cadaver organs and the demand for them, guidelines would have to be much more aggressive than under the Uniform Anatomical Gift Act. Great pressure would have to brought to bear on people when they are least able to resist or cope with it, namely when a loved one's life (usually a young person) has been lost. At best such pressure would be morally suspect. At worst it would be outright exploitation.

A second method for obtaining cadaver organs is exchange. The idea here is to induce people to allow their organs to be used upon their deaths. As an incentive potential donors would be offered something in return for this permission. One sort of incentive would be financial compensation. Individuals could be paid before death by the state for permission to use their organs when they die. Another sort of incentive would be the right to use any available organs needed while alive. Persons who give permission to use their organs when they die would have the right to available organs for transplantation should they be needed. All persons who refuse permission would be denied this right (McConnell 1982). Although either version of the exchange approach would probably generate more cadaver organs for transplantation, both would be somewhat morally problematic.

For cash payment to induce enough people to contribute organs, the payment would have to be reasonably large. Because many of the organs donated would not be usable for various reasons (many donors will not die until they are very old and their organs are no longer suitable for donation, the organs of others will be made unusable by disease, the organs of yet others will be unusable because of congenital defects, and so on), the pool of potential donors would have to very large. The total cost then of this version of the trading approach could prove to be too burdensome to bear.

This problem could avoided by allowing the development of a market for cadaver organs. The idea of a market for cadaver organs is not purely speculative. As journalist Fern Schumer Chapman (1984) reports:

One entrepreneur has already been attracted. . . . Lance Weddell, 41, who heads a Frankfort, Maine, firm that manufactures survival equipment for hunters, created a company a year ago to procure cadaver organs for sale to patients in need of transplants. He invited potential donors and recipients to register with his firm, Medical Lifeline, Inc., intending to charge each $40. In addition, would-be recipients were told they had to deposit $10,000, to be paid to the estate of the eventual donor. The company was to make its fee from interest earned on the funds held in trust for the donors and the recipient. In the company's five-month life, Weddell says, 267 potential recipients called, though none sent money. But in September, Weddell went out of business, and he refunded the money he had collected.

He said the venture failed because he "couldn't get the co-operation of the physicians," who refused to remove organs purchased from donors.

A staunch advocate of free markets for most goods and services, economist Milton Friedman notes that the adverse effects of a cadaver organ market could be quite dangerous. "[If] the price is right, the unscrupulous might commit murder to sell organs" (Chapman 1984). History supports this worry. As the demand for cadavers increased in the late eighteenth century, due to increased demand by medical schools, some individuals resorted to grave robbing, whereas others even committed murder to capitalize on this situation (Chapman 1984).

The notion of a market in human organs has not been limited to cadaver organs. Some have advocated supplementing the supply of cadaver organs with a market in organs obtained from living donors. A defrocked physician, Barry Jacobs, has recently sought to become a broker for human organs. Jacobs proposed to offer a bounty to kidney donors, which would be paid to living donors on the spot. There is reason to believe that there are many who would be induced to sell their organs if a proposal such as Jacobs's were allowed to be put into operation. Consider the following two cases (Chapman 1984):

Larry Ernest Carter of Surrey, British Columbia, wrote to [Congressional Representative] Gore that he sees his kidney as a means of acquiring a graduate education in land-use planning and environmental management. Carter estimated that the training will cost about $22,000. Said Carter's letter: "For the poor, the sale of bodily tissue may offer the only . . . opportunity of breaking out of the poverty cycle. My kidney is the only capital resource I still possess that can be marketed in order to provide me with a chance of gaining access to educational and employment opportunities. . . . I am far more able to sacrifice one kidney, rather than the continued sacrifice of what should be the most productive years of my life."

. . . Bob Reina of Ocala, Florida, unemployed except for occasional gardening, placed a classified ad in the Ocala *Star-Banner*: "Kidney for sale." Reina, 40, asked $20,000 for his kidney—below the middle price asked in dozens of ads that have appeared in newspapers nationwide, offering kidneys, at $10,000 to $50,000 each. "You get after 35 years old or 40 and you see you've never accomplished anything," said Reina in an appearance on ABC-TV's "Nightline," "and you know you're really never going to have $5,000 or anything in the bank or any kind of security, you kind of get afraid. . . . But I'm a nice person. I'm not a callous person that's just looking to take advantage of a person who has a bad kidney. I think in a way it's a nice thing to do, even if you do get some money for it."

Advocates of this method of supplementing the supply of cadaver organs point out that it helps to reduce the shortage of organs for transplantation in a way that is respectful of individual autonomy. No one is forced to sell an organ who does not want to do so. Though a donor does significant harm to himself by selling an organ, as a society we have traditionally allowed individuals to jeopardize their well-being for money and even for sheer pleasure. If society allows us to play football, join the armed forces, skydive, and climb mountains, doesn't simple consistency demand that it allow competent adults to choose to sell certain of their organs?

Critics of the notion of a market in human organs obtained from live donors argue that this sort of system would exploit the poor. Those who would stand to gain the most by selling their organs are the poor. For many of them the income from the sale of a kidney— from $10,000 to $50,000—could be a means of escaping poverty, at least temporarily. Would the decision to sell an organ be a genuinely autonomous choice in these sorts of circumstances? Though not as coercive as holding a loaded gun to a person's head, isn't it extremely coercive to place the prospect of obtaining a large sum of money before a poor person? Few people could resist such a prospect even if they had serious misgivings about it. The poor, unlike their wealthier counterparts, may find that they do not have the luxury of being able to allow their misgivings to get the upper hand. Critics ask us to consider whether we are prepared to turn the poor into living organ banks for the wealthy.

Under the second version of the exchange policy, individuals would have the right to available organs for transplantation only if they had themselves given permission for their own organs to be used when they die. Suppose though that at a given time, there is but one person who needs an available organ for transplantation. No one else needs the organ, and this person would die without it. It would seem immoral to refuse to allow this person the organ simply because, before his need, he had not himself agreed to be an organ donor in the event of premature death.

A third approach to obtaining human cadaver organs for transplant is salvaging (also referred to as the presumed consent approach). Under this sort of policy a person's usable organs would be removed for transplantation upon his death unless he has explicitly requested that this not be done. "This method presumes that the deceased person has consented, and thus favored preserving life; the burden of objecting is put upon those who would deny life to another" (Dukeminier 1982). Because most members of our society favor the use of cadaver organs for transplantation, we have good reason to believe that few people would in fact object to this presumption of consent. Indeed "routine salvaging of cadaver organs for transplantation is practiced in many countries, including France, Italy, Sweden, Norway, Denmark, and Uruguay" (Dukeminier 1982). Of the three approaches salvaging is the one most likely to close the gap between the demand for organs for transplantation and the supply. Because organs would be obtained for free, the only cost of this approach would be the cost involved in setting up a system for determining when permission has been denied and for obtaining organs immediately upon the deaths of persons who have not denied permission (McConnell 1982).

Opponents of salvaging as a method of obtaining cadaver organs have expressed a number of objections to the morality of this approach. First, it has been argued that a taking policy puts an unfair burden on persons suffering from terminal illnesses. If such persons have objections to the use of their organs, such objections must be expressed at this most unfortunate time. Second, it has been argued that salvaging deprives people of the opportunity to exercise the virtue of generosity by freely donating their organs when they die. Third, it has been argued that it is very dangerous to allow the state to assume that its citizens' organs are its property (unless they explicitly declare otherwise) (McConnell 1982).

Whether any one of these three approaches to obtaining cadaver organs for transplantation can be successfully defended against objections is a matter that is not pursued here. It should be noted though that the moral issues raised by these approaches to obtaining cadaver organs, as well as those raised by obtaining organs from live donors, cannot be resolved solely on the basis of beneficence and nonmaleficence. For these traditional norms of Western medicine are quite narrow in their focus. They concern the therapeutic relationship, the relationship between patient and health care provider. Although most of the issues we have discussed in this section have some effect on that relationship, they go beyond it. Knowledge of what kinds of medical care will benefit or at least not harm a patient will not settle these kinds of concerns.

Allocation of Scarce Medical Resources

Another important instance of the kind of ethical concern that lies beyond the bounds of beneficence and nonmaleficence and their narrow focus on the relationship between the patient and health care provider is our concern with distributive justice—the fair distribution of benefits and harms. Even if what counts as a benefit or a harm in the use of modern medical technology were more certain, the question of distributive justice, the question of who gets what, would still have to be faced. The technologies of modern medicine are often scarce and expensive. They require costly hardware, expensive facilities, highly trained personnel, and in general large investments of society's all too limited resources. The simple and often painful fact of the matter is that not all of the benefits of modern medicine can be provided to all who need or desire them. Some deliberate determination of who is to get what must be made.

How to do this fairly lies at the heart of our concerns with distributive justice. And because medical resources are often crucial to health and continued life, these concerns are especially urgent.

Although the efficient use of medical resources is surely a relevant consideration in determining who gets what, it is primarily an empirical rather than a moral issue. The moral issue is one of fairly distributing the resources that go into the production, maintenance, and deployment of medical technology and of distributing the benefits (and burdens) that this technology engenders. The issue of fair allocation arises at two levels—the level of macroallocation and the level of microallocation. At the macroallocational level there are two closely related questions: (1) How much health care should be produced? and (2) What type of health care should be produced?

The first of these macroallocational questions concerns what portion of society's collectively controlled resources should be invested in the production of health care as opposed to other desired goods. Society's decision to produce a particular level of health care is always also a decision to forgo the opportunity to produce a larger quantity of some other valued good such as education, public parks, the elimination of poverty, pollution control, and so forth. Concern with fairness arises here because the valuation of these different goods is not uniform across society. Different sectors of society will rank these goods differently. Whereas the poorer members of society may regard the elimination of poverty and education as the highest priorities for the use of society's resources, better-off members may regard health care and pollution control as most important. Given such disparities in preferences and interests, how is a fair decision on what level of health care to produce to be made?

In posing the second macroallocational question, society's concern is to determine how much money, energy, skill, and other resources under its collective control are to be devoted to the various kinds of medical care. For example, medical care can be distinguished into two general types—rescue medicine and preventive medicine. In rescue medicine health care efforts focus on restoring the health and normal functioning of people who are already ill or disabled. (A further distinction can be made within the category of rescue care—crisis care. Crisis care is the form of rescue care in which the patient requires the sort of medical support typically found only the intensive care units of hospitals. Organ transplantation is an excellent example of crisis care.) In preventive medicine health care efforts focus on efforts to prevent people from becoming ill or disabled. At present society seems to give priority to rescue medicine. In the 1970s to early 1980s, only 1 percent of U.S. dollars devoted to health care were spent on prevention and control of illness (Menzel 1982). The majority of the major developments in medical technology over the past two to three decades have been in medical rescue devices and techniques (particularly devices and techniques related to crisis care); developments including human, animal, and artificial organ transplantation; sophisticated diagnostic technologies; and cardiac surgery to name only a few. Yet many commentators on the U.S. health scene agree that our society has arrived at a point where appreciable improvements in the health of our population will depend on improvements in preventive medicine rather than better or more rescue medicine.

Among the reasons these commentators conclude that improved health care in the United States requires giving less priority to rescue medicine is that many of the technologies of rescue medicine are "halfway technologies." When the underlying mechanisms of an ailment are not yet adequately understood, the best that medicine can do is to develop some complex technological manipulation or gadgetry that modifies the symptoms without curing the disease. Such technologies are labeled "halfway" for obvious reasons. Though often better than nothing, they fall short of definitive means of controlling conditions that they are intended to ameliorate. Many times halfway technologies are extremely expensive and must be used repeatedly during the course of a patient's ailment. Two of the more conspicuous instances of today's more expensive halfway technologies are renal dialysis and kidney transplantation. In most cases where these technologies are used, the underlying disease of the kidney is chronic nephritis. But because the underlying mechanisms of this disease are not adequately understood, therapies that are able to cure or prevent it are not available. Other conspicuous examples of expensive halfway technologies are the artificial heart and heart transplantation. Disease of the heart muscle, heart valves, and coronary arteries is not understood well enough to enable development of therapies that cure the sorts of disease that make the artificial heart and heart transplantations necessary. The diseases that are inadequately understood, and thus for which genuinely definitive technologies do not exist, include stroke, heart attack, congestive heart failure, most forms of cancer, arteriosclerosis, cirrhosis of the liver, emphysema, senile psychosis, and rheumatic arthritis (Bennett 1977).

The history of medicine shows that "each time a major disease has been controlled, the definitive technology has been much cheaper than the technologies devised before the disease was understood" (Bennett 1977). Consider polio, for example. Before this disease was adequately understood, treatment involved expensive special facilities, iron lungs, intensive care, hot packs, corrective orthopedic operations, braces, and other prosthetic and orthopedic devices (Bennett 1977). Today polio is controlled much less expensively by means of vaccination, a relatively simple and inexpensive procedure that prevents its occurrence. Furthermore, even in those cases where the underlying mechanisms of disease are not well understood, incidence of the diseases for which many halfway technologies are used could be significantly reduced by changes in individual behavior and in the natural and social environment. On the basis of these sorts of considerations, many commentators urge that greater priority be given to preventive medicine and less to rescue medicine. We may find that efforts to prevent certain ailments prove far more successful and less costly than efforts to rescue persons already in their grip (hence the old adage, "An ounce of prevention is worth a pound of cure").

Several considerations show this issue to be a matter of fairness. First, the benefits of rescue medicine are often expensive to produce. Accordingly they consume resources that could be put to uses that some regard as more important. Second, the benefits of crisis medicine are alleged to accrue to relatively few persons compared with the numbers who could benefit from advances in the technologies of preventive medicine. Third, whereas the beneficiaries of preventive medicine are primarily those who are young and not yet ill, the beneficiaries of rescue medicine are primarily the elderly and the sick. Fourth, preventive medicine benefits unknown persons in future peril, and rescue crisis medicine benefits known persons in present peril. Each of these considerations implies that a decision to favor the one form of medicine over the other is simultaneously a decision to give greater weight to the interests of some persons over those of others. Because this cannot be avoided, knowing how fairly to do it is important. Here beneficence and nonmaleficence are of particularly little use, for the issue is not one of what counts as a benefit or harm but rather how such are to be distributed equitably.

Before proceeding to discuss the topic of microallocation, we should note that a decision to shift resources from rescue medicine to preventive medicine is not necessarily antitechnological. Technology would still be important in the provision of health care. Among the uses to which technology could be put in preventive medicine are more frequent monitoring of significant physiological, nutritional, and daily living parameters for the general population; automatic dispensing of drugs; monitoring of newborns once they leave the hospital; and the prevention, early detection, and treatment of alcohol and drug addiction. Although the technologies required by preventive medicine might differ substantially from those useful for rescue medicine, in all likelihood their importance and efficacy will be at least as great.

At the microallocational level the central question is, Who should have access to the health care that is produced? This question is often tragic, for answering it often means deliberately denying resources to persons who will certainly die without them. On what sorts of standards should this sort of decision be based to be fair? Two different sorts of criteria seem necessary—rules of exclusion and rules of selection. Rules of exclusion are used to determine who should be considered as potential beneficiaries of some scarce medical good or service, and rules of selection are used to make the final choices from this pool of prospective patients.

Rules of exclusion distinguish patients according to considerations such as age, ability to pay, geographic location, presence or absence of complicating ailments, and likelihood of benefit. Is exclusion from access to some medical resource that could save one's life or improve one's health fair on any of these grounds? However this question is answered, most commentators agree that medical suitability should the crucial consideration. Scarce medical resources should not be squandered on patients who cannot benefit from them or who are likely to die from some other illnesses in the near future. Once a pool of prospective patients has been determined, rules of selection come into play. Two sorts of standards of selection have been proposed—social worth criteria and randomization.

Social worth criteria tend to include considerations such as economic productivity, years of productive life remaining, marital and family status, social relationships, and history of antisocial behavior, among others. Such criteria are formulated as means to determine which patients have the most value to society, under the utilitarian rationale that scarce resources should be used to generate maximum benefit for society. Advocates of social worth criteria argue that because medical resources are in large part produced with social resources, society has the right to get the maximum value for its investment, that because medical institutions and personnel are trustees of society's well-being, to use any other sort of criteria would be to act irresponsibly, and that by adopting a social-value–oriented approach to allocation, we create additional benefits, stemming from the contributions of those we save; the creation of those additional benefits increases our chances of bringing benefit to ourselves or to those for whom we care (Basson 1979).

Social worth criteria have met with much criticism. Critics have argued that unbiased judgments of the social worth of individuals simply cannot be formulated. These commentators argue that such judgments are extremely likely to be mere reflections of the arbitrary preferences of those who make them. And this is not so much because persons charged with formulating and applying social worth criteria are likely to prejudiced or unconscientious but rather because we have no way of arriving at objective measures of social worth.

This difficulty was well illustrated by the efforts of Swedish Hospital in Seattle during the mid-1960s to allocate access to its limited number of renal dialysis machines. (This particular micro-allocational problem has been relieved for the most part by the Federal government's decision to provides funds to make dialysis available to all who need it, thus showing how macroallocational decisions can affect microallocational ones.) The Artificial Kidney Center at Swedish Hospital formed an Admissions and Policy Committee composed of seven lay persons and doctors to devise criteria to determine which patients among those medically suitable would have access to dialysis. The committee wanted to know the following of each prospective patient: age, sex, marital status, number of dependents, income, emotional stability, nature of occupation, educational background, past performances, and future potential. Critics of the Seattle committee charged that its criteria surely favored solid middle class persons much like the committee members themselves. In the words of one critic, "[W]hen one examines the patient profiles as to who received dialysis in those early days, the image that emerges is one of young executive types, family members, and religious individuals, while old people, blacks, alcoholics and others are left in the ideal patient's wake" (Caplan 1981). And as another critic has noted, a needy Henry David Thoreau certainly would not have been selected by the Seattle committee. Yet quite often those persons ultimately regarded as having made the greatest contributions to society are, like Thoreau, persons who seemed to their contemporaries anything but socially valuable.

The inability to formulate unbiased criteria of social worth is not the only consideration that critics of social worth have noted. They have also argued that our judgments about who is valuable to society depend in part on society's needs, and these needs are constantly changing and difficult to predict; and, perhaps most compelling, they have argued that allocating scarce social resources on the basis of social worth criteria violates our belief in the moral equality of people (Childress 1981). The use of social worth criteria is like saying to a person that her worth is contingent on the contribution that saving her will make to the collective good. And given that different persons are differentially important to the collective well-being of society, different persons will have different amounts of moral value. But according to the critics, one of the fundamental commitments of Western morality is that people, unlike things and animals, are not mere instruments to be valued solely in terms of their usefulness. Rather they are entities with equal and intrinsic value, and this makes them special beings. If this is the case, to use social worth criteria in allocating scarce medical resources is to fail to respect this special status and to treat persons as if they were mere things—to dehumanize them.

For these sorts of reasons the critics of social worth criteria urge that the better way to determine which persons from a pool of prospective patients will get access to a scarce medical resource is to use some form of randomization such as a lottery or a queuing system, that is, first come, first served. This method of allocation is alleged by its proponents to have the advantages of eliminating the possibility of making choices—which are often life-and-death choices—in a biased and hence unfair fashion and of giving all medically suitable persons an equal opportunity to have access to the desired medical resource, thereby respecting the equal moral worth of all persons.

However society resolves the allocational issues facing it on both the micro- and macroallocational levels, medical technology is implicated, because the need to make these decisions is a product of the scarcity and expense of various therapies. To the extent that advances in medical technology can reduce this scarcity and expense, the necessity for and difficulty of such decisions will be lessened, which would indeed be a beneficent result.

Summary

In this chapter we have examined some of the many ethical issues posed by modern innovations in medical technology. We began with the background in ethical theory needed for full grasp and appreciation of the moral dimensions of the technology of contemporary medicine. Topics covered included distinctive characteristics of moral judgments, standards, and principles; the nature of metaethics, the study of moral concepts; and the strengths and weaknesses of the two main normative ethical theories—utilitarianism and Kantianism.

With this background the central norms of traditional medical morality—beneficence, the provision of benefits, and nonmaleficence, the avoidance of doing harm—were presented. One of modern technology's most important effects has been a shift in the relative importance of these two norms. Because of medicine's therapeutic importance and the grave dangers it often has posed to patients for much of its history, nonmaleficence was traditionally given priority over beneficence. Indeed avoidance of harm was often the best thing that a physician could do for his patients in a wide variety of circumstances. The innovations in medical technology that have occurred in the past fifty years have given medical practitioners unprecedented therapeutic, preventive, and rehabilitative powers. Hence today, at least according to some commentators, beneficence and its stress on benefiting the patient should be the dominant norm of medical morality.

Unfortunately along with the increase in ability to be beneficent that modern technology provides to today's health care professionals comes greater uncertainty as to what counts as a medical benefit or harm. In short, modern medical technology has made it more difficult than ever before in the history of medicine to know just what constitutes being beneficent or maleficent. This situation was illustrated through discussion of a number of moral problems facing contemporary health care professionals. These problems include defining death, deciding when the quality of a patient's life justifies euthanasia, and meeting the growing demand for transplantable human organs. These are problems that seem to shake the very foundations of traditional medical morality.

In the last section of this chapter we examined a moral issue that goes beyond traditional medical morality—the allocation of scarce medical resources. Not everyone can be provided with everything they need or want by way of medical care, and this is especially true of some of the highly sophisticated technologies that have recently become available. Here the issue is not what to count as a benefit or harm but rather how to distribute scarce benefits fairly. This issue is often one of great urgency, for it is often a matter of deciding who will live and who will die. Modern medical technology is crucial to this issue because of its potential to both exacerbate and ameliorate these sorts of allocational dilemmas.

References

Basson, M. 1979. Choosing among candidates for scarce medical resources. *The Journal of Medicine and Philosophy* 4(3).

Bennett, I. J. 1977. Technology as a shaping force. *Doing and Feeling Worse: Health Care in the United States*. Norton.

Caplan, A. L. 1981 (Fall). Kidneys, ethics, and politics: Policy lessons of the ERSD experience. *Journal of Health Politics, Policy & Law* 6(3):488-503.

Chapman, F. S. 1984. The life-and-death question of an organ market. *Fortune* June 11.

Childress, J. 1981. Priorities in the allocation of health care resources. In: *Justice and Health Care*. D. Reidel Publishing Co.

Dukeminier, J. 1978. Organ donation: Legal aspects. *The Encyclopedia of Bioethics*, vol. III. Free Press.

Menzel, P. 1982. *Medical Costs, Moral Choices: A Philosophy of Health Care Economics in America*. Yale University Press.

Mill, J. S., 1957. *Utilitarianism*. Bobs-Merrill.

McConnell, T. 1982. *Moral Issues in Health Care*. Wadsworth.

McCormick, R. 1978. Organ transplantation: Ethical principles. In: *The Encyclopedia of Bioethics*. Vol. III. Free Press.

Murphy, J., and Coleman, J. 1984. *The Philosophy of Law*.   Rowan and Allenheld.

Valesquez, M. 1988. *Business Ethics*. 2nd ed. Prentice-Hall.

4

Cardiac Technology

Despite the significant improvements in the diagnosis and treatment of cardiovascular disease, heart disease remains the leading cause of death and disability in the United States. In 1984, for example, it was reported that over 980,000 people in the United States died from diseases of the heart or blood vessels (American Heart Association 1984). The majority of them died as the result of a stroke or a heart attack (which alone accounted for over 540,000 deaths). An even larger number, more than 4.5 million people in the United States, suffered a heart attack or experienced chest pain (angina) because of heart disease. Clearly heart disease is a medical problem of major proportions.

The economic costs associated with heart disease are massive by anyone's accounting. The American Heart Association estimated that in 1987 more than $85 billion was spent on, or lost as a result of, heart disease. This sum represented about 1.5 percent of the U.S. gross national product. Certainly if people took better care of themselves, much could be done to ameliorate the incidence of heart disease and the associated mortality and morbidity. In fact recent reductions, although extremely modest, in cardiac-related deaths can be accounted for by life adjustments such as reductions in smoking and drinking, sensible exercise programs, and improved low-cholesterol diets.

Innovations in medical therapies that use sophisticated technologies have also been important in both the treatment of cardiac patients and the cost of their care. Some of these technologies include cardiac assist devices such as pacemakers and defibrillators that have enabled more than half a million people in the United States to enjoy normal or more normal and comfortable lives. More dramatic innovations in cardiac medical treatment such as heart transplants and artificial heart devices, including left ventricular assist devices (LVAD) and the complete artificial heart, have to date had only marginal effects on cardiac disease–associated mortality and morbidity. Since their introduction in 1960 artificial hearts of one kind or another, however, have been used to aid only a handful of patients (Bernhard et al. 1970). This technology is still considered to be in the development stage of clinical experimentation. Nevertheless some evidence suggests that relatively large numbers of patients (up to 66,000 per year) could benefit eventually from the artificial heart program, an endeavor that has excited the imagination of the world.

This chapter focuses on these various cardiac assist technologies (pacemakers, defibrillator, and partial and complete artificial hearts). First, some basic biological information about the functioning of the human heart is provided to permit the reader to understand the need for and role of each technology. Second, the evaluation of each technology from a historical perspective is presented to include the current state of the art. Third, the economic issues that surround each technology are examined with special attention given to the costs and benefits associated with the use of pacemakers and artificial heart research and development programs. Finally, some of the major ethical issues associated with human experimentation and the use of nonvalidated technologies such as the artificial heart are examined.

Cardiovascular Assist Devices–Artificial Heart

Fundamental Principles of Heart Action

Before exploring the development, current status, and future of many of the tools of cardiology, a knowledge of the basic anatomy and physiology or function of the heart and blood vessels is required. The heart consists of muscular tissue organized into four chambers with four one-way valves at various levels to separate the chambers and their outflow blood vessels (figure 4.1). The valves permit the development of different pressures within the heart to allow for low levels on the inflow site and higher levels for output.

The two upper cardiac chambers, called atria, receive blood from either the body or lungs and serve as booster pumps to maximize filling of the two lower chambers, named ventricles, which in turn eject the blood out of the heart. It is easier to consider the heart and cardiovascular system as a group of pumps and valves connected

Figure 4.1
Normal heart

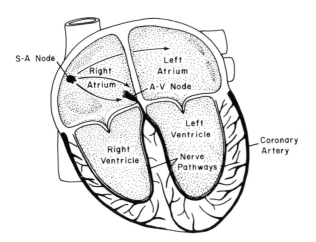

in a series with only one-way flow. The right atrium receives blood from the upper and lower portions of the body via the two major veins, called the superior vena cava and the inferior vena cava. This blood has had oxygen removed from it by the body and needs to have that essential element replenished by the lungs. The right atrium pumps the blood through the tricuspid valve into the right ventricle, which in turn develops a higher pressure and pumps the blood through the pulmonary valve into the lungs. In the lungs the blood comes into contact with a thin interface within the air sacs, or alveoli, where gas exchange takes place. The blood, now rich with oxygen, returns to the left atrium, which in turn boosts it into the left ventricle through the mitral valve. The left ventricle, the major pump for the body, then ejects the blood into the major artery— the aorta—through the aortic valve.

The left ventricle is the most essential part of the heart and is the site of much of the disease in the heart. The first branches of the aorta are the coronary arteries that bring oxygen-rich blood and nutrients to the heart muscle, or myocardium. It is disease of these blood vessels that causes many heart problems. After passing the coronary artery origins, the blood is distributed throughout the body, where oxygen is extracted and the entire cycle starts again with the return of blood to the right side of the heart.

The pressures produced by the various heart chambers are governed by the simple physical law that states that pressure equals the product of flow and resistance, $P = Q \cdot R$. Each cardiac chamber and great vessel normally produces its own characteristic pressure and wave form. For the atria these are determined by the flows into the chambers from either the body (the right atrium) or the lungs (the left atrium) and by the resistance offered by the respective ventricles that receive their output. A ventricle whose myocardium has been damaged either may not relax well during diastole or may not empty adequately. This in turn could cause the atria to work harder. The ventricles, which must generate enough flow to meet the needs of the body for cardiac output, develop pressure levels proportionate to the resistance offered by either the lungs or the body. The lungs usually offer very low resistance so that the right ventricular pressure is normally below 30 mmHg at peak systole. The left ventricle must overcome the resistance of the entire body so it will typically generate a peak systolic pressure of between 110 mmHg and 140 mmHg in most adults. The pressure during cardiac relaxation and filling is the determinate of the diastolic pressure. Systole is thus the period of the cardiac cycle when the heart is emptying and diastole is when it is filling.

The measurement of these pressures is quite important to the assessment of health and disease. The only pressure that can be measured even indirectly without inserting an instrument or needle into the patient is the arterial blood pressure in the body. This is most simply done by the use of a sphygmomanometer (Kirkendall et al. 1980). This device consists of a compression bladder, an inflating bulb or pump with a control valve to allow for slow deflation, and a manometer from which the pressure of the bladder is read. A stethoscope is used to listen for the sounds of blood flow. The bladder is inflated to a pressure sufficient to stop all flow in the limb and is gradually deflated. When the sounds of flow are first heard, the manometer pressure is noted as the systolic pressure. The cuff is further deflated until the sounds of disturbed flow are no longer heard and this is recorded as the diastolic pressure. This system is simple but not sufficiently accurate in critical situations.

For critically ill patients continuous measurement of blood pressure is needed; thus often a direct assessment is required. This is accomplished by the use of an arterial catheter line, a needle or plastic cannula inserted directly into a major artery, typically the radial artery at the wrist. A skilled physician can usually insert the cannula directly through the skin and into the vessel without having to make a large incision in the skin. These puncture sites must be watched carefully for infection, bleeding or compromise of the hand or other parts of the body that dependent on the artery for flow. The cannula is typically connected to a fluid-filled tube, which is in turn connected to a transducer for continuous measurement of blood pressure. The arterial line is also useful for the withdrawal of blood samples for the assessment of blood pH and oxygen and carbon dioxide content. The clinical setting for the use of this type of monitoring is further addressed in the section on intensive care units.

To function efficiently, the cardiac pumping action must proceed in a coordinated fashion. The coordination is carried out by specialized groups of cardiac muscle fibers that generate and conduct electrical pulses. Electrical activity in the normal heart begins with a voltage generated by a group of excitable cells located in the right atrium and called the sinoatrial (SA) node. This group of cells is the natural pacemaker for the entire heart. The electrical impulse generated by the SA node spreads across both atria, causing them to contract and pump blood into the two ventricles. The impulse then spreads to another group of specialized cells, the atrioventricular (AV) node. This group of cells delays the electrical signal slightly so that the mechanical event of atrial contraction, which is much slower than the electrical conduction, can take place before the electrical stimulation of the ventricles. The delay is essential because the atrial contraction occurs at a much slower rate than electrical conduction and it must take place to allow the atria to serve as booster pumps for the ventricles. After the atria have pumped, the electrical impulses spread into the ventricles through a special pathway called the bundle of His. The ventricles then contract vigorously in response to the electrical impulse and pump blood out of the heart.

In this process it is important to note that although all of the specialized cells in the heart can generate electrical stimuli, in a normal heart the SA node controls the overall pattern of heart activity. The frequency at which the SA node generates an electrical impulse is the frequency that the cardiologist detects as the heart rate. Moreover it is important to remember that all impulse formulation takes place in the heart itself.

Because the electrical control of cardiac activity is essential to its function, a means of analysis of this function is needed to permit the clinician to evaluate a patient for heart disease. This analysis is based on the recording of an electrocardiogram (ECG) (figure 4.2). The electrical signal is detected by a series of electrodes placed on

Figure 4.2
Electrocardiographic recording (ECG) obtained from normal heart

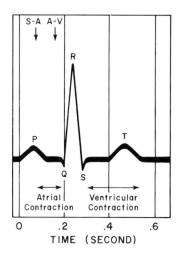

the body and displayed either on an oscilloscope or recorded on paper. The electrocardiograph, the instrument used for the detection and recording of the ECG, was the first biomedical device to have widespread use and was introduced into hospitals around 1910. To make these tracings universally useful, standards have been established. As a result a tracing obtained in one place—say, California—can be interpreted by a cardiologist in another place— say, New York—without the New York physician having to know the specific manufacturer of the machine used to make the tracing.

Each portion of an ECG represents electrical activity in various parts of the heart. It is well known that electrical depolarization of the atria is represented by the P wave in the ECG trace. The QRS complex comes from the depolarization of the ventricles, and the T wave segment represents repolarization of the ventricles. All electrical activity is initiated by the pacemaker cells of the SA node, but this voltage is so small that it cannot be detected in itself by a standard ECG machine.

Standards for the size and duration of these ECG components have been established for both adults and children. A larger amplitude of the QRS, for instance, reflects an increased muscle mass, or hypertrophy, of a ventricle. If the electrodes are varied in their position so that different areas of the heart can be analyzed, then specific chambers can be separately studied. In this way if cardiologists note that the voltage from the left ventricle is reduced and there are alterations in depolarization and repolarization, they can make an ECG diagnosis of an infarction or death of a portion of the ventricular muscle from a heart attack.

The heart rate is documented by the ECG, and the normal sequence of P-QRS-T can be analyzed. Heart rates slower than normal are called bradycardia, and faster rates are called tachycardia. The extremes of these variations are asystole, or cessation of all cardiac electrical activity due to a lack of impulse formation, and ventricular fibrillation, which is very rapid, chaotic electrical activity conducted through the ventricles in a random and uncoordinated fashion. Because there is no effective mechanical cardiac action in either of these states, death would occur if neither were treated.

Fortunately medicine and biomedical engineering have developed devices—the artificial pacemaker for bradycardia and the defibrillator for ventricular fibrillation—to treat these situations. Prompt recognition of these conditions leads to appropriate therapy, and lives are regularly saved through the use of these devices.

Cardiac Pacemakers

The use of artificial cardiac pacemakers is well established in medical practice (Judson et al. 1967). More than 500,000 pacemakers have been implanted into patients in the United States alone, making this the most common of all cardiovascular therapeutic procedures (American Heart Association 1984). The most common indication for a pacemaker is a very slow heart rate or bradycardia due to a failure of the natural pacemaker or a component of the conduction system. Patients with bradycardia are prone to loss of consciousness because their hearts cannot speed in response to stress and their brains do not receive adequate blood flow.

The first animal experiments with electrical stimulation of the heart took place over 100 years ago, but the first human use of a pacemaker did not take place until 1932 (Hays 1964, Hyman 1932). On that occasion a needle was passed through the chest wall into the heart of a dying patient and an electrical stimulation applied.

This crude approach was unsuccessful, but led to more scientific study. In 1952 Paul M. Zoll, a cardiologist working with the engineers of the Electrodyne Company, developed an external pacemaker that pulsed energy to the heart through large electrodes placed on the chest wall (Falk 1983). The system was not practical because it required such high voltages to stimulate the heart that other muscle groups also contracted and patients developed skin burns. In 1957 C. Walton Lillehei paced the heart directly during an open-heart operation by sewing wire electrodes into the heart muscle and using an external pulse generator (Thevenet et al. 1958, Weirich et al. 1957). This effort was successful only for a brief period. What was needed was a self-contained, battery-powered unit that could be implanted in the body and connected by reliable electrodes directly to the heart. The technological development of more effective power sources and reliable solid state electronics led to the development in 1960 of the first implantable cardiac pacemaker by William Chardack, a physician, and Wilson Greatbatch, an electrical engineer (Chardack et al. 1960). Their device was placed into the body, and at a predetermined rate it delivered an electrical impulse that was conveyed to the heart's ventricles by a wire electrode. Its major shortcoming was that it could not sense the heart's own electrical activity. Some patients receiving this unit therefore developed a competition between their own slow rhythm and that of the pacemaker. This led to the development of ventricular fibrillation and, in some patients, death.

The next generation of pacemakers was developed to sense the heart's own rhythm and to suppress the artificial pacemaker's own output when necessary to prevent the occurrence of ventricular fibrillation. These "demand" pacemakers had electronic circuits capable of sensing the heart's own native rhythm through electrodes. If the natural heart beat exceeded a predetermined rate, typically 60 beats per minute, then the pacemaker would not stimulate the heart. When the natural heart rate fell below the preset rate, the pacemaker would be instructed to send an electrical impulse to cause the heart to beat. These early units did not address the need for coordinated atrial booster function because they could only sense and pace electrical impulses in the ventricles.

The current generation of cardiac pacemakers incorporates three major components: the electrodes, the power source, and the pulse generator and its control mechanism. All of the electrodes used today are placed directly in contact with the heart. Two types of electrode systems are used: (1) bipolar systems in which two leads, a positive (+) and a negative (−) electrode, are used, and (2) unipolar systems in which only the negative (−) electrode is positioned in or on the heart with the positive (+) electrode as one surface of the pulse generator, which is placed under the skin of either the chest or abdomen. The electrodes are either sewn directly into the heart muscle or placed via a vein into the atrium or ventricle and into close contact with the inner lining of these chambers, which is called the endocardium. If myocardial electrodes are used, they can be placed on any surface of the heart, but endocardial leads can reach only the right-side cardiac chambers. Myocardial sewn-in electrodes are shaped like corkscrews or barbs to permit firm attachment, and the endocardial leads have tines to snare any rough surface in the right-side chambers to achieve a firm anchor.

Presently most patients have pacemakers that are implanted via a large vein and that use an endocardial electrode system. In this technique a chest incision into the thoracic cavity is not needed, and the postimplantation hospital stay can be quite brief. The cardiologist makes a small incision in the skin over the subclavian vein, isolates that blood vessel, and inserts a catheter with the tined electrode at its tip (figure 4.3). She then uses a fluoroscope, an X-ray viewing device, to pass the catheter into the heart to the precise location where pacing or sensing would be most useful. The pulse generator is then connected to the catheter leads, tested, placed under the skin, and the incision is closed. Many patients leave the hospital on the day after such a procedure.* The electrodes and connecting wires themselves must be very strong to avoid breakage, but also flexible enough to permit catheter passage into the heart. The metals most commonly used today are a stainless steel alloy, Elgiloy, for the wires, and iridium or platinum for the tips of the electrodes.

The power source for pacemakers has also undergone several technological transformations. Initally batteries for the units had working lives of one year, and patients had to be subjected to reoperations to be provided with a new generator. As a result three separate avenues of biomedical research were explored to develop a solution to this problem. First, rechargeable units that were implanted and charged weekly through the skin by radio frequency stimulation were developed (Holcomb et al. 1969). These units worked quite well, but were inconvenient and are no longer in widespread use because the useful life of the rechargeable batteries was only ten years. Second, a nuclear-powered pulse generator was introduced. This type of pulse generator used the heat generated by

*In Europe many patients receive implants on an ambulatory or outpatient basis.

Figure 4.3
Implanting the electronic pacemaker requires intravenous catheterization. Usually the unit is implanted in a convenient body cavity, and a catheter containing the electrodes is inserted into the heart through a vein (in this case, the right cephalic vein). (Courtesy of Medical Systems Division, General Electric Co., Milwaukee, WI)

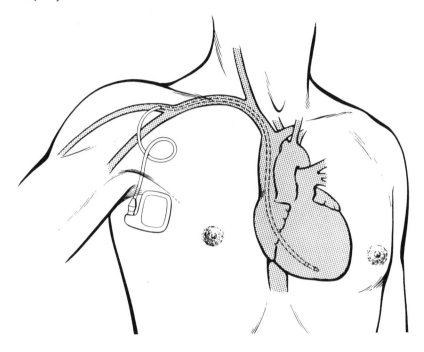

the decay of radioactive plutonium to produce electrical energy. These units were quite heavy because of the shielding needed and had a life expectancy of fifteen years. The third and most successful approach was the development of better conventional power sources. Most units now implanted use a lithium-iodine battery that has a life expectancy of 15 years. These units have proved very reliable and can be monitored externally for signs of power source depletion (Ludmer and Goldschlager 1984, Owens 1986, Tarjan and Gold 1988).

The modern pacemaker is also a far cry from the early impulse generators with respect to its pulse generator. The latest units available can sense the heart's own electrical activity and pace appropriately. For patients with SA node dysfunction, the pacemaker can pace the atrium directly through an atrial electrode and allow the heart's own conduction system to function normally. For patients with a blockage of impulse conduction between the atria and ventricles, a condition called heart block, the heart's own atrial activity can be sensed by the pacemaker and a stimulus sent directly to the ventricle. This type of physiological pacing allows the patient's own SA node to set the rate. Many patients have widespread conduction system disease. At various points in their conduction system, problems may occur and flexibility is needed to control the situation. These patients may experience initially SA node dysfunction but subsequently heart block. To deal with this problem of change in the status of the heart, cardiologists requested and engineers developed a dual-chamber programmable pacing system. After implantation the functions of these units can be controlled and reprogrammed by an external device using radio frequency commands. The functions that can be controlled are the selection of the chambers to be paced, the chambers to be sensed, and the mode of response. All of these functions plus battery status are presently monitored by telemetry (Sutton et al. 1980).

Typically a pacemaker unit will be installed in a patient and programmed by a cardiologist. The status of the unit will be monitored by telemetry, increasingly via the telephone, on a monthly basis, and periodic reprogramming of the pacemaker will take place as needed (figure 4.4). Present pacemakers are so durable they can be reused (Amikam et al. 1983). In the future it is almost certain that even better batteries and pulse generators will be developed to provide even more flexibility.

Defibrillation Techniques

The normal heart contracts rhythmically and in a set sequence. In the diseased heart, especially in the case of coronary artery disease, when the heart muscle does not receive adequate blood flow through the blocked coronary arteries, rhythm disturbances occur. These are usually limited to occasional extra beats, called ventricular extrasystoles, but can progress to the potentially lethal rhythm called ventricular fibrillation. In this state there is no orderly beat, and the ventricle is overstimulated electrically to the point where mechanical activity ceases and there is no effective cardiac pumping. Death occurs in a few minutes if the condition persists (figure 4.5).

Ventricular fibrillation is the primary cause of death from heart disease in the United States. It is usually related to coronary artery insufficiency and is called a "heart attack" by most laypersons. It is, however, a condition that can respond to treatment if promptly detected before the heart muscle or other vital organs are further damaged by the lack of blood (DeSilva et al. 1980).

The first instrument to treat this condition was produced in the 1930s by William B. Kouwenhoven, an electrical engineer (Geddes and Hamlin 1983). This "defibrillator" was designed to pass alternating current through the chest of a person with ventricular fibrillation. Significant refinements have been made since that time, but the basic principle of causing all the cells of the entire heart to depolarize at the same time is involved. All cardiac cells would be made to enter their electrically silent or refractory period at the same time. The heart cells that would recover first would be ideally the normal pacemaker cells of the SA node so that normal heart rhythm would return (Kerber et al. 1983, Tacker et al. 1974).

In 1962 Bernard Lown, a Harvard cardiologist, developed the direct current (dc) defibrillator, in which a capacitor is charged to

Figure 4.4
Complete Teletrace system. (A) Patient connected to transmitter. (B) Display of all components, including receiver, transmitter, and electrodes. (C) Telephone ECG receiver being monitored by nurse.

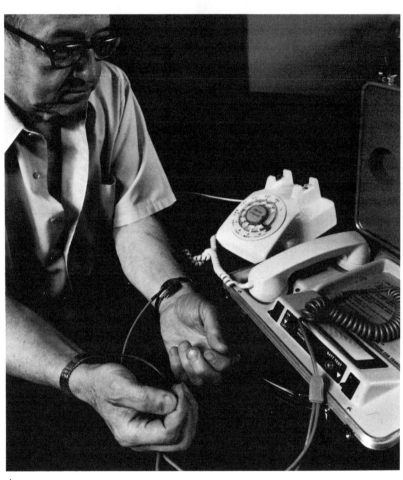

A

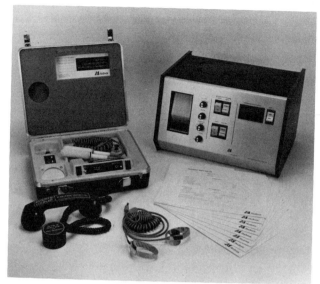

B

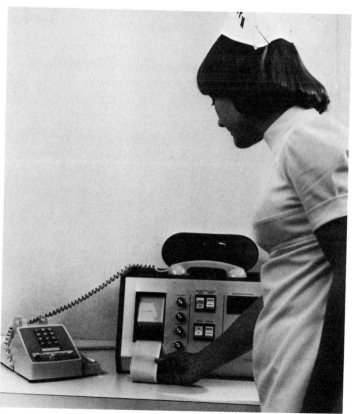

C

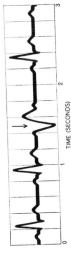

FIRST SIGN OF ARRHYTHMIA in the heartbeat, following a coronary heart attack, usually takes the form of a premature ventricular beat (arrow).

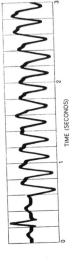

MORE SERIOUS ARRHYTHMIA is tachycardia, or fast heartbeat, in which ventricular impulses occur at two or three times the normal rate. If not halted, they can cause death.

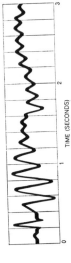

FATAL ARRHYTHMIA known as fibrillation can develop from tachycardia or when premature beats fall within a critical part of the T wave.

ELECTRICAL ACTIVITY OF DAMAGED HEART is erratic because nerve impulses can no longer flow smoothly from the atrioventricular node through the pathways that feed impulses to the ventricular muscles. Most heart-attack patients develop one or more of the arrhythmias shown below.

Figure 4.5
Electrical activity of damaged heart (Redrawn from Lown, B. 1968 (July). *Scientific American* 219:19-27.)

the desired dc voltage and then discharged through large metal paddles passed on the patient's chest (figure 4.6). This method is still being used today. During ventricular fibrillation this voltage may be discharged at any time because there is no existing organized rhythm to be coordinated.

The defibrillator can also be used to correct other serious rhythm disturbances of the heart. If a patient experiences a rapid but still electrically organized rhythm, the ventricles may not have time to fill before they are required to beat again. Two common conditions, atrial and ventricular tachycardia, can cause this clinical problem. A random electrical shock in this situation is not a desirable therapy because it could overstimulate the heart and cause ventricular fibrillation. To avoid this possibility, the defibrillator can be programmed to sense the heart's own rhythm and allow an electrical charge to be delivered only at the right moment in the ECG cycle. The charge is never allowed to occur during the critical T wave phase of the cycle so as to avoid ventricular fibrillation.

In practice the defibrillator is now carried as part of ambulance equipment in most areas. The paramedical team, upon arrival at the location of a patient who is thought to be suffering from a heart attack, promptly attaches ECG leads to the patient, observes the ECG, and sends a copy of the signal by radio to the physician at the hospital base. If ventricular fibrillation is seen, cardiopulmonary resuscitation is started, and by remote direction of the physician, the defibrillator paddles are placed on the patient's chest. Ventricular fibrillation is again confirmed by analysis of the ECG signal sensed by the paddles. Good skin contact is essential to overcome the approximately 2000 ohms resistance of the skin. This is usually accomplished by use of an electrode gel or paste. One paddle is placed over the upper chest and the other over the lower chest, taking care not to allow the electrode gel to form a bridge between the two paddles. The discharge switches are located in the handles of the paddles and both must be pressed at the same time to allow the defibrillator to discharge. Care is taken to ensure that no one is touching the patient so as to avoid unwanted and dangerous shocks to the medical care team and that the patient is

Figure 4.6

A simple defibrillator circuit. Simple defibrillator consisting of direct current power supply such as a battery, stores electrical energy in capacitor "C" when switches S_1 and S_2 are connected to it. This energy is discharged through an electrical circuit containing resistance "R" and inductance "L" into the patient only when operator-activated switches S_1 and S_2 are connecting the charged capacitor directly to paddles applied to chest wall of the patient.

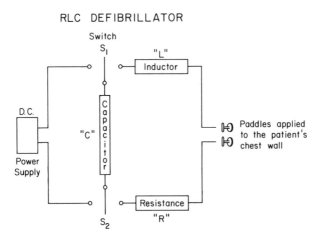

RLC DEFIBRILLATOR

defibrillated. Within seconds the ECG trace reveals, ideally, a return to a normal rhythm. The patient can then be safely transported to the hospital for further care. This system of portable monitoring, telemetry, and defibrillation is a major reason that patients who have heart attacks now survive in greater numbers.

A recent development in the area of defibrillation has been the automatic, implantable device (Mirowski et al. 1980). In patients with a very unstable rhythm that is only partially controlled by medication, this type of device can be used to allow them to leave the hospital and resume a more normal life. This new electronic device is designed to have two electrodes placed on the heart. The device is programmed to monitor cardiac electrical activity and to recognize ventricular fibrillation. If ventricular fibrillation is sensed, an internal discharge of the device can be triggered. Less energy is required for these pulses because the electrodes are in direct contact with the myocardium. Early experience with this type of device in selected patients has been encouraging, and further work is underway. Particular emphasis is being placed on accurate sensing of the cardiac rhythm and energy storage to allow repeated discharges if needed.

The Artificial Heart and Assist Devices
Over twenty years ago it became clear to many researchers that many patients with heart disease could not be aided by medication or surgical repair of their damaged hearts. The heart valves could be replaced with artificial devices, but the heart muscle or myocardium was often so damaged that it required replacement or the patient would die. Two separate avenues of research to the problem were initiated.

First, research on cardiac transplantation began. Kidney and cornea transplantations from other humans had proved so successful that cardiac transplantation was studied. Much of the pioneering work was done by Norman Shumway of Stanford University, but the first human heart transplantation was undertaken by Christian Barnard in South Africa on December 3, 1967. Many other centers quickly began to develop their own programs. The most successful center is still the one operated by Shumway at Stanford. Many problems with rejection, donor organ procurement, and infection remain, but considerable progress has been made, and this approach has now become an accepted part of cardiac therapy. In 1983, 172 heart transplantations were done in the United States, with a one-year survival rate of 80 percent (American Heart Association 1984).

The second area of research in the area of cardiac replacement was the development of the artificial heart. This effort was subdivided into two areas: the left ventricular assist device (LVAD) and total cardiac replacement. Both approaches require close co-operation between the biomedical engineer and the physician and surgeon, and the successes that have been achieved are testament to the great accomplishments of those disciplines (Akutsu and Kolff 1958, Akutsu 1975, Altieri et al. 1986, Bryson 1974, Gott and Klopp 1974, Harmison 1972).

The first temporary use of a mechanical device to sustain life while the heart was unable to function was the pioneering effort of J. H. Gibbon, Jr. and C. W. Lillehei at the University of Minnesota (Allen and Lillehei 1957). In 1953 their group began using the techniques of cardiopulmonary bypass, enabling the current practice of open heart surgery to evolve. The device they developed consisted of a pump to maintain blood flow and an oxygenator to permit gas exchange when the heart and lungs were removed from the path of circulation.

The modern heart-lung machine typically consists of a double-roller pump that propels the blood forward by a squeezing action and either a bubble or membrane oxygenator. The roller pump is quite effective as a generator of flow, but prolonged use causes mechanical damage to the red blood cells (hemolysis). This damage can be tolerated for the short periods (several hours) that are typical in open heart operations but not for prolonged periods because the damaged red cells release hemoglobin into the blood-stream, which eventually causes kidney failure. To permit support for longer periods, pumps that compress the blood between the walls of flexible chambers similarly to the natural process were found to be more desirable. As appropriate pressures and materials are identified, the new types of pumps are being used in the mechanical heart programs. It is important to remember that the materials in these new devices are sensed by the body as foreign. Although they are not rejected as unacceptable biological tissue (a problem that occurs with heart transplants), the body does attempt to form clots on their surface (Hershgold et al. 1972). Such clotting must be avoided because it could cause the mechanism to fail. Patients with these devices must therefore be given drugs, called anticoagulants, to prevent clotting, for as long as the mechanical device is sustaining circulation. Care must be taken to avoid overuse of anticoagulants, however, because spontaneous bleeding could take place. Hemorrhaging could occur in vital organs such as the brain or the gastrointestinal tract.

Anticoagulants are given also to patients who have received artificial heart valves. These mechanical devices are usually made of stainless steel and pyrolized carbon and would cause clots if the patient did not receive anticoagulants. In other situations the materials selected are used to promote clots and a reaction by the body to completely coat the material with the body's own tissue. These materials are used for vascular grafts and for patches within the heart to close a hole caused by a birth defect. In these cases a knitted Dacron material is used to promote ingrowth by the body's own tissues. Within six weeks the body has completely encased the Dacron fibers, making the patch a framework for the body's own patch (McGoon 1982).

Both the artificial heart and the LVADs are extensions of previous efforts to develop the heart-lung machine for open-heart operations. These machines were developed both to pump blood and to oxygenate it during the short period (1 to 3 hours) typically needed to repair a cardiac defect. In the case of the artificial heart, the lungs do not have to be replaced, so that all that is needed is a "simple pump."

In many patients the left ventricle is damaged during an episode of coronary artery blockage. The left ventricle, which is the pump for the entire body, can be revascularized by coronary artery bypass, but often requires time to heal. Because the body must be served by the ventricle immediately after the operation, there can be no rest period. What many patients require therefore is temporary assistance for the left ventricle (Kantrowitz et al. 1965). The development of the LVAD was led by John C. Norman and colleagues (Norman 1975b, Norman and Huffman 1974). Such assist devices are designed to provide partial or total support of the circulation. It was found that most patients demonstrated improvement in myocardial function after 96 hours sufficient to permit withdrawal of the devices.

The LVAD is based on an initial design by William Bernhard of the Children's Hospital Medical Center in Boston in collaboration with the engineers of the Thermo-Electron Corporation (Bernhard and LaFarge 1969, Bernhard et al. 1970). The system is air-powered and has two one-way valves (figure 4.7). An inlet tube is sewn directly into the apex of the left ventricle, replacing a core of myocardium that is removed to make room for the device. The inlet tube is made of pyrolyte carbon, a durable material that does not induce rejection by the heart tissue. The inlet tube is coupled to a series of tubes flexible enough to permit optimum positioning of the LVAD in either the thoracic or abdominal cavities. When the LVAD is removed, the inlet tube is left in place in the apex of the left ventricle and the tubing is disconnected, thereby minimizing trauma to the heart tissues. The LVAD pumping chamber consists of a flexible bladder that deforms to displace a stroke volume of up to 75 ml. The pumping chamber is in turn connected to an outflow tube, which is in turn sewn into the side of the aorta as it passes through either the thoracic or abdominal cavities. The aortic connection is designed to be removed completely when the LVAD is no longer needed by the patient.

Patients receiving support from the LVAD in a carefully controlled study conducted by Norman at the Texas Heart Institute were initially judged to be beyond help with more conventional therapy. Although most patients eventually died from their underlying diseases, all were shown to have benefited from the use of the device in the sense that their cardiac output was substantially

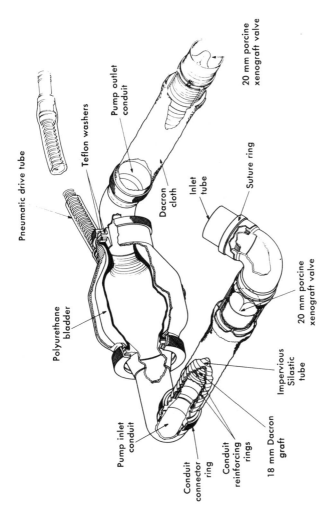

Pneumatic drive tube

Teflon washers

Pump outlet conduit

20 mm porcine xenograft valve

Dacron cloth

Inlet tube

Suture ring

Polyurethane bladder

20 mm porcine xenograft valve

Impervious Silastic tube

Pump inlet conduit

Conduit connector ring

Conduit reinforcing rings

18 mm Dacron graft

Figure 4.7
Design of model X left ventricular assist device
(Courtesy of Thermo-Electron Corp., Waltham, MA)

augmented. Moreover those patients who survived owed their lives to the use of this device (Pierce et al. 1981). Clinical trials continue with improved materials and better understanding of the appropriate role of these devices.

The LVAD has been investigated and developed also by a Stanford University team headed by Philip E. Oyler, who works in conjunction with the biomedical engineering team of the Novacor Medical Research Corporation of Oakland, California (LV assist 1985). Their device, which has FDA approval for experimental use, has already been used in prototype form in a human. The Stanford LVAD is a two-component system consisting of an implantable pump and an electronic controller. The pump is made of a seamless polyurethane bladder that is compressed by two plates. The inflow tube is connected to the apex of the left ventricle and the outflow tube into the descending aorta. The device is electrically powered so that the bulky pneumatic tubing needed for the Texas device is not needed.

In the prototype device that has already been used in one patient, both the power source and the control mechanism were external. The control mechanism is designed to sense the heart's own rhythm and directs the pump to function in a coordinated fashion. In the planned new Novacor LVAD, the power supply and electronic controller will be battery powered and implanted in the patient. The engineers are also developing a microprocessor-based unit that would control the pump unit just as the pacemaker controls the heart's rhythm.

The power source for the projected Novacor LVAD will do what previous units could not—it will provide the patient with the ability to move about freely without a bulky power console. The unit consists of two induction coils. The secondary coil is implanted under the skin and completely encircles the patient's waist. It is connected to implanted rechargeable batteries that in turn power the pump. The primary coil is worn outside the skin like a belt above the secondary coil and connects to either a battery pack during ambulatory activity or a 110-volt source for major charging. Work on the LVAD continues to overcome the remaining problems of clot formation and materials fatigue, but this avenue of research promises much for the future.

The second and more highly publicized effort has been in the development and use of the total artificial heart. These devices are designed to completely replace the damaged heart and to support the circulation on a permanent basis. Unlike the LVAD, which allows the patient's own heart to recover and the support device to be removed, the artificial heart is implanted in place of the patient's own heart, which is then discarded. The only alternative available at this point would be a heart transplantation.

The first animal heart replacement was performed in 1957 by T. Akutsu and W. J. Kolff at the Cleveland Clinic (J. Kolff 1975, W. J. Kolff et al. 1975). Although that animal survived for only 90 minutes, a new era of research was born. In 1963 the National Advisory Heart Council recommended giving priority as well as the necessary federal funding to this type of research, focusing on such areas as materials, driving mechanisms, and control systems (Kusserow 1958, Lindgren 1965, Norman 1975a, Roe 1969).

In 1969 sufficient progress had been made to permit Denton Cooley of the Texas Heart Institute in Houston to implant an artificial heart in a patient awaiting a heart transplant whose own heart had failed. The device sustained the patient for 64 hours until a suitable donor heart was found and the heart transplantation took place. A similar type of temporary artificial heart was again used in a similar circumstance by Cooley in 1981 to sustain a patient for 54 hours. In both these cases the use of the artificial heart was described as temporary, that is, providing assistance until a heart transplantation could be performed.

The work of William C. De Vries and Robert K. Jarvik, supported initially by W. J. Kolff at the University of Utah and subsequently by the Humana Heart Institute in Louisville, Kentucky, has led to the use of a *permanent artificial heart* to sustain a patient (De Vries et al. 1984, DeVries 1986).

The artificial heart used by this group, the Jarvik 7, consists of two pneumatically powered spherical ventricles with attached atria and great vessels that are sewn to the remnants of the patient's own atria, aorta, and pulmonary artery (figure 4.8). Air is pulsed into the ventricular air chambers, which in turn compress the blood-filled ventricles. These chambers, which contact the blood, are constructed of polyurethane and have a stroke volume of 100 ml. The first heart used in Utah had four tilting-disc valves of pyrolytic carbon of the Bjork-Shiley design (Shiley Laboratories), whereas the subsequent device used in Kentucky used the monostrut Medtronic-Hall valve, which was believed to be more durable. The connections to the atria are made of Dacron felt, and connections to the great vessels are made of Dacron vascular prosthetic graft. The air drive lines are made of reinforced polyurethane tubing, which at the skin level were encased in Dacron velour to enhance tissue ingrowth to better secure the lines during patient movement.

The air drive unit is designed to allow for the different pressures and frequencies needed for the artificial left and right ventricles. In practice the unit has typically been set to deliver up to seven liters of blood flow per minute from each ventricle. The ejection fraction, or percentage of blood emptied from the ventricle with each beat, is usually about 65 percent. A higher ejection fraction is possible but could cause damage to the blood.

During the first clinical use of the Jarvik 7 heart (figure 4.9), one of the artificial valves suffered a break in one of its support struts, and the patient was returned to the operating room where DeVries replaced the entire left artificial heart. This was facilitated by the use of a modular concept with quick connect-disconnect coupling.

The pneumatic drive unit is controlled electrically and is backed up by rechargeable batteries that allow the unit to continue to function even if regular electricity fails. A second, smaller portable drive unit has been used in Kentucky. This unit can be carried by the patient to permit him considerable mobility for up to four hours, the safe limit for the portable unit's batteries.

While the implantation of the Jarvik 7 heart continues in Louisville, the design team of Symbion, the firm in Salt Lake City headed by Jarvik, works on the successor model, the Jarvik 8. The major problems of the Jarvik 7—clotting, size, durability, and "mass" production—are being addressed in the redesign (Lavin 1985).

Figure 4.8
The Jarvik 7 Artificial Heart—how it works
(©1984 Time Inc. Reprinted by permission.)

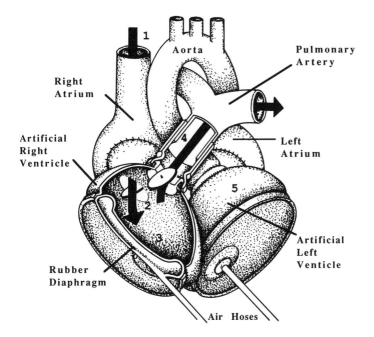

1. Blood flows into atrium.   2. It enters artificial right ventricle through one-way valve.   3.   Air, pumped through hose attached to external power supply, inflates rubber diaphragm, forcing blood out.   4. It passes through second one-way valve and into the pulmonary arteries.   5. Oxygenated blood returns from the lungs to the artificial left ventricle and is pumped into the aorta for distribution to the rest of the body.

Figure 4.9
Mechanical replacement used by the Humana surgical team

The clotting problem can be approached two ways. The first is to use anticoagulants in the patient, but this makes the patient susceptible to bleeding at other sites in the body. The second is the use of materials that do less to stimulate clot formation. This approach is the subject of much research in materials for implantation and is by no means a problem with an easy solution. The development of new, less reactive, more durable materials will be crucial to the future of the artificial heart program.

The size of the Jarvik 7 artificial heart limited its use to adults weighing at least 150 pounds. Its shape was determined by engineering considerations and required atrial cuffs of a relatively large size. The Jarvik 8 heart will come in three sizes, the smallest suitable for a person of about 100 pounds. This new size will allow the use of the artificial heart for the first time in women and smaller men. There are still no plans for an artificial heart for children.

The future of the artificial heart program depends not only on the success of the current trials in Louisville but also the production of the artificial heart in numbers sufficient for demand. The Jarvik 7 model is presently a hand-built device not designed for production in large numbers. The Jarvik 8 and other similar devices are capable of being produced in an almost assembly-line fashion to allow for sufficient numbers of artificial hearts to be available as more implantation teams are trained in its use. Plans are now being developed to train teams in ten centers around the country to select patients, implant the artificial hearts, and provide the needed postoperative care.

Both the total artificial heart and the LVAD face similar engineering problems (Galioto 1986). Materials for their construction must be durable because they must truly sustain life. Their mechanisms must safely pump blood, but should not damage the delicate cellular elements of the red blood cells. Most important these devices should allow for a reasonable and tolerable life. Blood can be protected by nonocclusive pumps and anticoagulants, but the patient's own long-term responses to such devices must be borne in mind at all times. Further development should address not only the engineering aspects of these devices but their effect on daily activity as well.

Economic Issues

Cardiac assist devices such as pacemakers and defibrillators provide a dramatic contrast to cardiac replacement devices (artificial
hearts), such as the Jarvik 7, when it comes to assessment of their
expected costs and benefits. Pacemakers and defibrillators are
proven technologies that have been used extensively in medical
therapies during the past three decades to control bradycardia and
extreme tachycardia. They are relatively inexpensive, and patients
can expect to receive significant benefits in the form of extended
life and improved quality of life during the additional years the
technologies provide to them. In contrast the artificial heart is a
relatively nonvalidated, unproven technology still in the earliest
stages of its "product life cycle."* Although proponents have
argued that each year as many as 66,000 people in the United States
could benefit from artificial heart transplants, the technology has
not yet proved to be a satisfactory therapy that extends both
quantity and quality of the patient's life. Pacemaker and defibrillator technologies have clearly been economically successful from
a benefit-cost perspective. It is too soon to make any economic
benefit-cost call about the artificial heart research and development
program, but, to say the least, the outlook is murky.

Benefit-Cost Analysis of New Cardiac Technologies
The economic issues associated with these technologies are examined in some detail. In each case questions concerning economic
benefit-cost analyses are central to the concerns that are raised.
The principles that underlie benefit-cost analysis are very simple.
If a given project—say, the development of a reliable artificial
heart—provides a stream of benefits that exceeds the costs associated with it, then the project should be adopted. On the other
hand, if the costs exceed the benefits, then the project should not
be adopted. The benefit-cost criterion is very straightforward;
problems with its use arise, however, because benefits and costs are

---

*The precise meaning of *nonvalidated* is examined in detail in the discussion of
ethical issues associated with the artificial heart program later in this chapter.

difficult to measure and to compare in terms of some common measuring rod.

Economists usually advocate the use of a monetary yardstick for the comparison of different types of benefits and costs and therein lies a major problem. Often decision makers such as health care providers and administrators believe that certain types of benefits cannot be measured solely in monetary terms. Human lives are one class of benefit (or cost) that often falls into the category of immeasurables. Quality of life is another. The reluctance of decision makers to place monetary values on benefits that accrue in the forms of increased life expectancies and adjustments to quality of life has a special significance in relation to the use of benefit-cost analysis in assessing health care technologies because those types of benefits form most, if not all, of the benefits to be obtained from many types of medical technologies, especially cardiac assist devices.

Economists are much more willing to place monetary values on human lives. They argue that society does make decisions about the allocation of resources that carry implications for life expectancies and quality of life. For example, expenditures on road repair speed up traffic, but also increase safety. Increasing the speed limit also speeds up travel (a benefit), but increases the rate at which automobile accidents occur and the severity of injuries and number of deaths associated with those accidents (costs). The latter example is interesting because policymakers have no difficulties with the notion that the value of time saved in travel can be measured in monetary terms. Implicitly, therefore, when policymakers trade off lives for reduced travel time, they are placing a monetary value on human life and quality of life.

Despite the fact that policymakers do make implicit judgments about the monetary value of human lives, several concerns lead them to avoid using explicit monetary values to assess benefits that occur in the form of changes in life expectancies. One such concern is that once a dollar value is placed on a human life, it provides an immediate and vulnerable target in the political arenas in which many of the debates over the development of medical technologies take place. Suppose, for example, it is asserted that the value of a statistical life is $2 million, and on that basis the expected benefits of a new medical technology do not exceed its costs to society. Those people who believe they are most likely to benefit from the technology will immediately claim that human life is being undervalued and that in fact the technology should be developed.

A second concern is that in truth the lives of different categories of people have different (monetary) values from a societal perspective in terms of their economic productivity. A 40-year-old physician is likely to be more highly valued in terms of his expected contribution to the economic welfare of his community than a 40-year-old resident of skid row. From a pragmatic view the political consequences of acknowledging that different lives are valued differently could be quite significant. At a minimum that view violates a moral rule embedded in what is essentially a Judeo-Christian culture that asserts that although individuals may differ in importance with respect to their class, status, and wealth, their lives (or deaths) are all equally important. Thus despite the existence of substantial economic inequities in American society, including the distribution of health care, the explicit use of different values for different lives in the assessment of health care technologies is not a process with which our society is comfortable.

Benefit-cost analysis also has its difficulties because although economists can estimate the values society places on many categories of benefit, they cannot always be very precise in making those estimates. Moreover, in the case of some benefits, they can do nothing at all with respect to "monetizing" their size. An example of such a benefit associated with the artificial heart program concerns the value to U.S. citizens as a whole of knowing that American health care scientists are at the forefront of medical research. In fact this type of benefit may provide the most persuasive social justification for the artificial heart program.

Just as benefits often are difficult to estimate, so too are the costs of particular projects, especially projects such as the artificial heart that involve uncertain outcomes. The uncertainty about the nature of the costs involved has three major components: First, the research and development costs are difficult to forecast. For example, no one is sure how much it will cost to develop a reliable implantable artificial heart. Second, the costs of the therapies associated with the technology (once it has been developed) also may not be easy to predict. When pacemaker technologies were in the initial phases of their development, the expected working life of a pacemaker was less than two years. Innovations in the development of electrical power sources that almost no one predicted have significantly extended the life of pacemaker power sources and reduced dramatically the annual per-patient costs of pacemaker therapies. Third, because costs and benefits associated with projects usually take place over extended periods of time, adjustments must be made to compare costs that are incurred, say, this year with benefits that will be received, say, in ten years. There is considerable uncertainty as to how these adjustments should be made.

Almost all economists agree that benefits and costs should be "discounted" to a common basis. The process of discounting is designed to take a benefit or cost with a specific monetary value that will occur ten years from now and use the "correct" rate of interest to calculate the sum of money that would have to be invested today in a perfectly safe financial investment (for example, a government bond) to provide that amount of benefit or cost ten years from now. Applying these procedures (described in the appendix in chapter 2) to all benefits and costs associated with any project, a decision maker can compare the benefits and costs of the project when those benefits and costs occur at different times. The uncertainty arises with respect to the choice of discount rate. No one is sure what the correct discount rate should be. Some economists have argued that the correct interest rate for discounting benefits and costs should be very high, equal to the interest rates that are charged on very risky loans to not very reliable customers by banks. Others suggest that the discount rate should be very low, perhaps even zero. A high discount rate makes projects whose benefits occur in the future appear to be relatively unattractive. A low discount rate makes such projects appear much more attractive. Proponents of public investment in the artificial heart program, whose benefits are not likely to be realized for many years, are likely to argue for the use of low discount rates. Advocates of other programs that yield large immediate and somewhat smaller long-term benefits (for example, dental care programs in schools) might well prefer to use a higher discount rate.

This discussion indicates some of the complexities that surround the use of cost-benefit analysis in the assessment of health care technologies. We now examine the specific economic issues surrounding the use of pacemaker therapies and the artificial heart program.

Cardiac Pacemakers

In their product development stages, both the pacemaker and defibrillator technologies were regarded as exotic. The first implantable pacemaker consisted of wire electrodes placed within the heart, but driven by an external power unit. The power unit was large, about the size of a two-drawer file cabinet, and the patient had to push it around the hospital corridors on a dolly when he went for a walk. The first pacemaker recipient lived for six months subsequent to the implantation of the pacemaker, but was in considerable discomfort and confined to the hospital. In 1988 dollars the cost of providing care to that patient amounted to well over $100,000. That level of expenditure would be difficult to justify from the perspective of social economic welfare if the only benefit that resulted was a gain of six months of low-quality life for one person when the monies, if allocated to other health care activities, could yield significantly larger benefits for larger numbers of patients. In fact it is quite possible that even the patient himself and his family would not have been willing to pay that amount for the additional months of low-quality life, either because they did not have the resources or because the patient did not value the additional life more than the cost required.

In fact hindsight clearly indicates that the resources invested in the development of pacemakers, including those devoted to clinical trials, have been more than justified in terms of subsequent benefits to patients. Today, in the United States alone, more than half a million people have pacemaker implants. Each year in the United

States over 200,000 new patients receive implants (American Heart Association 1984). More than 50 percent of the recipients will live for at least eight years after receiving the implant, even though over 30 percent will die within 18 months.* The average patient therefore has a life expectancy of more than four years subsequent to receiving a pacemaker implant.

Recent data provided by a study of the costs of implants in a New York hospital since 1965 indicate that the average cost of an implant (in terms of 1980 dollars) amounted to $11,259 (Lopman 1981). If, very naively, this cost is spread over four years,† the annual cost to an average patient (given an expected life of more than four years) would be less than $2,814. Economic analyses of the value that persons place on their own lives indicate that a statistical life is valued at between $400,000 and $4 million.§ Using the lower figure and assuming that a statistical life consists of 75 years, the value of an additional year of good-quality life to an average person is at least $5,333 and probably a good deal more than that. The average benefit-cost ratio for pacemakers is therefore clearly attractive. Additionally over the product life cycle of pacemaker therapy, the benefit-cost ratio has improved, in part because the cost of the therapy has declined. In 1965 the annual per-patient average cost of pacemaker therapy at Montifiore Hospital in New York was $2,590. By 1980 it had fallen to $1,407. The reason for the decline in annual average costs was twofold: (1) increases in the working lives of pacemakers and (2) a sharp reduction in the number of days the patient had to spend in the hospital for pacemaker implantation (from 29 days to about 6 days over an 8-year period of therapy).

*Today deaths shortly after pacemaker implantation are not the result of the pacemaker implant except in extremely rare cases. This was not the situation in the 1960s when fibrillation sometimes occurred because pacemakers could not self-adjust to changes in the patient's heartbeat.

†The procedure of simply averaging the cost over a given period is incorrect from a technical point of view. Ideally an appropriate interest rate should be used to annualize any lump sum outlays (for example, the purchase and implantation of the pacemaker), taking proper account of the effects of the tax structures.

§The science of economics is not always as precise as its practitioners wish.

We assume that the continued survival of each patient depends on the implantation of a pacemaker. In many but perhaps not all situations, the assumption is reasonable. If such is the case, in an average year (given that more than 500,000 people at any one time have pacemaker implants) pacemaker technology provides U.S. residents annually with over $1.5 billion in net benefits, more than enough to pay back any of the research and clinical trial costs associated with the technology's development. On a worldwide scale the benefits are even more substantial. Pacemaker implantation has become commonplace in all countries with sophisticated health care systems. Literally millions of people enjoy additional years of good-quality life through the use of pacemakers. However, all of the above do not preclude the existence of difficulties associated with the use of the technologies. In the United States two major policy issues that have economic ramifications have arisen in recent years. The first concerns the reuse of pacemakers (Amikan et al. 1983, Fried 1982, Mugica et al. 1986, Pringle et al. 1986), and the second concerns allegations of excessive implantation of pacemakers (Scherlis and Dembo 1982). A third issue concerns the question of what constitutes the most efficient system for providing patients with pacemakers and associated monitoring services and is related to the general economic environment in which health care services are provided (Faivre and Dodinot 1983, Zegelman et al. 1986).

The reuse of pacemakers has become feasible subsequent to innovations with respect to the power packs that significantly increased the durability of pacemakers. In 1965, for example, the expected life of a mercury cell pacemaker was one and a half years. By 1985 the expected life of a lithium cell pacemaker had increased to at least ten years. More than 30 percent of all patients who receive pacemakers die from one cause or another within two years* (Amikan et al. 1983). The pacemakers implanted in those patients can be recovered, gas sterilized, and reused. Protocols for

---

*This is not too surprising because, in the United States, the average age of a recipient of a pacemaker implant is 75 years. The cause of death in virtually all pacemaker recipients is related not to pacemaker malfunctions but to other health problems.

the reuse of pacemakers have been developed in several countries, although the procedure has not yet been approved formally by the Food and Drug Administration (FDA) in the United States. All of the corporations that market pacemakers in the United States have lobbied Congressional representatives and the FDA to prevent pacemaker recycling largely on the grounds that product quality control cannot be maintained for reused machines (Meyers 1986). However, a statistical analysis in Europe of a sample of over 2,000 patient records in which recycled pacemakers were used indicated that there is no difference in the performance of recycled pacemakers and new pacemakers (Mugica 1986). If anything the data suggest that recycled pacemakers perform better.

Assuming that manufacturers are aware of the clinical evidence, it is probable that their position is based on concern about the effects of pacemaker reuse on the size of the U.S. market. Currently over 200,000 new pacemakers are implanted annually, each at an approximate cost of $3,000. The sale of pacemakers thus constitutes a $600 million business. Conceivably recycling could reduce the demand for new pacemakers by between 10 percent and 20 percent,* cutting the size of the U.S. market by between $60 million and $120 million per year. Such revenues are a matter of serious concern to any industry.

The benefits that would accrue from recycling would be enjoyed by those who pay the bills for health care; in some cases the patients themselves (through lower direct payments or insurance premiums), and in other cases by taxpayers who underwrite the Medicaid and Medicare programs and in others by charitable institutions. The size of the gains from recycling in any given case would not be the full cost of a new pacemaker. Recycling involves certain costs; for example, resources are needed to obtain and sterilize the pacemaker, to check its performance capabilities, and to organize its distribution to an appropriate patient. Nevertheless the recycling costs appear to be considerably less than the price of a new unit. The

*Approximately 35 percent of patients receiving a new pacemaker die within 18 months of the procedure. Thus about 35,000 pacemakers are potential candidates for reuse. Not all of these candidates could be accessed, but even if "wastage" were as high as 70 percent, 10,000 pacemakers could be reused each year.

recycling issue illustrates the dynamic nature of medical technology. It also suggests that the development and rate of adoption of technological innovations depend not only on the costs and benefits of the innovations but also on the broader regulatory and market environment in which they are developed.

The second problem, an apparent excessive number of implants, also reflects the interplay between technological innovation and economic behavior. In 1979 a public interest group in Maryland published a report asserting that 35 percent of all pacemaker implantations that took place in Maryland between 1978 and 1979 were unnecessary from a clinical perspective. The argument put forward was that hospitals were using the procedure to generate revenues when drug-based therapies that were less expensive and less debilitating could have been used. The report generated such a furor that the state commissioned an independent review of all the patients identified in the Public Interest Report as having received implants unnecessarily (Scherlisand 1982). The independent report concluded on the basis of its review that only 1.5 percent of all patients received a pacemaker they did not need.

The economic issue at hand concerns risk. Pacemakers require less patient participation than a rigorous drug-based therapy. Because they are a (relatively) low-cost option with very low surgical risks for the patient,* it may often be beneficial from an economic perspective to invest in them as a risk reducing procedure. This is not to deny the legitimacy of public interest group concerns with respect to patient abuse. However, in this case there may be good economic reasons for the extensive use of pacemaker implants. Individuals, either acting on their own behalf or in social groups, have always been willing to use resources to reduce the risk of catastrophic events (for example, through the provision of fire and police services, and insurance against personal injury and property damage). Their approach to the provision of health care through the use of medical technology also incorporates such concerns.

---

*The implantation operation can be performed on an ambulatory or outpatient basis. In one pacemaker center in Germany, for example, over 75 percent of all implantations are carried out that way (Zegelman 1986).

The third problem concerns alternative systems for the provision of pacemaker implants and subsequent patient care. The economic structure of national health care systems affects the delivery and the costs of health care services. In Britain and Germany, both countries with national health care services, in areas with relatively heavy population densities special pacemaker centers have been developed to handle all pacemaker implantations (Watkins 1983, Zegelman 1986). The objectives are to reduce implantation costs and to improve quality of service by spreading the overhead costs associated with expensive monitoring and programming equipment, specialized personal, and the like, over large numbers of patients, thereby providing the patients with access to low-cost state-of-the-art care. In the United States most hospitals offer implant services, often relying on pacemaker manufacturers for monitoring and reprogramming services because the patient load with respect to such services is much smaller than in the German and British centers. In part the more dispersed delivery system for implant services in the United States is a function of smaller population densities. However, it is also the outgrowth of a system in which hospitals explicitly compete for patients and therefore need to offer the widest possible range of services. Thus pacemaker implantation procedures may represent one therapy with respect to which some degree of central planning works by improving the delivery and reducing the costs of health care benefits associated with a technological innovation.

The Artificial Heart
Artificial heart technologies are a completely different kettle of fish. The potential benefits to be derived from the successful development of artificial hearts for the purpose of partial or total cardiac replacement are significant, although less extensive than those associated with pacemaker implants. If the technology were available, as many as 66,000 patients per year in the United States although a more likely number is approximately 35,000 people (Blumenthat 1984). Moreover many of the agonizing situations that arise in the case of heart transplantations because of the relative unavailability of donor hearts could be avoided.

Research efforts related to the development of the artificial heart have been carried out since the early 1950s. In relation to other areas of health research, the program has received considerable support. Since 1965 the Federal government alone has provided over $200 million, and funding continues at the rate of approximately $10 million per year. Private corporate support also has been substantial in recent years. For example, in 1983 the Humana Corporation made a commitment to underwrite all costs associated with up to 100 artificial heart transplantations at an anticipated cost of over $200,000 per patient (a commitment of over $20 million) (Altieri 1986).

Despite the considerable advances that have been achieved in artificial heart technology, a reliable, long-term artificial heart therapy has not yet been developed. Further significant research and development (R & D) efforts will be required if the needs of potential beneficiaries are to be met. Ideally, in assessing the potential of a technology, decision makers would like to know what will be the costs as well as the benefits associated with the future development and use of that technology. In the case of the artificial heart, there is no clear cut guidance as to what future investments in R & D will be required to identify a successful therapy. We do know, at least approximately, how much has already been spent on the project, but, paradoxically, from an economic perspective that information is of no interest in evaluating whether support for artificial heart R & D should continue. Every economic textbook points out that historical expenditures should not count in relation to economic decisions about the future; bygones are bygones. In the past, resources that could have been allocated to other areas of medical research were instead allocated to artificial heart research. Those resources cannot be reallocated today. By the same token the fact that resources were allocated to a project in the past does not necessarily make that project an unattractive candidate for further support. Thus just because very few patients have benefited directly from artificial heart research expenditures up until now does not automatically imply that putting more resources into the program is ill advised. Nor does the argument that too much has been spent on the project already to let it die hold water. The correct concern is whether the expected future benefits from the project will exceed its expected future costs.

The expected future costs associated with the artificial heart include both R & D expenditures and the per-patient costs of installing and maintaining the device. We simply do not know what the R & D costs will be, but on a per-patient scale they are likely to be much smaller than the per-patient costs of using the technology. To speculate for a moment, if R & D expenditures totalled an additional $200 million over the next ten years, and as a result an average of 10,000 patients per year were to benefit from the technology, the future per patient R & D costs would be $2,000.* The per-patient costs of implanting the artificial heart (once it has been fully developed) have been estimated to be $28,000, with additional support costs of approximately $2000 per year (according to the Office of Technology Assessment). When set against these per-patient costs, the R & D costs do not loom large.

A more fundamental set of economic questions arises in relation to the artificial heart research program. The first and most important of these is whether the program is worthwhile in terms of expected costs versus expected benefits to the patient once the R & D problems have been overcome. The answer may be No. Current studies indicate that the recipient of an artificial heart would live on average an additional 1.6 to 3.6 years, extra years whose quality is likely to be quite poor. How one concludes whether the benefit-cost ratio is acceptable depends on one's evaluation of the value of extra years of life and explicitly ethical concerns, but other considerations are also relevant.

The social opportunity costs of the artificial heart program, as with any other activities, consist of the value society places on the next best alternative use to which the resources could have been put. Often a simple accounting of the market value of the resources

---

*Of course the analysis presented here ignores the original issues that involve discounting future costs. It is therefore an illustrative rather than a technically precise assessment of the costs of R & D relative to the costs of providing patient care at the time of implantation.

provides a reasonable approximation of the value to society of those resources in other uses. In the case of health care, particularly health care resources, this is not necessarily the appropriate approach. In the short to medium term, the total federal health care research budget is fixed. A dollar spent on R & D in the artificial heart programs represents a dollar lost to other forms of medical research. These other forms of research may yield the prospect of higher paybacks in terms of improvements in human health. Similarly a dollar spent on health care for an artificial heart patient may represent a dollar lost to the support of some other form of health care that yields higher benefits for its recipients. As an aside it is perhaps worth noting that although, as some have argued, it might well be a better world if resources were given to the artificial heart program instead of being used to manufacture luxury sports cars or nuclear weapons, if such trade-offs are not possible, then the argument is irrelevant. Moreover it might still be the case that even if such a trade-off could take place, resources should be allocated to other health care research and operational programs because they would yield greater benefits to society in terms of human welfare.

None of the above holds if in fact the artificial heart R & D program generates its own revenues. It is conceivable that the resources allocated to the artificial heart program would have been put to another, more worthwhile use if the program not been developed. There is no guarantee, however, that such would be the case. For example, perhaps the Humana Corporation would have allocated the monies now underwriting artificial heart implantations to mindless moments of advertising during the Superbowl. The federal research budget allocations might have been diverted to a tax refund that would be spent by citizens on unhealthy foods and drinks, thereby creating the possibility of future health care problems. More serious it has been argued that, of the many high-tech health care research programs, the artificial heart program has done more to focus attention on the need for research in health care and, as a result, stimulated the provision of resources for other health care research projects that in themselves hold the prospect of much higher social paybacks.

Any economic assessment of the artificial heart program is be-deviled by uncertainties. We do not even know whether a viable artificial heart will be developed. We do not know the extent of the benefits that would result to a typical patient or the number of patients likely to benefit from the technology, even though we do have some idea of the per-patient cost of the technology, once it has been developed. In that context there is one important concern: if the artificial heart is developed, the political economy of health care may require that the implant be made available to patients through Medicare (as is the case with renal dialysis). People who require an artificial heart transplant are, for the most part, unlikely to have either the personal resources or access to sufficient private insur-ance to meet the costs of the therapy. If then heart transplants become, like dialysis, a right to which the desperately sick are entitled, approximately 30,000 patients a year would qualify at a per-patient cost of somewhere in the region of $30,000, and the burden on the U.S. treasury would amount to over $1 billion. This sum is considerable, especially in light of the fact that other social welfare programs (for example, prenatal and early postnatal nutri-tion programs for the poor) that cost a lot less have recently been cut out of the U.S. federal budget. Thus we have a policy decision to make that is not trivial.

In this section we have been concerned with the economic issues that surround the benefits and costs of pacemaker therapies and the artificial heart program. We now turn to their ethical dimensions.

Ethical Issues Related to the Artificial Heart Program

Of the cardiac technologies discussed in this chapter, the artificial heart is the most controversial. This is due in some small measure to the fact that the heart has been viewed not only as a bodily part performing functions crucial to health and well-being, but also as the seat of love and commitment. Although most people are well aware that the heart is basically a pump essential to human exis-tence, these romantic associations still linger. Much of the contro-versy, however, arises from the fact that the artificial heart is not a validated technology, and its use may therefore constitute unjuati-fied research on humans. Although there is much hope that this technology will ultimately bring significant benefits to many patients, presently insufficient data exist for this hope to be well founded. Therefore, in an effort to gain such information, the FDA has approved implantation of the Jarvik 7 in several patients.

Experimental status is not unique to the artificial heart by any means. All the cardiac technologies discussed in this chapter were nonvalidated at some point, even though some, such as the pacemaker, are now widely accepted tools of medical practice. No matter how much laboratory research has been conducted, no matter how extensively a technology has been tested on animals, if its use in medical practice is to be well founded, if its use is to be based on reasonable comprehension of its risks and benefits, then it must ultimately be tested on humans. Research on humans is therefore the critical link between laboratory research and animal studies on the one side and scientifically well-founded clinical practice on the other side. This is expressly true of the artificial heart more than any other modern medical technology. Because this issue is so crucial, much of the remainder of this chapter focuses on the issues posed by conducting research on humans. First, the central ethical issues posed by human research are examined, with particular attention given to the artificial heart, especially William DeVries's use of the Jarvik 7. Then, FDA regulation of human experimentation is discussed along with regulation of nonvalidated technologies outside the context of research. Here particular attention is given to Jack Copeland's "emergency" use of the Phoenix heart. Finally, the ethical concerns raised by the temporary use of the artificial heart, that is, to keep patients with end-stage heart disease alive until a human heart transplantation can be performed, are discussed.

Human Experimentation: Basic Notions
Discussions of human experimentation usually begin by distinguishing experimentation from therapy in terms of their respective goals. Whereas the overriding aim of therapy is to benefit the

patient, the goal of experimentation is to acquire knowledge that will be useful to medical science. Thus the clinician qua therapist looks primarily to the well-being of the patient, and the clinician qua researcher is concerned immediately with gaining information and ultimately with the benefit to humankind that such information can yield. The distinction between therapy and experimentation is predicated on intention rather than outcome. Even when a course of action does not benefit a patient, it is ranked as therapy when it is *intended to improve* the patient's condition in some way. This remains true even when it has the unintended consequence of providing new and useful knowledge as well. On the other hand, even if a particular course of action does yield definite medical benefits for a subject, it is considered experimentation if it is primarily motivated by pursuit of knowledge. This remains the case even when it fails to achieve this aim.

The next logical step in discussions of human experimentation is usually to distinguish therapeutic from nontherapeutic experimentation (Fried 1982):

In therapeutic experimentation a course of action (or studied inaction) is undertaken in respect to the subject for the purpose of determining how best to procure a medical benefit to the subject. In nontherapeutic experimentation, by contrast, the sole end in view is the acquisition of new information.

Once again the crucial difference is one of intention rather than actual outcome. An experiment is deemed therapeutic if its purpose is to determine how best to benefit the patient, even if it actually results in no such benefit. It is deemed nontherapeutic if it is undertaken in order to gain new knowledge regardless of whether this will benefit the subject. And it remains nontherapeutic even if unintended medical benefits for the subject in fact are generated.

The distinction between therapy and experimentation, along with the distinction between therapeutic and nontherapeutic experimentation, can be misleading in ways that make a related but more adequate set of distinctions preferable. The use of the term *therapy* in opposition to *experimentation* may lead some to fail to appreciate that therapy is but one kind of clinical practice aimed at benefiting patients and therefore connotes that all nontherapeutic practices are experimental. Clearly this is incorrect. In addition to attempting to heal patients or palliate the effects of disease, medical caregivers also seek, where possible, to prevent disease. Furthermore therapeutic efforts are generally preceded and directed by the results of diagnosis, efforts to determine the source and nature of the patient's affliction. Therapy, prevention, and diagnosis are all conducted with the aim of benefiting the patient. This also applies to the distinction between therapeutic and nontherapeutic experimentation. Just as a course of action may be taken to determine what form of therapy will best benefit the patient, so can a course of action be taken to determine what will yield the best diagnosis or best prevent the patient from becoming ill in the first place. And although attempts to determine the best form of prevention or diagnosis are nontherapeutic, they are nonetheless aimed at benefiting the patient.

For these reasons, in place of the experiment/therapy distinction, a distinction drawn in the mid-1970s by the National Commission for the Protection of Human Subjects of Biomedical and Behavioral Research—that is, *the distinction between practice and research*—is more adequate. Quoting the commission, Alexander Capron (1986) wrote,

"[T]he term 'practice' refers to interventions that are designed solely to enhance the well-being of an individual patient or client and that have a reasonable expectation of success." In the medical sphere, practices usually involve diagnosis, preventive treatment, or therapy; in the social sphere, practices include governmental programs such as transfer payments, education, and the like.

"By contrast, the term 'research' designates an activity designed to test a hypothesis, to permit conclusions to be drawn, and thereby to develop or contribute to generalizable knowledge (expressed, for example, in theories, principles, and statements of relationships)." In the polar cases, then, practice uses a proven technique in an attempt to benefit one or more individuals, while research studies a technique in an attempt to increase knowledge.

Although the practice/research dichotomy has the advantage of not implying that therapeutic activities are the only clinical activities intended to benefit patients, like the therapy/experiment distinction it is based on intention rather than outcome. Interventions qualify as practice when they are proved techniques intended to benefit the patient even if they fail utterly sometimes.* Likewise interventions that are used to increase generalizable knowledge constitute research even when they result in direct benefits to the subjects.† What of the interventions that constitute neither pure research nor pure practice? As Capron (1986) has pointed out,

... not every use of unproven technique amounts to research; the way in which the technique is used must be designed so as to permit generalizable knowledge to be gained—something more than just seeing what results from using the technique in one particular instance. Conversely, not every intervention carried out with the intent of benefiting the patient (or other subject of the intervention) thereby qualifies as an instance of practice; prior study of the technique must have provided adequate information about it, to give rise to a reasonable basis for believing that it will achieve the intended result.

To refer to these interventions that are neither pure therapy nor pure research, instead of the standard distinction between therapeutic and nontherapeutic experimentation, the notion of nonvalidated practice is used here. Like the term *therapeutic experiment*, *nonvalidated practice* connotes that the course of action taken aims to benefit the patient. Unlike *therapeutic experiment* it explicitly encompasses prevention and diagnosis along with therapy. Addi-

---

*Even a proved intervention may fail in some cases. A validated intervention is not thereby infallible.

†In the following pages, although the term *practice* is consistently used instead of *therapy*, the terms *research* and *experiment* are used interchangeably with the understanding that both refer only to interventions that are intended to provide generalizable knowledge.

tionally the term seems to be at war with itself by implying that an intervention can be primarily intended to benefit the patient yet also primarily directed at generating useful scientific information. The limits that must be placed on an intervention for it to contribute to a scientifically sound research effort may be inconsistent with what would most benefit the patient. *Nonvalidated practice*, on the other hand, clearly connotes that the intervention's primary purpose is to benefit the patient while making clear that it has not been shown by rigorous prior study to be safe and efficacious.

An important difference between the therapeutic/nontherapeutic experimentation dichotomy and the concept of nonvalidated practice is that whereas the former limits attention to new and innovative interventions, the latter encompasses some interventions which are neither new nor innovative. Robert Levine, who coined the term *nonvalidated practice*, wrote (Capron 1986),

A practice might be nonvalidated because it is new; i.e. it has not been tested sufficiently often or sufficiently well to permit a satisfactory prediction of its safety or efficacy in a patient population. An equally common way for a practice to merit the designation "nonvalidated" is that in the course of its use in the practice of medicine there arises some legitimate cause to question previously held assumptions about its safety or efficacy.

Whether an intervention's safety and efficacy are in question because it is new or because clinical experience has raised questions challenging its status as a validated practice, the central moral issue is the same. Patients are subjected to a course of action for which there is no well-founded basis from which to assess its risks and benefits. Consequently there is no well-founded basis on which to judge whether the benefits make the risks acceptable. This is a crucial issue that warrants attention.

A discussion of the moral concerns raised by nonvalidated practice, however, requires that attention be first directed to concerns raised by the general topic of research on human subjects (Capron 1978):

The basic ethical problem . . . is the use of one person by another to gather knowledge or other benefits that may be only partly good, if at all, for the first person; that is to say, the . . . subject [of the research] is not simply a means but is in danger of being treated as a mere token to be manipulated for research purposes.

As these remarks indicate, the central ethical concern is posed by the fact that research on humans is not intended to benefit research subjects themselves directly. Instead it is intended to yield generalizable knowledge that will be useful to medical science and thereby to humankind in general. Whether one finds this to be an important ethical concern depends on one's view of the proper relationship between the well-being of distinct individuals and the well-being of the social whole to which they belong (Capron 1978):

Those who regard individuals as part of a collective whole, which allocates rights and protects them, can justify a wide scope for experimentation that serves to advance the interests of the collectivity. Conversely, for those who emphasize the inviolability of human persons and the moral obligation not to treat them simply as parts related to a social whole, any human experimentation whose purpose is to benefit persons other than the subjects of the experimentation requires very strong justifying reasons.

From a perspective that regards the common good (however that is conceived) as greater than the good of particular individuals—in short an essentially utilitarian perspective—the fact that a program of research sacrifices the interests of particular individuals for the good of others need not be problematic. From such a perspective what justifies the research in question is that it indeed poses a genuine prospect of benefit for others and that the probability and level of benefit are sufficiently high to outweigh any loss to the subjects of the research. The important consideration from this perspective is that once risks, costs, benefits, and probabilities are assessed, the research promises a net gain, even if that gain does not accrue to the individual subjects of the research. Accordingly the worry that research always poses the danger of turning some persons into mere instruments of benefit to others, or mere resources for the good of others, has little significance from this essentially utilitarian point of view.

From a perspective that considers the individual's dignity to be paramount, an essentially Kantian perspective, there are clear limits to the pursuit of the common good. While not denying the importance of attempting to achieve "the greatest good for the greatest number," this perspective insists that this pursuit must be conducted only in ways that do not demean persons. This in turn entails that what is done on behalf of the larger society must only be what can be done without reducing individuals to mere resources for the good of others. Adherence to this imperative clearly would limit what can be done in the name of medical research.

The importance of failing to give adequate regard to the wellbeing of research subjects is well illustrated by the Nazi experiments of the 1930s and 1940s. Although it was not the first situation in which atrocities had been committed upon persons in the name of the greater good, the atrocities perpetrated by Nazi physicians shocked the world (Redlich 1982):

... unparalleled in history were the inhuman experiments primarily in the concentration camps Auschwitz, Buchenwald, Dachau, and Ravensbrueck. Members of the Medical Corps of the SS, among them university professors and high dignitaries of the medical profession, carried out experiments on human beings with the authorization of the SS leadership. These experiments may be divided into three categories: (1) experiments that served primarily military purposes, such as experiments exposing human beings to very low water temperatures ... , experiments on the effects of high altitude in simulated parachute jumps ... and experiments with the ingestion of salt water ... ; (2) experiments that served both military and general medical objectives such as experimental approaches to the etiology and therapy of infectious diseases, of cancer, and of fractures, and the transplantation of extremities; and (3) experiments that served only the SS. These latter included experiments in exterminating large populations, or what Dr. Leo Alexander calls "ktenology," the science of killing; experiments on twins by the notorious Dr. Joseph Mengele in Auschwitz; the absurd experiments with mass sterilization by hidden radiation equipment and subsequent castration; and the most macabre experiments of all, performed by Prof. A. Hirth, who collected and examined the decapitated skulls of Jewish Soviet commissars and other ethnic groups expected to become extinct.

The subjects of this "research" were coerced participants who had no choice in the matter and most of this "research" was conducted in an unscientific fashion that yielded little useful scientific knowledge. Probably more than anything else, these atrocities made the second of the two ethical perspectives outlined, the Kantian perspective, dominant in Western thinking about research on human subjects and made protection of persons who are research subjects a paramount priority.* As a result numerous guidelines and codes of ethics for research on human subjects have been formulated since World War II in an effort to limit treatment of human subjects to conduct that is consistent with and respectful of human dignity. (Indeed the Nuremburg Code, one of the first formal codes of ethics for research on humans, was developed by the Allies as a basis on which to judge Nazi treatment of human research subjects and as a basis for judging future research.)

These various codes differ in detail, but generally agree that several requirements must be met before research can properly be conducted on human beings (Capron 1986). First, research on humans must itself be based on laboratory research and research on animals as well as on established scientific fact "so that the point

*A proponent of utilitarianism might argue that the Nazi experience does not weigh against this ethical theory because the Nazis were badly mistaken about what in fact would promote the greatest good and may even have had no genuine concern with this end in any case. On this view, if the Nazis had been concerned with doing the greatest good and had correctly identified what would have had this result, they would not have engaged in such atrocities. But the crucial point that the world took from the Nazi "research" atrocities was that such practices would remain immoral even if the overall good they produced were overwhelmingly large.

under inquiry is well focused and has been advanced as far as possible by nonhuman means" (Capron 1986). Second, research on humans should use tests and means of observations that are reasonably believed to be able to provide the information at which the research is directed. Research methods that are not suited for providing the knowledge sought are pointless and rob the research of its scientific value. Third, research should be conducted only by persons with the relevant scientific competence. Fourth, "all foreseeable risks and reasonably probable benefits, to the subject of the investigation and to science or more broadly to society, must be carefully assessed, and . . . the comparison of these projected risks and benefits must indicate that the latter clearly outweighs the former. Moreover, the probable benefits must not be obtained through other less risky means" (Capron 1986). Fifth, participation in research should be based on informed and voluntary consent. Sixth, "participation by a subject in an experiment should be halted immediately if the subject finds continued participation undesirable or a prudent investigator has cause to believe that the experiment is likely to result in injury, disability, or death to the subject" (Capron 1986). Imposition of these conditions on research on humans probably does limit the pace and extent of medical progress, but society's insistence on them is its way of saying that the only medical progress truly worth having is that which is consistent with a high level of respect for human dignity. Of these requirements the requirement to obtain informed and voluntary consent from research subjects is widely regarded as one of the most important protections. But before discussing this requirement further, let us turn to a second important ethical issue regarding research on humans.

This issue concerns the most ancient of Western medicine's traditional moral norms, beneficence and nonmaleficence, the duty to benefit the patient and the duty to avoid harming the patient. The aim of research is to provide new knowledge, and although this does not mean that research has no benefits, it does mean that its benefits are not aimed primarily at the research subject. It also means that any harm that the subject suffers is justified, if at all, not by benefits to the subject himself but by benefits to others. So when the researcher is a physician and the subject is a patient, then the physician qua researcher is not discharging his duty to concern himself primarily with his patient's well-being. If the patient benefits, these benefits are incidental to the research and unintended, and if the patient is harmed, the harm cannot be justified on the grounds that it was an unavoidable concomitant event to providing some compensating benefit to the patient, for no such benefit is intended. Indeed if the intervention in question is genuinely an instance of research, the physician cannot have well-founded beliefs about what benefits or harms it will pose for the patient, for the testing needed to have such beliefs is precisely the substance of the research. How then can physicians or researchers reasonably intend that their actions benefit or not harm their patients? Clinicians may hope for this but cannot reasonably intend it. Can the aims of research be reconciled with the traditional moral obligations of physicians? Or is the researcher/physician in an ethically untenable position?

Answering these questions leads back to the requirement of informed and voluntary consent. The leading idea behind this requirement is that if the subject of an experiment agrees to participate in the research, then what happens to him during and because of the experiment is a product of his own decision. It is not something that is imposed on him, but rather, in a very real sense, something that he elects to have done to him. Because his autonomy is thus respected, he is not made a mere resource for the benefit of others. Although he may suffer harm for the benefit of others, he does so of his own volition, as a result of the exercise of his autonomy rather than as a result of having his autonomy limited. An extremely important difficulty here is knowing when consent is genuine, when it is truly voluntary and not the product of coercion. Not all sources of coercion are as obvious and as easy to recognize as physical violence. A subject may be coerced by fear—by the fear that she has no recourse for treatment of her ailment (even when she knows any benefit the intervention brings her will be incidental to its purpose), by the fear that she will alienate the physician on whom she depends, or even by fear of the disapproval of others. This sort of coercion, if it truly ranks as such, is often difficult to detect and remedy.

Given that genuinely voluntary consent must be uncoerced consent, an admission made by William DeVries, the surgeon who has implanted more permanent artificial hearts than anyone else and who is by far the leading defender of this technology, concerning the nature of the consent given by his patients is extremely worrisome. Although Robert Jarvik, the developer of the Jarvik 7, and others believed that the device was not ready for use in humans because it was neither totally implantable nor easily portable, DeVries was of a different view (Annas 1988). DeVries felt that use of the device was acceptable as long as the consent of the patient was obtained. Yet he described the consent he obtained from the four patients in whom he had implanted the device as coerced. George Annas (1988) quoted Lawrence Altman, a medical reporter for the *New York Times*, as reporting that

Dr. DeVries has repeatedly said that the four men in whom he has implanted artificial hearts were so coerced by their diseases that they felt death was their only alternative. In signing the 17-page consent form, each recipient, Dr. DeVries has said, "told me in their own way they didn't care" if they read it or not, and had signed it primarily because they had to [in order] to get the device.

If DeVries's truly permanent implantation of the artificial heart was to be justified on the basis of informed consent, and if consent is not genuine unless it is voluntary and hence uncoerced, DeVries's own words stand as condemnation of his work. His own account of the nature of the consent obtained from his implant patients indicates that they were coerced, although not by any human source, and thus not acting in a truly autonomous fashion.

In any case absence of coercion is not enough. Voluntariness can be compromised by ignorance and misunderstanding as well. One's consent to something is not autonomous unless one understands what he is consenting to. Accordingly a prospective research subject must be given information sufficient to arrive at an intelligent opinion concerning whether to participate in the research or to continue to participate once the research has commenced. Without such information the subject may consent to something quite different from what she believes, and her consent may, as a consequence, not reflect her own values and priorities. Two difficulties arise here. The first, and probably less worrisome, is knowing how much information is enough and what information is relevant, knowing when the subject has been rendered able to arrive at an intelligent opinion. Although a subject need not be provided with all the information that the researcher has, how much should be provided and what can be left out without depriving the subject of something relevant? The second difficulty lies in knowing whether the subject is competent to understand the information given and to render an intelligent opinion based on it. This makes research on children and the mentally handicapped especially suspect. It also makes research suspect when its subjects include patients whose ailments compromise their competence. In any case efforts must be made to ensure that sufficient relevant information is given and that the subject is sufficiently competent. These are matters of judgment that probably cannot be rendered with absolute precision (the most important ethical issues rarely can) but unless rigorous efforts are made in good faith, experimentation on humans runs a large risk of undermining human dignity.

How does the requirement of informed and voluntary consent speak to the tension between the goals of research and the traditional medical ethical obligations of physicians? Being concerned about the well-being of the patient is often crucial to prevent him from being dehumanized, that is, from being treated as a mere thing. In recent years Western societies have tended to regard respect for a person's autonomy as the most crucial aspect of respecting her humanity. That is, the Kantian notion that what makes human beings morally special is their capacity for autonomy seems to have become a widely accepted value, expressed in a strong and widespread rejection of paternalism. To treat a person paternalistically is to limit his liberty for his own good. Although many paternalistic measures would benefit those subject to them, Western societies, especially the United States, have been exceedingly reluctant to take such measures. In medicine this respect for individual autonomy has taken the form of limiting the traditional obligations placed on physicians. Increasingly patients' rights advocates and others have persuasively argued that there is no valid justification for not allowing patients to do as they please as long as they do not impose harm on innocent others. Acceptance of this Kantian point of view has meant that physicians are expected to discharge their duties of beneficence and nonmaleficence within the limits set by the autonomy of the patient.* And it has also meant that when their own consciences allow, they can legitimately set aside concern with the patient's well-being if the patient agrees. Accordingly the patient who voluntarily and with adequate information chooses to become a research subject releases the physician from her duty to make the patient's well-being her primary concern. The physician can then take on the role of researcher and attend to the goal that justifies research in the first place—benefit to humanity in general. Of course this does not mean that anything goes. The physician is still obligated to pursue the ends of the research with as little risk to the subject as possible.

Nonvalidated Practices: The Artificial Heart
In this section issues raised by nonvalidated practice are discussed. Capron (1986) has examined this topic:

What should one's ethical response be to activities that fall within this region of nonvalidated practices? Since they are not carried out according to a research plan (or "protocol"), they cannot be justified by their benefit to science; since they lack a valid basis in science, they cannot be defended for the benefit they will provide to those with whom they are used. The solu-

---

*Indeed this is why it is widely accepted that before any intervention is undertaken, no matter how well validated, the informed and voluntary consent of the patient must first be secured.

tion, of course, is to employ the procedure in the context of an appropriate research plan or to substitute a proven treatment. Yet this will not always be possible, especially when attempting to cure problems (such as life-threatening conditions) for which no satisfactory therapy exists and when the need to intervene is too urgent to allow a formal research plan to be adopted.

A nonvalidated practice thus lacks the justification available to ethically sound research, namely, benefit to science and thereby to society as a whole. Furthermore it lacks the justification available to practice, namely, benefit to the patient, precisely because there has not been sufficient study to show that such benefit can reasonably be expected. Is nonvalidated practice therefore always unjustified? Not necessarily. In circumstances where life itself is at stake or where severe debilitation is highly probable and where validated practices have failed or cannot be used, it may be morally acceptable to turn to nonvalidated practice. Of course the requirement of informed consent must be met here as well. Indeed given the desperate nature of the circumstances, informed consent may well be even more important here than in the circumstances of normal research. The issue is, Why should a patient who is in desperate circumstances, who is well apprised of the uncertainties of the intervention being contemplated, and who is competent to make his own choices be deprived of the opportunity to have his life or health saved, however uncertain that opportunity might be? (As we discuss in the following, the FDA allows nonvalidated practices in just these sorts of circumstances.) Of course the great difficulty here is knowing whether the patient in desperate circumstances really is well apprised of the relevant information and competent to decide on the basis thereof.

None of the cardiac technologies discussed in this chapter is as controversial as the artificial heart. Much of this controversy derives from the fact that the artificial heart is a technology in the process of being tested on humans. Many have high hopes for it, but at this stage data sufficient to show whether these hopes are in vain have not been obtained.

In light of the preceding discussion, it can be seen that even DeVries is confused about the status of his use of the artificial heart. On the one hand DeVries and others concerned with the artificial heart recognize that the use of this device is experimental, that is that it is primarily intended to provide information about the safety and efficacy of the device, information that will allow it to be soundly assessed and further developed, if warranted. Before implanting the Jarvik 7 artificial heart, DeVries had to comply with the same federal regulations that anyone wishing to conduct medical experiments on humans within the United States must meet. (More is said about these regulations in the following.) If nothing else this process surely must have made DeVries aware that others regarded the proposed use of the artificial heart in humans as justified not on the basis of reasonable beliefs about its value to those who would receive the implant but rather on the basis of the scientific information that would be generated. Yet in defending his implantation of the artificial heart, DeVries has consistently appealed to the dire circumstances of his subjects and has characterized the artificial heart as the only hope that was available to them.

Why is this problematic? Can't an intervention be part of a research plan and a nonvalidated practice at the same time? Clearly it can in certain circumstances. An intervention may be part of a research plan and thereby intended to provide generalizable knowledge, and yet be used outside the plan and thus on a patient who is not one of the research subjects, if that patient is in a medical crisis where validated practices are not available or ineffective. It is important to note that this sort of use of the intervention as a nonvalidated practice occurs outside the research plan and so on someone who is not a research subject. And this use has a very different sort of purpose than the use of the intervention in the context of the research plan. Whereas the latter use aims at providing generalizable medical knowledge, with benefits accruing to the research subjects being incidental to this purpose, the former use aims at providing a possibility of benefit to a patient for whom all other options are closed, and any information thereby obtained is incidental to this purpose. In this case these different primary goals can be kept separate even when the physician conducting the research is also the physician conducting the nonvalidated practice because the interventions are applied to different persons—a research subject in the former case and a patient in the latter case.

But what about the possibility of an intervention being a nonvalidated practice and part of a research plan in a meaningfully different kind of situation? Can one and the same intervention be applied to one and the same person and also be both nonvalidated practice and research? This seems to be how DeVries conceives of his uses of the artificial heart. On his view the first recipient of a Jarvik 7 artificial heart, Barney Clarke, was a patient and a research subject at the same time. In one sense this situation is unproblematic. Suppose that some interventions performed on Clarke were designed primarily to benefit him and not to yield scientific information. With respect to these interventions, whether they constituted validated or nonvalidated practice, Clarke was a patient. Suppose that other interventions used along with these were designed primarily to generate generalizable knowledge rather than to benefit Clarke. With respect to these Clarke was a research subject. The problematic case is where one and the same intervention is supposed to have two equally primary aims, benefiting the patient *and* generating scientific information. And this seems to be how DeVries has regarded his use of the artificial heart as a permanent implant.

This case is problematic for at least two reasons: First, the limits that research must observe to be scientifically sound, that is, to have a reasonable possibility of yielding generalizable knowledge, might not be compatible with providing maximal benefit to the patient. When this is the case, if the physician observes the limits required of legitimate research, he fails to do what is required of him as a physician. He does his duty as a researcher, but not as a physician. If he does his duty as a physician and violates the canons of legitimate research, he fails to do his duty as a researcher. His situation is thus ethically untenable. But if he uses the intervention only as a research instrument, treats the patient only as a subject with respect to this intervention, and forgoes any benefit to the patient that might be attainable by violating the research protocol, the informed consent of the subject releases him from any obligation to attend primarily to the subject's benefit and any obligation to limit harm to the subject to what is necessary to benefit him.

If he uses the intervention as nonvalidated practice, then his main concern must be doing what promises greatest benefit to his patient at the least risk, even though the nonvalidated nature of the intervention means that he doesn't have good grounds on which to assess the effect of the intervention. He must do whatever is necessary to serve this concern even if it means that little or no generalizable knowledge will result. If a physician cannot in good conscience stay within the requirements of the research protocol, then he ought to withdraw his patient from the research and treat him as a patient rather than as a research subject (assuming of course that the patient gives informed and voluntary consent).

The second reason this case is problematic is that it may undermine the soundness of the patient's consent. If the intervention is portrayed as research, the patient's consent must be based on the understanding that her well-being is secondary to the aims of the research. Presumably then the patient will take this risk only because she regards the benefits to science or some other nonmedical benefit as sufficiently worthwhile. The patient may hope that she will benefit medically, but cannot reasonably have the expectation of such benefit as her rationale for participating as a subject. Nor can she expect that the research protocol will be violated to secure her maximum benefit. If the intervention is portrayed as nonvalidated practice, the patient's consent must be based on the understanding that, although everything feasible will be done on her behalf, she will be subjected to an unproved technique because nothing else available has been effective and her situation is critical. In the latter case the patient consents with the understanding that her well-being is the highest priority, whereas in the former she consents with the understanding that her well-being is secondary to other considerations. What could the patient possibly think she is consenting to if she is lead to believe both that her well-being is the primary consideration and that other considerations have greater priority? Chances are that a patient who gives consent will have one or the other of these beliefs rather than both. That is, she will think that the risks to which she will be exposed are either on behalf of others or on behalf of her own welfare. The danger here is that the patient will agree to the intervention with false or confused beliefs and expectations and thus that her consent will be radically compromised.

At least two reasons can be given to explain why someone like DeVries would have trouble portraying his use of the artificial heart as research and research alone. First, as a trained surgeon all of his professional socialization inclines him to be concerned exclusively with benefit to his patients. And setting this inclination aside has to be exceedingly difficult. Second, public support for the use of the artificial heart in any given case—especially where the intervention leaves the patient very ill, subjects him to numerous unforeseen complications, and extends his life only briefly without saving it—is likely to be far greater if it is portrayed as a heroic effort in desperate circumstances than if it is portrayed as research. The benefits of research lie mainly in the future and accrue to presently unknown persons, whereas the risks and harms of research occur in the present and to identifiable persons. The public is more likely to approve of the intervention if these risks and harms are characterized as the only recourse to save a patient in desperate circumstances rather than as a means to future benefits for presently unknown persons. Thus the temptation to garner such support will likely (wittingly or unwittingly) cause researchers to want to characterize their interventions as both research and practice (albeit nonvalidated practice.)

FDA Regulations of Medical Technology: The Artificial Heart

The FDA is the primary federal mechanism for regulation of research on humans and thus for ensuring compliance with the requirements such research must meet to be scientifically sound and morally licit. It is charged with reviewing new drugs and medical devices and ensuring that they are not brought to market until their safety and efficacy have been scientifically established. First, we discuss the development of the FDA's role in the regulation of drugs and then its role in the regulation of medical devices. In 1906 Congress gave the FDA the authority to act against manufacturers and vendors of drugs that were shown to be misbranded or adulterated. "Before 1906, medicines containing narcotics such as morphine and cocaine were sold without restriction. Quack remedies—some claiming to cure cancer—and contaminated drugs were largely uncontrolled by the federal government" (Flannery 1986). But the authority given to FDA by the 1906 Food and Drugs Act was limited to taking action only after such drugs had been placed on the market. Thus the FDA was empowered to act in many cases only after harm had already come to innocent consumers. According to Flannery (1986),

The inadequacy of this approach became clear in 1937. One manufacturer mixed sulfanilamide, considered a valuable drug, with diethylglycol, a highly toxic solvent, to make a liquid preparation. The resulting "Elixir Sulfanilamide" was promptly marketed, and it caused almost 100 deaths. The manufacturer had done no safety tests before marketing. Responding to this incident, Congress strengthened the drug law in 1938 by requiring manufacturers to submit to FDA proof of a new drug's safety before putting the drug on the market.

A further tragedy, in 1962, moved Congress, through the Drug Amendments of 1962, to give the FDA even greater control over new drugs than had been given to it with the Federal Food, Drug, and Cosmetic Act of 1938 (Flannery 1986):

Thalidomide was marketed in Europe as a new sleeping pill, but it caused gross deformities in the children of women who had taken the drug while pregnant. Thalidomide was not commercially marketed in the United States because FDA blocked the new drug on grounds of inadequate evidence of safety. Nevertheless, thalidomide had been distributed to over one thousand

U.S. doctors as an investigational drug; as a result, seventeen cases of birth defects were reported in this country.*

Among the requirements placed on manufacturers of investigational drugs by the Drug Amendments of 1962 were a requirement that informed consent be obtained before engaging in human research, a requirement that controlled clinical studies be undertaken to prove the safety and efficacy of new drugs, and a requirement that no new drug be marketed without prior FDA approval.

The Federal Food, Drug, and Cosmetic Act of 1938 also gave the FDA authority to prevent adulteration and misbranding of medical devices. Congress gave the FDA this power because of the great variety of fraudulent and often dangerous medical devices being placed on the market in the late 1930s. "... for example, lead nipple shields caused lead poisoning in nursing infants and contraceptives caused genital infections. Some devices were economic frauds, such as simple boxes with colored lights that claimed to treat and cure virtually every disease" (Flannery 1986). But the 1938 Act limited the FDA to act only after devices had actually been marketed, and thus to when harm had already been done.

Not until 1976, with passage of the Medical Device Amendments, did Congress provide the FDA with both premarket review and premarket approval authority over medical devices. Again harm to innocent persons motivated Congressional action (Flannery 1986):

In 1970, the Dalkon Shield was introduced as a supposedly safe and effective contraceptive device. By mid-1975, sixteen deaths

---

*"An example of the lack of safeguards for human subjects before 1962 is found in the research involving diethylstilbestrol (DES). From 1950 through 1952, approximately 1,000 women were given DES on a medical experiment conducted by the University of Chicago and a manufacturer of the drug. The drug was administered to the women during their prenatal care in a study to determine the effectiveness of DES in preventing miscarriages. The women were not told that they were part of a medical experiment or that the drug they received was DES. Years later, in 1975 or 1976, the university notified the women that they had received DES and that cancer and reproductive tract abnormalities had been associated with it in pregnant women" (Flannery 1986).

and twenty-five miscarriages were attributed to the Shield, as well as numerous infections and removals for medical reasons. Congress cited the Dalkon Shield as one example where "[i]n the search to expand medical knowledge, new experimental approaches have sometimes been tried without adequate premarket clinical testing, quality control in material selected, or patient consent."

In response to this situation the FDA, under the Medical Device Amendments of 1976, was given regulatory powers over new medical devices according to the degree of risk. The FDA's greatest powers are with respect to implanted devices and life-supportive or life-sustaining devices. Before placing a device of this nature on the market, the manufacturer must provide the FDA with research reports establishing that the device is safe and efficacious for use on humans. This means of course that before marketing, the FDA requires that such devices be assessed in research using humans.

To gain FDA approval to conduct the research on humans required to obtain agency permission to market a new device or drug, the manufacturer or other sponsor of the research must submit a formal application to the agency (Flannery 1986):

The application must include: a description of the investigational plan; a list and description of components of any drug; a statement of methods and controls for manufacturing the drug or device; information about all prior investigations involving the drug or device, including tests on laboratory animals; a list of the investigators and a summary of their training and experience; a list designating each institution at which a part of the investigation may be conducted; copies of all labeling; copies of the forms and informational materials relating to informed consent; and assurance that the investigation has been, or will be approved by an institutional review board (IRB). The IRB must be constituted in accordance with FDA regulations and must conduct its activities in compliance with the regulations.

Submission of this application with all the required materials does not of course ensure that the proposed research will be allowed. But by acquiring an application with all the required information, the FDA can make a reasonable determination concerning the value, scientific validity, and ethical soundness of the proposed research. An important aspect of the FDA's requirements for an application to conduct research on human subjects, and to protection of those subjects, is the demand for assurance of IRB review of the proposed research. An IRB is a committee constituted by the institution authorized to conduct the research. On a typical IRB two-thirds to three-quarters of the members are biomedical and behavioral scientists.* Most of the remaining members are persons associated with the institution in some other respect. On some IRBs one or two people who are not associated with the institution are included as representatives of the community. According to Flannery (1986),

The purpose of IRB review is to ensure that risks to subjects are minimized, risks are reasonable in relation to anticipated benefits to the subject and the importance of the knowledge expected to result from the investigation, informed consent will be sought, selection of subjects is equitable, the research plans call for monitoring the data collected to assure the safety of subjects (where applicable), and mechanisms exist to protect the privacy of subjects and confidentiality of records.

In addition to approving use of nonvalidated drugs and devices for research purposes, provided that all its conditions are met, the FDA also will allow nonresearch uses of such drugs and devices, the sorts of uses designated as nonvalidated practices. The FDA will approve of nonvalidated practices in two sorts of situations: First, the FDA allows physicians to treat their patients with investigational drugs or devices when those patients have serious diseases that have not responded to validated practices. This kind of nonvalidated practice must meet two conditions: the physician in

*A very important criticism of IRBs focuses on this preponderance of professional researchers, who are usually biased toward making scientific discoveries and advancing scientific knowledge rather than toward protecting research subjects.

question must successfully apply to the FDA or join an FDA-approved research plan, and preliminary evidence must show that the intervention may be safe and effective for human use. Whether the physician applies to the FDA directly or joins an FDA approved research protocol, IRB review is involved before the use of the intervention, and so some explicit mechanism exists to protect the very vulnerable persons who will be subjected to it.*

A second sort of situation in which the FDA will allow nonvalidated practice is an unanticipated emergency (Francis 1988):

Unanticipated emergency situations may not wait for the deliberations of IRBs. To handle such crises, the FDA regulations allow a second possibility: use of an experimental article without an approved protocol on a one-time emergency basis. This use must be reported to the IRB within five days. No further use of the experimental article in question is permitted until the investigation undergoes IRB review. At this point, the investigator is aware that the experimental article may be needed for emergency use again; therefore he or she should seek IRB approval beforehand.

The FDA requires informed consent to be obtained in both these situations unless (Francis 1988)

. . . a human subject confronts a life-threatening situation necessitating use of the test article, he or she cannot give the consent,

---

*That the FDA requires the physician in question to submit an application to the agency or join an approved protocol indicates that it suffers from the same confusion attributed to DeVries. That is, it seems to want to view itself as allowing research in emergency situations that is also nonvalidated therapy in emergency situations. Research in emergency situations would test the safety and efficacy of an intervention with subjects who are in a grave condition and have not responded to available conventional treatments. The aim here would be to yield generalizable knowledge about the intervention in such circumstances. Nonvalidated practice in emergency situations would not aim at providing such generalizable knowledge but rather at benefiting the patient by means of a nonvalidated intervention because such an intervention is the only hope left. In the former case the demands of science are primary, whereas in the latter case the demands of benefit to the patient are primary. Which does the FDA regard as primary when the two sets of demands are not compatible, when what serves the demands of science in an emergency situation does not also serve the demands of patient benefit?

and there is no available alternative therapy with an equal or greater chance of saving his or her life. These conditions must be attested to by the investigator and evaluated by an independent physician if possible before the article is used and otherwise within five days.

An implantation of an artificial heart that was ultimately treated by the FDA as a legitimate case of the second situation was the implantation of the Phoenix heart on March 6, 1985 by Jack Copeland. Professor of Health Law, George Annas (1988), described what happened:

On Tuesday morning, March 5, 1985, Dr. Jack Copeland, Chief of University Medical Center's Heart Transplant Team in Tucson, Arizona, performed a human heart transplant on Thomas Creighton, a 33-year-old, divorced father of two. The procedure was not a success, as Mr. Creighton's body rejected the heart. At 3:00 A.M. Wednesday morning a search for another human heart began, and Dr. Copeland placed Mr. Creighton on a heart-lung machine. At 5:30 A.M. the medical team placed a call to Dr. Cecil Vaughn of Phoenix, asking if he had an artificial heart ready for human use. Dr. Vaughn was scheduled to implant an experimental model developed by dentist Kevin Cheng into a calf later that day, and had never considered use of the device in a human. Nonetheless, he called Dr. Cheng. Dr. Cheng told him, "It's designed for a calf and not ready for a human yet." Asked to think about it for ten minutes, Dr. Cheng recalls, "I knelt and prayed." When Vaughn called him back, he said, "The pump is sterile, ready to go." The two helicoptered from the hospital to the airport, chartered a jet to Tucson, and then took another helicopter to the Tucson hospital. They arrived at 9:30 A.M. Wednesday morning. The implant procedure began at noon. Designed for a calf, it was too large, and surgeons could not close the chest around the device. The implant maintained circulation until 11:00 P.M. that night when, in preparation for a second heart transplant, doctors turned it off and put Mr. Creighton back on the heart-lung machine. By 3:00 A.M. Thursday, Dr. Copeland completed a second human heart transplant. The next day Mr. Creighton died.

It should be noted that Copeland did discuss the possibility of using the artificial heart as a temporary measure with Creighton's mother and sister and obtained their consent before performing the implantation. He also alerted his chief of surgery and had Creighton examined by the physician transporting the artificial heart. Beyond this examination "no additional or independent assessment of the patient's condition was performed" (Francis 1988). The Phoenix heart was used because when Copeland contacted sources for artificial hearts after Creighton had suffered rejection of the donor organ, it was found to be the most readily available device.

Copeland's first line of justification for his action was that he simply had no choice if he was not simply to let his patient die. In other words Copeland felt that the choice was between life and death. Annas (1985a) offered a dissenting account:

. . . the choices were not just "live or die." The reality was closer to: "Accept the implant and you'll almost certainly die anyway; and if you do live, you could spend the rest of your life severely disabled, mentally and physically." When the possibility of a "halfway success," survival in a severely impaired state, is added to the equation, the patient has much to lose, including his self-determination and his dignity as a human being.

If Annas is correct, then Copeland's action did not merely offer his patient the possibility of life in the face of sure death but also exposed him to substantial risk of life in a severely impaired condition, a condition that the patient might find a greater evil than death. Does a physician or anyone else, including his closest relatives, have the right to impose this sort of risk on a person even in attempt to save his life? Or is this the sort of decision that only the patient should have the right to make? In any case if Annas is correct, Copeland's first line of defense may prove uncompelling.

Copeland's second line of defense was that, to save the life of his patient, no limits apply to what a physician can legitimately do in an emergency. Again Annas (1985a) offered a dissent worth notice:

The emergency rule is "treat first, and ask legal questions later." Thus an emergency situation may justify a physician's decision not to review federal regulations prior to acting, but it can never justify a physician in not considering medical data before acting. The medical "reasonableness" of using the artificial heart in a true emergency is debatable. But was this a true, unanticipated, emergency?

Organ rejection is a known risk of all transplant procedures. Mr. Creighton was Dr. Copeland's third patient to suffer immediate heart rejection. He vowed after the second to do all he could to save the next such patient. Thus, not only is organ rejection a "reasonably foreseeable risk" of transplantation, Dr. Copeland knew of this risk firsthand and had ample opportunity to develop a plan to deal with the next case he encountered. Under these circumstances organ rejection is not an unanticipated "emergency" that justifies unthought-out, extreme interventions.

Annas's comment directs two main criticisms against Copeland's appeal to emergency circumstances in defense of his use of the Phoenix heart. The first is that although emergency circumstances may relieve the physician of any obligation to be sure that his efforts to save his patient comply with the relevant legal dictates, it does not relieve him of an obligation to consider existing medical information concerning the intervention being contemplated. Annas seems to be quite correct on this. For although a nonvalidated practice is one for which sufficient testing has not been done to lead to the reasonable belief that it is safe and efficacious, nonetheless a physician who turns to such a practice, even in emergency circumstances, must base his hope that the intervention will offer some slim chance of saving the patient on some evidence. Otherwise the physician acts arbitrarily altogether and, even in an emergency, arbitrariness is hardly a morally sound course to take. If this is correct, then Copeland's claim that in an emergency a physician may do anything he believes is appropriate is simply false. Even in an emergency the physician's beliefs must be founded on consideration of available relevant evidence regardless of how far short of definitive that evidence might be.

Annas's second point is that Creighton's rejection of the first heart transplant was not a genuine case of an unanticipated emergency. Dr. Copeland surely knew, just as many laypersons now know, that organ rejection is a risk always carried by transplantation operations. Further, immediate heart rejection had occurred in the two patients Copeland had performed transplantation on before Creighton. Copeland's own vow to take every possible step to save his patient the next time immediate rejection occurred indicates that he in fact did anticipate the likelihood of this emergency occurring yet again. This indicates that Copeland could have thought through the possibility of using the artificial heart as a temporary device in case of immediate heart rejection and made the appropriate application to the FDA for permission to use an investigational device in emergency circumstances.* If he had done so and received FDA approval, he could have discussed the possibility of a temporary implant of the artificial heart with Creighton before the initial heart transplantation operation and thus have given Creighton the opportunity to offer his informed agreement or refusal.† Denying Creighton the opportunity to consent to or refuse the temporary implantation of an artificial heart in the event of the failure of transplantation is probably the most serious ethical shortcoming of Copeland's behavior.

When the FDA initially learned of this case, Copeland's conduct was termed a violation of the law. But only a few days after the implantation had occurred, according to Francis (1988), the FDA

... recommended taking no action against Dr. Copeland or the University of Arizona Medical Center. The recommendation was based on findings that the emergency was genuine and that Dr. Copeland had followed basic FDA procedures for unanticipated emergency use by notifying the IRB, consulting other clinicians, and obtaining consent from the patient's mother and sister. Of particular importance to ... [FDA's] investigatory

---

*Of course by not doing so, he avoided the possibility that his application would have been denied.

†In avoiding the possibility of FDA rejection of his use of the artificial heart, Copeland also avoided the possibility that his patient would not have consented.

findings was that "there was no evidence that the use of the artificial heart was premeditated." Also noted in the report were efforts at the University of Arizona Medical Center to establish a Bioethics Committee and formulate guidelines for use of the artificial heart in subsequent emergencies. Based on this report, the FDA decided to withhold regulatory action against those involved in the Phoenix heart case. The FDA's decision, however, cautioned the University of Arizona hospital that the FDA has procedures for the approval of protocols allowing experimentation in emergency situations and that any further use by the institution of an artificial heart must be under such an approved protocol.

The most morally troublesome aspect of FDA's position in this case is its contention that Copeland had been faced with a genuine and unique emergency. The authenticity of the emergency is beyond question. But its alleged uniqueness is highly doubtful. To hold that the emergency was unique is to hold that it was not reasonably foreseeable. But as Annas persuasively argues, much factual evidence exists that strongly suggests both that the emergency was foreseeable, indeed it was the kind of thing that can occur in any instance of organ transplantation, and that Copeland himself was well aware of it and highly motivated to do something in the event that it did in fact occur in his practice again.* Given the dangerous precedent this establishes and communicates to other physicians dealing with critically ill patients, it is very difficult to comprehend why FDA ruled as it did.†

To its credit the FDA published, in October 1985, a guidance document clarifying the regulations that cover the use of nonvalidated medical devices in emergencies. According to Francis (1988),

*This does not imply that Copeland had actually planned to use the Phoenix heart in this way, although as Leslie Francis writes, "There is evidence of earlier inquiries [before Creighton's operation] by Dr. Copeland about cardiac assist devices including artificial hearts" (Francis 1988).

†A possible reason for FDA's ruling was the media portrayal of Copeland as a heroic physician fighting desperate odds and governmental regulation to save his patient's life. If the agency unwittingly removed itself from the role of "bad guy" because of this portrayal by the media, then its effectiveness in protecting patients from misuse of nonvalidated practices must be questioned.

The document defines an emergency as a situation that is immediately life threatening for the patient, without acceptable treatment alternatives, and too imminent to allow FDA approval of
use of the device. Physicians are cautioned to use "reasonable
foresight with respect to potential emergencies" and to make
arrangements to use the regular FDA approval process whenever
possible. Patient protection is mandated, including informed
consent, review by an independent physician, institutional approval and concurrence by the IRB chair, and sponsor authorization. Under these conditions unapproved devices may be
shipped—as was the artificial heart in Arizona—without FDA
objection, provided the FDA is immediately notified. Emergency use of devices the FDA has disapproved for investigation
is prohibited entirely.

Temporary Implantation of the Artificial Heart
The use of the artificial heart as a permanent implant has been much
less successful than initially expected. In each case, although the
device did prolong the life of the recipient (some would say that a
more accurate description is that the device prolonged the dying of
each recipient), devastating adverse effects were experienced, and
the quality of life was very low.* Although DeVries has FDA
approval to make other permanent implants, the artificial heart is
now most often used as a temporary device to keep the patient alive
until a suitable donor heart can be obtained. Although many who
are skeptical about the artificial heart as a permanent implant nonetheless hail its use as a temporary device, that use too is morally
problematic in its own right. Annas has argued that use of the
artificial heart as a temporary device is both useless and unfair.

What is Annas's reason for concluding that this use of the
artificial heart has no value? Temporary use of the artificial heart
occurs in the context of a shortage of transplantable organs. If every

---

*Barney Clarke's experience was representative. "Clarke survived 112 days on
the artificial heart. During that time he underwent two major operations in
addition to the implantation and suffered from a number of other difficulties. An
artificial heart ventricle broke on the operating table, bubbles appeared in his
lungs, and a heart valve broke. Clarke had serious nosebleeds, contracted
pneumonia, suffered from kidney failure, and experienced prolonged episodes
of psychological disorientation" (Jamieson 1988).

organ medically suitable for donation were donated, supply might well exceed demand. But in fact very few people make provisions for donation in the event of premature death, and families are often reluctant to sanction donation of the organs of loved ones. Furthermore some physicians are reluctant to broach the topic of organ donation with the family so soon after the (usually) unexpected death of a loved one, typically a young person, a time when the family's trauma is greatest and, to prevent deterioration of the organs, the very time when donation would be optimal. In light of this context Annas (1985b) reached the conclusion that

. . . temporary artificial hearts cannot increase the total number of human heart transplants performed; they can only change the identity of the individuals who obtain them. For example, assume hypothetically that we have 1,000 individuals annually qualified for human heart transplants in the U.S. and that 600 human hearts are available. The remaining 400 will not receive a heart and will therefore die. If before their deaths these 400 are put on artificial hearts, the "waiting pool" will increase to 1400 the next year, although only 600 hearts will be available. If priority is given to those on artificial hearts, these 400 will get hearts, but only 200 of the 1000 remaining on the list will get them. If the remaining 800 are not to die, they will get temporary artificial hearts. But this will already outstrip the next year's supply of human hearts. Six hundred will get them, 200 will wait another year, and the new group of 1000 potential recipients will either all die, or will have to have "temporary" hearts that will actually be "permanent" for most of them.

If Annas's logic is correct (although his hypothetical numbers may not match reality), in a context where the supply of donor hearts exceeds the demand and where this shortage is likely to continue, there are compelling utilitarian objections to temporary use of the artificial heart. The temporary artificial heart cannot increase the number of heart transplantations that can be performed because it cannot increase the number of donor hearts available for transplantation. Its use does result in "jumping the queue" though and thereby changes the identity of the patients who are given transplants. That is, patients who are given temporary artificial hearts are automatically placed at the head of the waiting list for human hearts. Patients who were originally at the head of that list are thereby pushed further down the list. This of course means that everyone behind them is pushed further down the list as well. The net result is that because the next available donor heart goes to the person with a temporary implant, everyone else has to wait longer than otherwise, and inevitably someone dies who would have gotten a heart. To the fact that the total number of transplantations is unaffected by temporary implants add the fact that each temporary implant is very costly, and it becomes clear that resources are not used to maximum benefit. The same number of transplantations are done as if no temporary implants were performed, but at greater total cost. This is the substance of Annas's charge that temporary implants are useless. They fail to increase the number of lives saved and increase the total costs of performing transplantations.

His charge that the use of temporary implants is unfair is based on the fact that it changes the identity of who gets a transplant. Whatever criteria are used for determining positions on the waiting list for donor hearts, patients who are given temporary artificial hearts are allowed to escape those criteria and go to the head of the queue. If there is one point on which writers on justice agree, it is that justice means treating like cases alike and unlike cases not alike on the basis of their relevant differences and similarities. This means that whatever criteria are used for distributing donor hearts to individuals with end-stage heart disease, all must be subject to those criteria equally. To allow recipients of temporary implants to bypass these criteria is to treat like cases not alike, and that is the most fundamental form of injustice.

What conclusions should be drawn from the preceding discussion? Should one conclude that the artificial heart program ought to be opposed? Not necessarily! The considerations adduced in the preceding pages provide compelling reasons to be very skeptical of the value of the artificial heart, *in its present form*, as a therapeutic technology. Its value as a permanent implant is brought into question by the extremely disappointing results so far. To see this, one need only to compare the condition of Barney Clarke, William Schroeder, and the other recipients of the Jarvik 7 to the optimistic predictions that DeVries and others offered when the FDA first approved permanent implantation of the device. As the preceding discussion also shows, as a temporary implant the artificial heart exacerbates rather than reduces the problem its use is intended to address, problems posed by the scarcity of donor hearts. Nothing said here entails that research use of the artificial heart, aimed at discovering and realizing its ultimate therapeutic potential, should not occur. Research use of this device should be clearly distinguished, however, both in the minds of research subjects and researchers themselves, from use of the device as a form of clinical practice. The goals that justify research are different in important ways from those that justify clinical use of a technology. Furthermore, as with all research on humans, research use of the artifical heart should conform to the standards that are definitive of ethical experimentation. Precisely because adequate research on this device has not yet occurred, representation of its use as validated practice would be misrepresentation and could cause potential patients to consent to receiving implants under false pretenses with false hopes and expectations. Its use as nonvalidated practice, however, would be acceptable as long as the conditions outlined in the previous discussion of nonvalidated practice and the Phoenix heart are met.

Summary

In this chaper we have examined the development of modern cardiac technology. Particular attention has been given to the fundamental principles by which the heart operates and the ways in which medical technologies have been developed to compensate for operational abnormalities. These technologies include pacemakers (which correct for bradycardia or tachycardia), defibrillators, left ventricular assist devices (originally developed to facilitate coronary bypass operations and more recently used for long-term life support), and the permanent artificial heart (intended to replace the patient's heart). Many of these technologies have proved extremely successful in terms of medical therapies they have made possible. Moreover, in most of their uses the economic benefits they provide are considerably greater than the associated costs (as has been demonstrated in some detail in the case of pacemakers).

One exception has been the artificial heart. Although the potential benefits from the development of a validated artificial heart technology are considerable, as yet no one has succeeded in producing such a technology. Moreover, although the potential benefits associated with the development of validated artificial heart technology are large, they do not appear at this time to exceed the associated R & D and treatment costs (although a technological breakthrough could alter that calculus). These concerns raise question about the degree to which the artificial heart R & D programs should be supported with public funds or encouraged at all. The most compelling reasons to support them may be that they provide hope of a better future for many patients suffering from chronic heart disease and that they have resulted in a larger basic research program for cardiac diseases (although the latter argument is subject to challenge).

In addition to these economic considerations, difficult ethical questions have been raised by the artificial heart program. Particularly the artificial heart program shows just how important it is to clearly distinguish research from practice and validated practice from nonvalidated practice. It is vitally important that the patient/ subject know whether the intervention to which he is consenting is for research purposes or for his direct benefit. And if it is for his benefit, he should know whether the technology to be used is validated. Without such knowledge his consent cannot be regarded as genuinely informed. Similarly the physician/researcher must be aware of the differences between research, validated practice, and nonvalidated practices. The aims that justify these different types of interventions are quite different and so are the moral obligations that each imposes on the physician/researcher. Failure to fully appreciate these distinctions could result in unethical conduct by the physician/researcher and inhumane treatment of the patient/ subject.

It is also important to note that some problems have arisen with respect to the ease with which recent innovations in cardiac technologies have been disseminated (for example, in relation to the use of pacemakers). These difficulties are not inherent to the technologies themselves, as the discussion of economic issues associated with pacemakers indicates. Rather they arise because of the interplay between regulatory agencies and social structures that permits vested interests to delay the advent or use of innovations that are potentially damaging to those vested interests.

References

Akutsu, T. 1975. *Artificial Heart: Total Replacement and Partial Support.* American Elsevier Publishing Company, Inc.

Akutsu, T., and Kolff, W. J. 1958. Permanent substitutes for valves and hearts. *Transactions of the American Society for Artificial Internal Organs* 4:230-232.

Allen, P., and Lillehei, C. W. 1957. Use of induced cardiac arrest in open heart surgery: Results in 70 patients. *Minnesota Medicine* 40:672-676.

Altieri, F. D., Watson, J. T., and Taylor, K. D. 1986. Mechanical support for the failing heart. *Journal of Biomaterials and Applications* 1:106-156.

American Heart Association: Heart Facts. 1984. American Heart Association.

Amikan, S., Feldman, S., Boal, B., Riss, E., and Neufeld, H. N. 1983. Long term follow-up of patients with reused implanted pacemakers. *Pace* 6:A88.

Annas, G. 1985a (June). The Phoenix heart: What we have to lose. *Hastings Center Report* 15:15-16.

Annas, G. 1985b (Oct). No cheers for temporary artificial hearts. *Hastings Center Report* 15:17-28.

Annas, G. 1988. Death and the magic machine. In: *Organ Substitution Technology: Ethical, Legal, and Public Policy Issues.* Westview Press.

Bernhard, W. F., and LaFarge, C. G. 1969. Development and evolution of a left ventricular aortic assist device. In Hastings, F. W. (ed.): *Artificial Heart Program Conference.* U.S. Government Printing Office.

Bernhard, W. F., LaFarge, C. G., Husain, M., Yamamura, N., and Robinson, T.D. 1970. Physiologic observations during partial and total left heart bypass. *Journal of Thoracic and Cardiovascular Surgery* 60:807.

Blumenthat, D., and Zeckhauser, R. J. 1984. The artificial heart: An economic issue. In Shaw, M. W.: *After Barney Clark*. University of Texas Press.

Bryson, F. E. 1974. Countdown for the artificial heart. *Machine Design* 46:34-42.

Capron, A. 1978. Human experimentation: Basic issues. In *The Encyclopedia of Bioethics*. Vol. II. The Free Press.

Capron, A. 1986. Human experimentation. In Childress, J. F., King, P. A., Rothenberg, K. H., Wadlington, W. J., and Gaare, R. D. (eds): *BioLaw*. University Publications of America.

Chardack, W. M., Gage, A. A., and Greatbatch, W. 1960. A transistorized self-contained implantable pacemaker for the long term correction of complete heart block. *Surgery* 48:643-654.

DeSilva, R. A., Graboys, T. B., Podrid, P. J., and Lown, B. 1980. Cardioversion and defibrillation. *American Heart Journal* 100:881-895.

DeVries, W. 1986. The artificial heart. In: *Technology and Medicine*. Trinity College Press, pp. 50-57.

DeVries, W. C., Anderson, J. L., Joyce, L., Anderson, F. L., Hammond, E. H., Jarvik, R. K., and Kolff, W. J. 1984. Clinical use of the total artificial heart. *New England Journal of Medicine* 310:273-278.

Falk, R. H., Zoll, P. M., and Zoll, R. H. 1983. Safety and efficacy of noninvasive cardiac pacing. *New England Journal of Medicine* 309:1166-1168.

Flannery, E. J. 1986 (Feb). Should it be easier or harder to use unapproved drugs and devices? *Hastings Center Report* 16:17-23.

Francis, L. P. 1988. Legitimate emergencies, experimentation, and scientific civil disobedience. In *Organ Substitution Technology: Ethical, Legal, and Public Policy Issues*. Westview Press.

Fried, C. 1978. Human experimentation: Philosophical issues. In *The Encyclopedia of Bioethics*. Vol. II. The Free Press.

Faivre, G., and Dodinot, B. 1983. Prevention of fraud and abuse in pacemaker therapy. *Pace* 6:A89.

Galioto, F. M. 1986. Cardiovascular assist and monitoring devices. In *Biomedical Engineering: Basic Concepts and Instrumentation*. PWS, pp. 79-134.

Geddes, L. A., and Hamlin, R. 1983. The first human heart defibrillation. *American Journal of Cardiology* 52:403-405.

Gott, V. L., and Klopp, E. H. 1974. Cardiovascular prosthetics and mechanical assistance. In Ray, C. D. (ed.): *Medical Engineering*. Year Book Medical Publishers, Inc.

Harmison, L. T. 1972. The totally implantable artificial heart. *DHEW*. Pub. No. NIH 74-191. U.S. Government Printing Office.

Hays, C. V. 1964. The development of the heart pacemaker. *Rose Technic Magazine* 10 and 32-33.

Hershgold, E. J., Kwan-Gett, C. S., Kawai, J., and Rowley, K. 1972. Hemostasis, coagulation and the total artificial heart. *Transactions of the American Society for Artificial Internal Organs* 18:181-185.

Holcomb, W. G., Glenn, W. L., and Sato, G. 1969. A demand radiofrequency cardiac pacemaker. *Medical and Biological Engineering* 7:493-499.

Hyman, A. S. 1932. Resuscitation of the stopped heart by cardiac surgery. II. Experimental use of an artificial pacemaker. *Archives Internal Medicine* 50:283-305.

Jamieson, D. 1988. The artificial heart: Reevaluating the investment. In *Organ Substitution Technology: Ethical, Legal, and Public Policy Issues*. Westview Press.

Judson, J. P., Glenn, W. L., and Holcomb, W. G. 1967. Cardiac pacemakers: Principles and practices. *Journal of Surgical Review* 7:527-544.

Kantrowitz, A. F., Gradel, O., and Akutsu, T. 1965. The auxiliary ventricle. *IEEE International Convention Record* 12.

Kerber, R. E., Jensen, S. R., Gascho, J. A., Grayzel, J., Hoy, R., and Kennedy, J. 1983. Determinants of defibrillation: Prospective analysis of 183 patients. *American Journal of Cardiology* 53:739-749.

Kirkendall, W. M., Feinleib, M., Freis, E. D., and Mark, A. L. 1980. Recommendations for human blood pressure determination by sphygmomanometers. *Circulation* 66:1146A-1155A.

Kolff, J., Olsen, D. B., and Kolff, W. J. 1975. The mechanical heart on the medical horizon. *Cardiovascular Diseases, Bulletin of the Texas Heart Institute* 2:265-272.

Kolff, W. J. 1975. Update on artificial organs. *Cardiovascular Diseases, Bulletin of the Texas Heart Institute* 2:273-284.

Kusserow, B. K. 1958. A permanently indwelling intracorporeal blood pump to substitute for cardiac function. *Transactions of the American Society for Artificial Internal Organs* 4:227-230.

Lavin, J. H. 1985. Almost here: The Jarvik-8. *Cardiology World News* 1:22-23.

Lindgren, N. 1965 (Sept). The artificial heart—Exemplar of medical engineering enterprise. *IEEE Spectrum* 2:67-83.

Lopman, A., Langer, C. L., Furman, S., and Escher, D. J. W. 1981. A fifteen-year study of cardiac pacing costs. *Pace* 4:241.

Ludmer, P. L., and Goldschlager, N. 1984. Cardiac pacing in the 1980s. *New England Journal of Medicine* 311:1671-1680.

LV assist systems progressing toward full portable device. 1985. *Cardiology World News* 1:14.

McGoon, D. C. 1982. Long term effects of prosthetic materials. *American Journal of Cardiology* 50:621-630.

Meyers, G. H. 1986 (Nov-Dec). Is reuse financially worthwhile? *Pace* 9(part II):1288-1294.

Mirowski, M., Reid, P. R., Mower, M. M., et al. 1980. Termination of malignant ventricular arrhythmias with an implanted automatic defibrillator in human beings. *New England Journal of Medicine* 303:322-324.

Mujica, J., DuCorge, R., and Henry, L. 1986. Survival and mortality in 3701 pacemaker patients: Arguments in favor of pacemaker reuse. *Pace* 9:1282-1287.

Norman, J. C. 1975a. The artificial heart: Perspectives, prospects and problems of a high applied technology. *Cardiovascular Diseases, Bulletin of the Texas Heart Institute* 2:259-264.

Norman, J. C. 1975b. An intracorporeal (abdominal) left ventricular assist device XXII: Precis and state of the art. *Cardiovascular Diseases, Bulletin of the Texas Heart Institute* 2:425-437.

Norman, J. C., and Huffman, F. N. 1974. Nuclear-fueled circulatory support systems III. In Russek, H. I. (ed.): *Cardiac Disease: New Concepts in Diagnoses and Therapy*. University Park Press.

Owens, B. B. 1986. *Batteries for Implanted Biomedical Devices*. Plenum.

Pierce, W. S., Parr, G. U. S., Myers, J. L., Dae, W. E., Jr., Bull, A. P., and Waldhausen, J. A. 1981. Ventricular-assist pumping in patients with cardiogenic shock after cardiac operations. *New England Journal of Medicine* 305:1606-1610.

Pringle, R. A., Leman, R. B., Dratz, J. N., and Gillette, P. A. 1986. An argument for pacemaker reuse: Pacemaker mortality in 169 patients over ten years. *Pace* 9:1295-1298.

Redlich, F. C. 1978. Medical ethics under national socialism. In *The Encyclopedia of Bioethics*. Vol. III. The Free Press.

Roe, B. 1969. Whole-body perfusion with heart-lung machines. In Burford, T. H., and Ferguson, T. B. (eds.): *Cardiovascuhar Surgery: Current Practice*. Vol. 1. C. V. Mosby.

Scherlis, L., and Dembo, D. H. 1988. Long-term follow-ups of patients with reused implanted pacemakers. *Pace* 6:A88

Sutton, R., Perrins, J., and Citron, P. 1980. Physiological cardiac pacing. *PACE* 3:207-219.

Tacker, W. A., Galioto, W. M., Jr., Giuliani, E., Geddes, L. A., and McNamara, D. G. 1974. Energy dose for human transchest electrical defibrillation. *New England Journal of Medicine* 290:214-215.

Tarjan, P. P., and Gold, R. D. 1988. Implantable medical electrical devices. In Kline, J. (ed.): *Handbook of Biomedical Engineering*. Academic Press, pp. 123-152.

Thevenet, R. P., Hodges, C., and Lillehei, C. W. 1958. The use of a myocardial electrode inserted Percutaneously for control of complete Atrioventricular block by an artificial pacemaker. *Diseases of the Chest* 34:621-631.

Watkins, J., Richings, M., Watkins, F., and Gribbin, B. 1983. Cost analysis of a regional pacemaking service. *Pace* 6:A88.

Weirich, W. L., Gott, V. L., and Lillehei, C. W. 1957. The treatment of complete heart block by the combined use of a myocardial electrode and an artificial pacemaker. *Surgical Forum* 8:360-363.

Zegelman, M., Kreuzer, J., and Wagner, R. 1986. Ambulatory pacemaker surgery: Medical and economical advantages. *Pace* 9: 1299-1303.

5

Critical Care Medicine

Many of the technological innovations that enable health care professionals to sustain life are highly visible in the intensive care units (ICUs) of contemporary hospitals. What may be less obvious, however, is that the presence of this technology has precipitated a revolution in the training of nurses and other staff who use these devices to attend to the needs of the critically ill. Consequently the availability of new medical technologies used by more highly skilled personnel in these critical care units has led both to improvements in the care of the critically ill and sharp increases in the costs of that care. Furthermore the ability to sustain life has improved to such an extent that now it is possible to support a patient's vital cardiac and pulmonary functions for extremely long periods of time. Even a patient classified as "brain dead" can be kept "alive," in a clinical sense, for many years. The capabilities provided to clinicians by these technical innovations have thereby forced our society to reconsider the very nature of life and death and to confront seriously the ethical dilemmas that surround questions about the quality of life and the definition of death

The various technologies found in modern ICUs and the economic and ethical issues associated with their use constitute the subject matter of this chapter. Although ICUs are devoted to a wide variety of physical disorders, our primary concern here is with cardiopulmonary ICUs which are designed for the monitoring and support of patients with insufficient heart and lung capacity. Because the functioning of the heart has been described in some detail in chapter 4, the first objective in this chapter is to provide a description of the physiology of the pulmonary system to permit a better understanding of the purpose of different technologies used to monitor and sustain cardiopulmonary function. These technologies include respiratory therapy, which involves devices such as the ventilator and extracorporeal membrane oxygenator. The second objective is to examine the historical development and current status of cardiac and pulmonary monitoring and resuscitation techniques, as well as the computer systems that integrate cardio-pulmonary data in the ICU. The third objective is to explore the various economic issues surrounding the allocation of resources to care for patients in the ICU. Finally, we discuss the ethical issues concerning the definition of death and euthanasia.

Pulmonary Physiology

The reduction in mortality of patients with cardiovascular disease is related directly to the availability of modern ICUs and resuscitation techniques (Callahan and Bahn 1974). Although these units were initially established for patients with heart disease, today many specialized units exist to meet the special needs of critically ill patients with a variety of problems. This chapter is largely concerned with those units in which monitoring of patients with cardiovascular problems—that is, cardiac and pulmonary function disorders—occurs.

As functioning units the lung and heart are usually considered a single complex organ; but because they essentially contain two compartments (one for blood and one for air), the tests conducted to evaluate heart function and pulmonary ventilation are usually separated. In this section we present the basic principles of operation of the pulmonary system itself and the biomedical devices that are available to evaluate lung performance and provide the necessary respiratory functions required when patients can no longer carry out these functions by themselves.

The air we breathe (inspiration) travels through a treelike arrangement of bronchial passages that end in the minute sacs called alveoli. A gaseous exchange takes place at the alveolar surface, which consists of a mesh of fine blood capillaries. Oxygen is exchanged for some of the blood's carbon dioxide. For this exchange to occur, the partial pressure of oxygen (pO_2) in the alveoli must be higher than in the venous blood of the alveolar capillaries, and the partial pressure of carbon dioxide (pCO_2) in the alveoli must also be less than in venous blood.

The gases present in the lungs are oxygen, carbon dioxide, nitrogen, and water vapor. Their partial pressures are normally 104, 40,

549, and 47 mmHg, respectively. These pressures represent average values for a resting healthy man at sea level, and the sum of all the partial pressures must be equal to the total pressure at sea level (760 mmHg). To meet the demands of the body, feedback mechanisms regulate ventilation to keep each of the partial pressures at or near the standard levels. Thus alveolar gas can be thought of as a compartment of gas lying between atmospheric air and alveolar capillary blood. Oxygen and carbon dioxide are continually exchanged with the blood flowing through the alveolar capillaries. Oxygen is supplied and carbon dioxide removed by the cyclic process of ventilation—the inspiration of fresh air followed by the expiration of some alveolar gas.

The gaseous exchange occurs in a surprisingly short period of time (less than 0.3 s) because the capillary network has so many branches that it offers remarkably little resistance to the blood. In fact the 80 ml of blood that enters the lung from the right ventricle during each heartbeat are already returning to the left atrium by the time the heart is ready to experience the next contraction. The 80-ml volume of blood, however, is distributed over approximately 50 to 100 square meters of respiratory surface in 300 million alveoli and terminal bronchiole.

The lungs, viewed as an entire unit, consist of elastic sacs within the airtight barrel (thorax) of the chest. The thorax itself is bounded by the ribs and the diaphragm, and any movement of these two boundaries will usually alter the volume of the lungs. The normal breathing cycle in humans is accomplished by the active contractions of the inspiratory muscles, which enlarge the thorax. This lowers intrathoracic and intrapleural pressure even further, pulls on the lungs, and enlarges the alveolar ducts and bronchiole, which in turn expands the alveolar gas and decreases its pressure below atmospheric levels. As a result air at atmospheric pressure flows easily into the nose, mouth, and trachea.

Tidal volume (or total ventilation) is normally considered to be the volume of air entering the nose and mouth with each breath (or each minute) when a person is at rest, but can be much larger. On the other hand alveolar ventilation is the useful volume of fresh air that actually enters the alveoli during this time and is therefore always less than total ventilation. The extent of the difference in the two volumes depends primarily on the anatomical dead space. Anatomical dead space refers to the 150- to 160-ml internal volume of the conducting airway passages of the trachea, bronchi, and bronchiole. The term *dead* is quite appropriate because there is no gas exchange across the thick walls of those passages. Because normal tidal volume is usually about 500 ml of air per breath, the presence of this dead space allows only about 340 to 350 ml to actually penetrate the alveoli and become involved in the gas exchange process. The lungs always contain some gas; that is, there is always a specific volume present. Residual volume, for example, is the amount of gas remaining at the end of maximal expiration. There is also a volume of gas that can be added to the normal inspiratory or expiratory cycle. For example, inspiratory reserve volume is the maximum volume of gas that can be inspired from the end of inspiration position, whereas expiratory reserve volume is the maximum amount expired from the end of expiration position. The relations between all of these volumes are shown in figure 5.1.

Because the lungs are elastic, they are capable of processing much larger volumes of air; hence the term *lung capacity* (or *compliance*) has been used to define specific inspiratory or expiratory events. All these events represent the storage capability of the lungs. Consider the following terms as a guideline (Comroe 1965, West 1974):

1. *Total lung capacity* (TLC)—the amount of gas contained in the lung at the end of maximal inspiration
2. *Forced vital capacity* (FVC)—the maximal volume of gas that can be forcefully expelled after maximal expiration
3. *Inspiratory capacity* (IC)—the maximal volume of gas that can be inspired from the resting expiratory level
4. *Functional residual capacity* (FRC)—the volume of gas remaining after a normal expiration

These volumes and specific capacities represent characteristics that enable us to quantify the status of the pulmonary system.

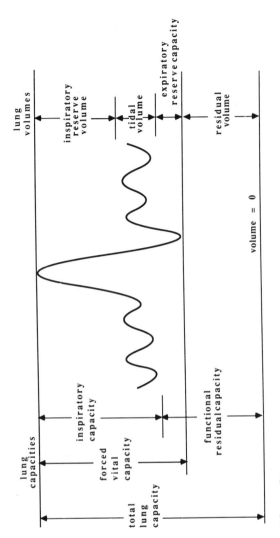

Figure 5.1
Lung volumes and lung capacities

Once these characteristics of the lungs are defined, tests can be designed to provide clinicians with information to help diagnose the presence and nature of any respiratory dysfunction (for example, inadequate air flow). These tests provide clues that enable the physician to determine if there is a physiologic reason for inadequate air flow (for example, when air passages are physically blocked with mucus or by bronchoconstriction), inadequate gas exchange (for example, when the alveoli are affected as in emphysema), or inadequate blood supply. These efficiencies may occur singly or in any combination. In addition some of these conditions affect the total quantity as well as the distribution of the air a person breathes. Others may affect the frequency of breathing and even blood pressure and chemistry. Thus to establish a complete diagnostic protocol regarding the presence or, one hopes, absence of these conditions, it is necessary to conduct a number of tests (Ray 1974) to investigate the status of both the air and blood components of the cardiopulmonary system. Furthermore if patients are unable to sustain respiration, then biomedical devices must be used to assist them.

Respiratory Therapy

The increasing incidence of respiratory and cardiopulmonary diseases has led to the evolution of the technical specialty called respiratory therapy (Leggitt 1984). At least 20 percent of all patients seen within the general hospital require respiratory treatment or support (Egan 1977, Burford and George 1984). Because new instrumentation and treatment modalities are continually being developed, physicians and nurses have come to rely on the assistance of the respiratory therapist, a skilled technical specialist (Burford 1984). Typical services provided by a respiratory therapy department include

1. Administration of oxygen to patients who cannot maintain adequate oxygen levels in their blood when breathing air
2. Administration of humidified air or oxygen to alleviate a variety

of respiratory symptoms and to maintain adequate moisture levels in the patient's airways

3. Administration of bronchodilator medication in aerosol form to reverse breathing difficulty resulting from airway obstruction caused by bronchitis, bronchospasm, or excessive secretion in the airways

4. Performance of chest physical therapy, or postural drainage, to break up and remove secretions and mucus from the lungs. This is usually done in conjunction with bronchodilator therapy.

5. Development of pulmonary rehabilitation programs for disabled patients. This includes training patients in breathing exercises and providing supplemental oxygen to allow ambulation.

6. Performance of pulmonary function testing. In many hospitals the pulmonary function laboratory is part of the respiratory therapy department.

7. Mechanical ventilation of patients who are unable to breathe on their own

To accomplish these tasks, a wide variety of instrumentation is required. For example, when patients are mechanically ventilated, the air they receive is blended with oxygen to deliver the prescribed oxygen level. This gas is then heated to body temperature and 100-percent humidified before being delivered to the patient. At the same time the patient may also be receiving aerosol medication through the ventilator circuit. Also, to provide an optimal response of the patient's physiologic systems, ventilator controls must be set properly for tidal volume, respiration rate, airway pressure limits, and the ratio of inspiratory time to expiratory time (I/E ratio). The patient's response is determined by monitoring such physiologic parameters as arterial blood gases, mean airway pressure, heart rate, arterial and central venous blood pressures, and cardiac output (McPherson 1977, Taykor 1978).

One of the most complex pieces of equipment used by the therapist is the ventilator. There are three categories of ventilators: the negative pressure ventilator, the positive pressure ventilator, and the high-frequency jet ventilator (Chatburn 1984, Egan 1977,

McPherson 1977, Alderson 1984). The negative pressure ventilator forces air into the patient's lungs by creating a negative pressure around the patient's chest. This negative external pressure forces the expansion of the thoracic cavity and air rushes into the stretched lungs. The negative pressure ventilator has been called a tank ventilator or the "iron lung" because of its physical characteristics. It was used extensively in the fifties to support polio victims, but presently is used only on a limited basis. For example, it is still used to ventilate patients suffering from neuromuscular diseases who cannot perform the muscular maneuvers required to expand their chest. Because patients suffering this type of nervous dysfunction usually have normal lungs, they do not require airway suctioning to remove secretions. Further because it is not necessary to "intubate" the patient, the lung tissues and airways are usually spared the trauma associated with other therapies (Alderson and Warren 1984). The major drawback to the negative pressure ventilator, however, is that access to the patient is very limited.

The positive pressure ventilator, on the other hand, provides ventilation by applying a high-pressure gas at the entrance to the patient's lungs. If the pressure gradient is sufficient to overcome the resistive and compliant forces of the lungs and thorax, the gas flows down the pressure gradient and into the patient. The high-frequency jet ventilator is still considered a research tool. This device delivers very rapid low volume bursts of air to the lungs. In the process the delivery of oxygen to the lungs and the removal of carbon dioxide is accomplished primarily by simple diffusion of molecules of gas. The advantage of using the jet ventilator is the overall reduction of positive pressure in the airways and the subsequent reduction of barotrauma to the lungs (Chatburn 1984).

The positive pressure ventilator is by far the most widely used. Its operation, however, requires that the patient be intubated (that is, a tube is inserted into the trachea, and the other end is attached to the ventilator breathing circuit). Despite this obvious drawback it does provide a greater degree of flexibility in ventilating the patient than other respiration techniques provide. Moreover it allows total access to the patient. Although a wide range of pressure ventilators is available, they all have certain common features. For example, most units provide for oxygen delivery in the range of 20 percent to 100 percent and include safety features such as high-pressure and low-pressure alarms, patient disconnect alarms, and power loss alarms. To quickly alert the staff of a problem, these alarms are usually both visual and auditory. Other common control features include permitting the therapist to set the tidal volume, the respiration rate, or the I/E ratio, the maximum airway pressure limit, and the baseline pressure (and expiratory pressure).

Most positive pressure ventilators operate in one of two modes, although in some instances a combination of the two modes is used. The first mode is the control mode, or time cycle mode, in which the ventilator is in total control and breathes for the patient at fixed time intervals that are determined by the control settings. Essentially the patient is not allowed to breathe on his own. The second mode is the assist mode, or the patient-cycled mode. In this mode the patient can initiate a breath by inhaling from the ventilator circuit. The negative pressure from the maneuver is sensed by the ventilator, and the breath cycle is begun. Thus although the ventilator is still breathing for the patient, it is the patient who initiates the breath cycle. Many ventilators will operate in an assist/control mode in which the patient is allowed to initiate the breath cycle. However, if the patient does not trigger the ventilator within a specified time interval, the ventilator will initiate the breath.

Once a breath is initiated by the ventilator, a method is needed for controlling the inspiratory phase of the cycle. Otherwise air could be forced into a patient until the lungs rupture. There are several methods of controlling, or limiting, the inspiratory phase. If the inspiration is terminated when a predetermined airway pressure is reached, the ventilator is pressure limited. If the inspiratory phase ends after a specific volume is delivered, it is volume limited. Finally, if the inspiratory cycle is determined strictly by timing, the unit is said to be time limited. Usually a time- or volume-limited ventilator will also provide pressure limiting as a safety feature.

Ventilators are powered either electrically or pneumatically. Typically pneumatically powered units require a source of compressed air or oxygen at a pressure of 35 to 55 psi to function properly. An electrically powered ventilator will deliver pressurized gas with a small rotary compressor or with a motorized piston. Also, independent of the power source, a ventilator can be controlled either electronically or pneumatically. The most recent generation of ventilators incorporates microprocessor control.

Modern ventilators require an external source of compressed air or oxygen to function. For this reason the use of these units is normally limited to the hospital. Recent concern about escalating hospital costs, however, has spurred the development of small, low-cost portable ventilators that can be used in the home. The ventilator illustrated in figure 5.2 is one such device. The unit is electronically powered and electronically controlled. An internal battery provides power when the unit is unplugged, and there is also a provision for use with a larger external battery. The small size of the ventilator and the battery capability make the unit portable because it can be easily mounted on a wheelchair. Although the practice of treating chronic ventilator patients at home has been gaining popularity, it requires a high degree of coordination between the hospital, the physician, the respiratory therapist, and the patient's family.

Extracorporeal Membrane Oxygenation

Positive pressure ventilation may have adverse effects on other physiologic systems. It can also produce permanent lung damage resulting from the prolonged use of high positive pressure. Consequently an alternative therapy was developed in the 1970s (Bartlett et al. 1977, Wetmore et al. 1979, Zapol et al. 1979). This therapy essentially used the technology developed for the heart/lung bypass during open heart operations. This procedure requires that blood be removed from the patient, passed through an artificial lung for oxygenization and removal of carbon dioxide, warmed to body temperature in a heat exchanger and then returned to the patient. As blood is pumped through the circuit the functions of the lungs and the heart are replaced by an artificial organ system. The procedure

Figure 5.2
Puritan-Bennett model 280 portable ventilator. (Courtesy of Children's Hospital National Medical Center, Washington, DC)

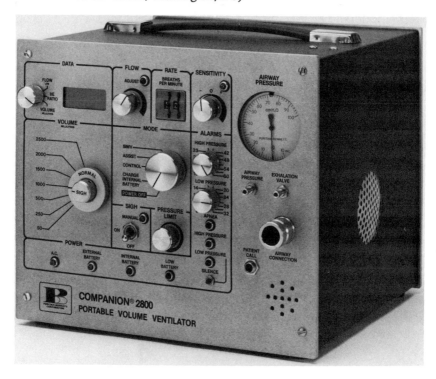

had been well defined and successful for short-term use, that is for the several hours required for an open heart operation. Further refinement of this system was needed to make it suitable for long-term cardiopulmonary support. Initial attempts at long-term support resulted in excessive damage to the blood cells in the bubbler oxygenator. With the development of a silicone membrane lung, the damage to blood cells was minimized, and long-term cardiopulmonary support became viable. In this artificial lung the blood is separated from the oxygen source by a thin silicone membrane (Zapol et al. 1977). As the blood flows through the membrane lung, the gases diffuse across the membrane. Following the partial pressure gradient, the blood is oxygenated and the carbon dioxide removed. With the use of the membrane lung, the procedure is now referred to as extracorporeal membrane oxygenation, or ECMO.

Initially ECMO was used in both adults and children with cardiopulmonary failure. Early criteria required that the patient have reversible cardiopulmonary disease and that the patient be unresponsive to conventional positive pressure ventilation and drug therapies. ECMO did not alter the probability of survival in adult patients (Zapol et al. 1977). This may be due to underlying sepsis in these patients that limited their response not only to conventional ventilator therapy but also to ECMO (Browdie et al. 1977). The most dramatic results have been obtained with newborn infants with reversible pulmonary disease (German et al. 1980, Hardesty et al. 1981, Kirkpatrick et al. 1983).

ECMO therapy oxygenates the patient's blood outside of the body, eliminating the need for pulmonary ventilation. While on ECMO the patient's lungs are allowed to rest and heal themselves. During this time ventilator settings are decreased to minimum settings to maintain open airways and to aid in the removal of secretions. Within two to three days the lungs will have healed, and the patient is then weaned from the ECMO system. Typically when infants are placed on ECMO, they have less than a 20-percent chance of survival with conventional therapy. If they did survive with conventional therapy, they would most likely have a chronic pulmonary disability. In contrast current survival rate for infants receiving ECMO is about 75 percent, and most of these patients leave the hospital as normal, healthy infants.

These results are quite dramatic, but it must be remembered that there is significant risk involved with the procedure. Complications such as intracranial hemorrhage, embolism formation, and sepsis can result in death. Also excessive bleeding may occur as a result of using anticoagulants in the system. The success of ECMO therapy can be attributed to the development of strict "entrance criteria" for the patient. To qualify for ECMO therapy, the infant must have reversible lung disease and must demonstrate little or no response to high-oxygen positive pressure ventilation for a specified time period. Also the patient must not have been on ventilator therapy long enough to sustain permanent lung damage. Therefore the patient must be identified early after birth and a timely decision to initiate ECMO must be made. Infants with sepsis, damage to the central nervous system, or other debilitating conditions are not considered for ECMO therapy because the presence of additional medical complications will guarantee a failure (Howard 1986).

A typical ECMO circuit is illustrated in figure 5.3. It consists of a rotary pump to push blood through the external circuit and into the patient, a membrane oxygenator for gas exchange, a heat exchanger and heater to warm the blood to body temperature, and a flowmeter to measure the blood flow through the circuit. The venous cannula empties into a small reservoir before entering the pump. Because the venous blood flow is determined only by a hydrostatic pressure gradient, it is necessary to provide servo-regulation for the pump. If the venous return cannot keep up with the pump flow, the blood in the reservoir will decrease. The servo mechanism (dubbed Venous Return Monitor at Children's Hospital) will then shut off the pump until the reservoir is filled. This eliminates the possibility of the pump pulling a negative pressure on the venous side of the circuit.

As of 1989 approximately ten medical centers in the United States have active ECMO programs. However, due to the demonstrated success with ECMO, many more medical centers are now looking at ECMO as a viable program.

Figure 5.3
Sketch of ECMO system used at Children's Hospital National Medical Center. (Courtesy of Children's Hospital National Medical Center, Washington, DC)

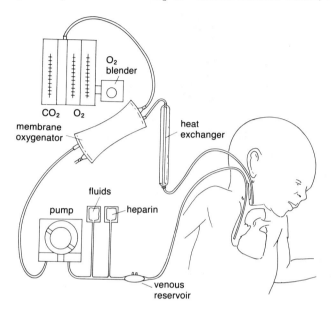

Figure 5.4
ECMO system used at Children's Hospital National Medical Center

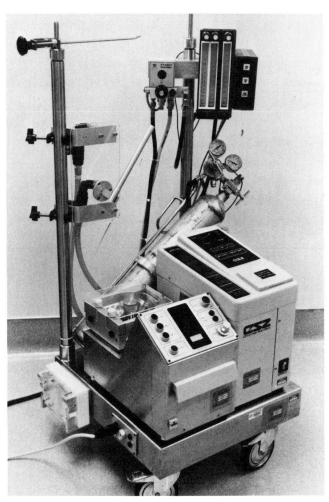

Essential Components of Patient Monitoring

Any patient monitoring system has a number of essential components, which may be divided into five categories (Terdiman 1974): sensor, signal conditioner, display device, logic module, and controller (figure 5.5).

Sensor The properties of various electrodes and transducers used to detect physiologic signals are well known (Geddes and Baker 1968). The ones most commonly used in physiologic monitoring systems include contact electrodes for electrocardiography and electroencephalography, intravascular and respiratory pressure transducers, ultrasonic and electromagnetic flow-rate detectors, and chemical and pH electrodes. The output of each of these sensors is an electrical signal that is directly related to the measured physiologic parameter.

Signal conditioner The electrical signals available at each sensor often must be amplified and shaped before they can be transmitted for review by the clinician. The signal conditioner extracts the needed information from the input signal and conditions it for display, recording, or analysis. The signal conditioner usually consists of a number of electrical devices such as detectors, amplifiers, filters, and other signal-transformation devices.

Display device The display unit receives an electrical signal from the signal-conditioning module and converts it into an appropriate display. Various types of displays include dial indicators, digital displays, cathode ray tubes, signal lights, audible alarm, and hard-copy displays. The latter may include strip-chart recorders, photographs, microfilm, and even computer printouts (figure 5.6).

Logic module (computer) The logic module may be used simply to compare a signal with a reference level or to perform more complex analyses. A common example of the former is a cardiac monitor that compares heart rate to preset upper and lower limits. In this case specialized digital electronic circuits can be used. However, if it is desired to include computation of specific parameters, such as cardiac output, or to analyze specific components of each electrocardiogram (ECG), then a digital computer is used.

Figure 5.5
Schematic of physiological monitoring system showing basic components and signal-low pathways. Pathway (a) contains a human interaction in the feedback loop. Pathway (b) contains an electromechanical controller in the feedback loop.

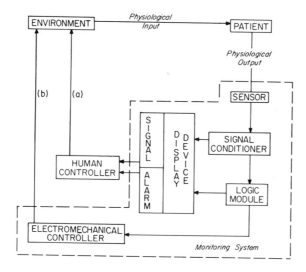

Figure 5.6
Computerized patient monitoring systems permit nursing staff at the central station to keep track of patients' clinical status at all times.

The extent of the hardware and software needed to perform a particular analysis depends on the nature and complexity of the analysis required.

Controller The type of control used, if any, also depends on the application. A passive control system, for example, provides for human interaction in the maintenance of a clinical state. It usually features visual or auditory alarms (or both) that are triggered whenever a physiologic signal falls outside a preset range. When an alarm occurs, the intended response is that a human observer will bring the physiologic signal back within its normal range by exerting some direct or indirect controlling action on the physiologic system, for example, by administering drugs or instituting other therapeutic measures.

With active control there is usually a direct feedback path between the logic module and a controller of the physiologic input. With an infusion pump it is possible to use a digital computer to convert the measured urinary output and central venous pressure to a signal that may be used to activate the motor of an infusion pump and thus maintain a specified fluid balance.

Any monitoring system can receive one or more inputs through its sensors and may produce a variety of outputs for display. A minimal monitoring system must contain a sensor, a signal conditioner, and a display device. Exactly how the patient monitoring systems are used in the hospital depends on the nature of the illness and the "hospital unit" that handles the particular condition (Sheppard 1979). In general patient monitoring equipment is used in one of the following three areas

1. *Diagnostic* Clinical area where unstable patient status require monitoring. Examples of this type of care include emergency room, special radiologic procedures area, and the cardiac catheterization lab.

2. *Treatment* Clinical areas such as the operating room or delivery room where specific invasive or noninvasive procedures designed to improve patient status are performed, but that carry the risk of status deterioration.

3. *Surveillance* Clinical areas where patients are monitored during stabilization and recuperative periods. Surgical, medical, coronary, and neonatal intensive care units are the most well-known areas. Intermediate care or "step-down" units also fall into that category because patients have progressed through critical stages and are not necessarily bedridden, but still require monitoring of the heart rate on an ambulatory telemetric basis (figure 5.7).

The patient monitoring equipment used in these critical areas involves primarily bedside monitoring stations and some sort of centralized facility, usually termed the nursing (or central) station. The standard bedside monitoring station is a compact and simply structured unit that provides a quick and comprehensive picture of monitored patient parameters, such as the heart rate and ECG (Sondak et al. 1979). The bedside monitors usually contain fairly small oscilloscope (CRT) screens that display the critical values and waveforms of each patient. In the nursing or central station the primary goal is to achieve multiple-patient supervision. As a result a number of bedside monitors are usually grouped together in an array to enable the nurse at the central station to monitor ECGs from several patients simultaneously.

In recent years the "modular concept" has been widely accepted and implemented in patient monitoring equipment, permitting clinicians to select a variety of physiologic monitoring units for each patient. In the process it has been possible for them to observe the dynamic characteristics of specific parameters for each patient in their care. Many modern units are compact, have responsive display capabilities, and permit printouts of each parameter. In some units alarm systems are available to alert the nurse or other observers by audible or visible signals in the event of a critical condition (for example, whenever the heart rate falls below 60 or exceeds 150 beats per minute). In others electric circuits have been provided to indicate that an electrode has become disconnected or that a mechanical failure has occurred somewhere else in the monitoring system. Although these systems are not without problems (noise and movement artifacts or false triggering of alarm circuits), they have been successful in detecting the presence of

Figure 5.7
Some computerized systems even provide data logging and trend displays at the bedside. (Courtesy of Hewlett-Packard, Waltham, Massachusetts)

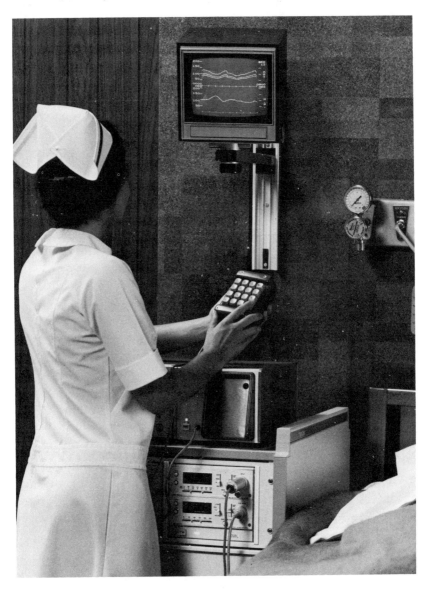

life-threatening events. With the information made available by these systems, better care can be given to critically ill patients. With this in mind let us examine the operation of automated patient monitoring systems in the ICU.

Intensive Care Units

ICUs have been designed for the management of patients experiencing a very broad range of medical problems, which may include trauma, respiratory insufficiency, coronary episodes, and renal failure. These units are characterized by specialized medical, nursing, and technical personnel, as well as appropriate monitoring devices, to increase both the efficiency and competence of medical care during periods of life-threatening illness. Any monitoring equipment must therefore provide responsible health professionals with relevant information regarding the patient's condition so that appropriate decisions can be made as quickly as possible.

Continuous monitoring of heart rate and rhythm, for example, is essential for all intensive care patients. Therefore disposable electrodes with adhesive on a foam backing are usually used and are intended to be left in place for days without causing skin irritation. These electrodes are interfaced with a cardiac monitoring system that has appropriate display and alarm capabilities.

In the ICU it is important for the physician and intensive care nurse to be able to monitor a series of parameters including (1) ECG, (2) blood pressure from an arterial catheter, (3) respiration from transthoracic impedance changes, (4) temperature from a probe placed either into the nose or on the skin, and (5) the partial pressure of carbon dioxide at the airway to assess ventilation and cardiac output. Cardiac output measurements are usually determined using pulmonary blood flow. This usually can be done without inserting any instrument into the body but by simply measuring the disappearance of an inert gas from a rebreathing bag. The removal of the gas corresponds to pulmonary perfusion and thus blood flow. Although the rebreathing approach is accurate, it requires some patient cooperation.

In most cases cardiac output is determined by use of an indwelling pulmonary arterial catheter and the use of thermodilution. In 1980 Swan and Ganz introduced balloon-tipped catheters for use at the bedside. These plastic catheters are usually introduced into a major vein directly. They are swept along by the blood into the right atrium, through the tricuspid valve, into the right ventricle, and out the pulmonary valve into the pulmonary artery. The catheters are very flexible, and as a result the manipulation required is minimal and X-ray visualization by fluoroscopy is not needed. The catheter will usually float out to a "wedge" position. Because the lumen of the catheter is open at its end and the balloon occludes the vessel further back, the pressure sensed in this position by the catheter is that of the left atrium as reflected in the pulmonary veins. This pressure can be displayed on the monitor and analyzed for signs of left heart failure. When the balloon is deflected, the pressure wave is then that of the pulmonary artery so that valuable information about the status of the right heart can be obtained (figure 5.8).

To measure cardiac output by thermodilution, that is, heat distribution, the Swan-Ganz catheter is adapted by placing a thermistor (which measures temperature) near the end so that it will be in the pulmonary artery in the patient. A second opening is created 20 cm from the tip so that it will be in the right atrium. The patient has a small amount of room temperature or iced saline or dextrose solution injected through the right atrial opening. As the cold fluid passes the thermistor, the temperature drops abruptly, then progressively rises as each heartbeat carries it away. This time/temperature curve is recorded, and the area under the curve is obtained and translated into flow of blood per minute, the standard way of stating cardiac output. In practice a microprocessor-controlled device is connected to the thermistor, and it computes and displays the cardiac output almost immediately after the injection has been completed. An important addition to the ICU is the central computer (the basic principles of which are discussed in detail in chapter 6). This computer is used for continuous surveillance of all parameters, interpretation of the data produced, and appropriate warning of potentially dangerous situations (Feldman et al. 1972).

Figure 5.8a
Three views of the Swan-Ganz catheter: the standard catheter with its four ports, a close-up of the inflated balloon, and a cross-section of the four lumens and the thermistor wires inside the catheter

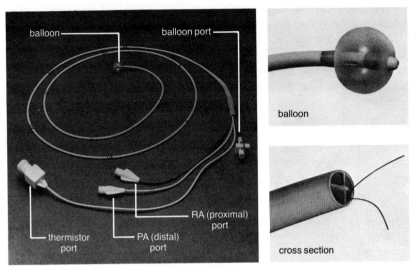

Figure 5.8b
Right atrial pressure. The catheter enters the right atrium (RA) of the heart through the superior or inferior vena cava. Normal RA pressures range from 1 to 6 mmHg. They reflect mean RA filling (diastolic) pressure (equivalent to central venous pressure) and right ventricular (RV) end-diastolic pressure (pressure at the end of the filling cycle, just before contraction).

A rise in RA pressure may signal RV failure, volume overload, tricuspid valve stenosis or regurgitation, constrictive pericarditis, left ventricular failure, pulmonary hypertension, cardiac tamponade, or RV infarction. (Redrawn from *RN Magazine*, September 1983.)

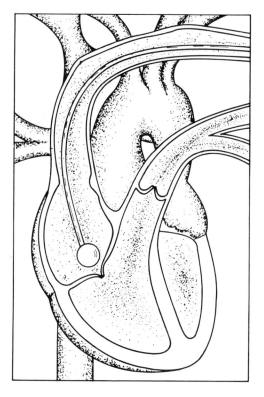

Figure 5.8c
Right ventricular pressure. After the catheter enters the right atrium, the balloon is inflated, and the catheter follows the blood flow through the tricuspid valve into the right ventricle. Right ventricular pressures normally range from 15 to 25 mmHg systolic and from 0 to 8 mmHg diastolic. (Redrawn from *RN Magazine*, September 1983.)

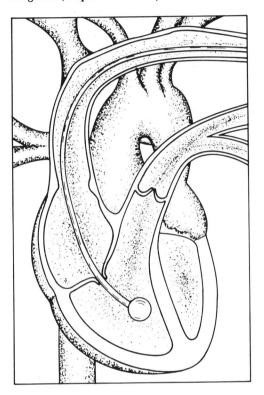

Figure 5.8d
Pulmonary arterial pressure. From the right ventricle the catheter passes through the pulmonary semilunar valve into the pulmonary artery (PA). Normal PA pressures approximate the venous pressure within the lungs and the mean filling pressure of the left atrium and ventricle. They also reflect right ventricular (RV) function because, in the absence of pulmonary stenosis, PA systolic pressure usually equals 1 RV systolic pressure. PA pressure will increase if there is left-heart dysfunction, increase pulmonary blood flow (left or right shunting, as in atrial or ventricular septal defects), or any increase in pulmonary arteriolar resistance (as in pulmonary hypertension, hypoxia, volume overload, or mitral stenosis). (Redrawn from *RN Magazine*, September 1983.)

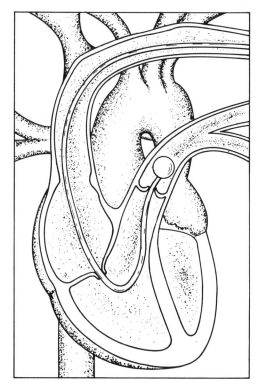

Figure 5.8e
Pulmonary capillary wedge pressure. When the balloon is inflated, the catheter floats to and wedges in a branch of the pulmonary artery, and a pulmonary capillary wedge (PCW) pressure is recorded. Once the PCW pressure has been recorded, the balloon must be deflated; the pulmonary arterial (PA) tracing then reappears.

Normal PCWP ranges from 6 to 12 mmHg. In most patients with no significant pulmonary vascular disease, PCW pressure equals PA diastolic pressure. Even nondiseased patients, however, commonly show a 3 to 5 mmHg gradient betwen PCW and PA diastolic pressures; a gradient of more than 5 mmHg is considered abnormal. Because PCW pressure is usually equal to left atrial pressures, it is a good index of left-heart function. The PCW value rises with left ventricular failure, mitral stenosis, or mitral insufficiency. (Redrawn from *RN Magazine*, September 1983.)

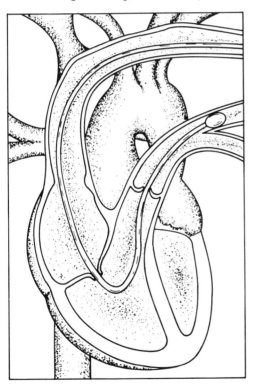

Noncomputerized monitoring systems have inherent problems. First of all, it is difficult for humans to continuously and reliably monitor. "Human monitors"—nursing and paramedic personnel whose primary assignment is to unblinkingly stare at ECG tracings on an oscilloscope—have been shown to miss a significant percentage of events (Romhilt et al. 1973, Lindsay and Bruckner 1975, Friedman and Gustafson 1977). Fatigue and distractions usually contribute to difficulties humans have in meeting multiple-task responsiblities, of which the detection of important events is but one.

Second, noncomputerized systems lack the capability to perform quantitative analyses and present the data in a variety of formats. Present-day hard-wired monitoring instruments, for example, have no storage capacity, thereby making trend analysis of any of the patient's physiologic parameters impossible. And yet such data are essential if the clinician is to ascertain what therapy to use and to evaluate the results of that therapy.

Third, automatic logging and data recall are not provided by noncomputerized monitoring systems. Significant events are recorded solely by manual charting. With such systems it is not possible to set priority alarm conditions according to previous events or recall and display past data conveniently and easily.

Finally, noncomputerized systems have routinely been found deficient, unable to detect and reject noisy signals. Artifacts—whether from misplaced electrodes, power line voltage functions, patient movement, or whatever—are at best a nuisance and at worst a severe deterrent to accurate monitoring. Even electronic devices that usually contain circuitry to detect and filter out some of these extraneous electrical signals are not completely successful. Human monitors need external aids to assist tham in assessing the presence of these electrical anomalies.

Automated monitoring systems overcome several of these problems and provide much assistance to the health professionals involved in the management of critical illness. The experience acquired in recent years through computerized monitoring systems, however, clearly indicates that effective automated systems

1. Provide continuous and accurate monitoring and display of relevant physiologic data

2. Provide rapid computerization of data for clinical review and action

3. Facilitate the documentation of clinical observations and the logging of events

4. Aid in the selection of treatments

Automated patient monitoring systems are now available that provide contiuous (or intermittent, depending on the rate of change of the physiologic parameter being monitored) monitoring of many input data channels. They are capable of providing patient-related data (such as temperature, blood pressure, ECG, and the like) quickly for clinical review. Because a computerized monitoring unit requires no sleep or coffee break, its use actually increases the attention paid to each patient. This implies not that these systems should replace the appropriate skilled health professionals but that they are able to extend the physical capabilities of the attending medical staff and thereby remove a significant burden from their shoulders.

One of the primary advantages of computerized systems is certainly their speed of computation. Once the data are supplied to the automated system, the computer can be programmed to perform the required calculations in a relatively short time—much faster than it would take the staff to go through the same computational process. With the calculations completed, the health professionals are provided with the results either in tabulated form or as graphic output, thereby enabling them to evaluate the relative status of the patient's condition. It must be understood, however, that these systems only do what they are instructed to do. Algorithms to accomplish a specific task must be developed and the systems properly programmed to ensure that the process of assimilating the input data is medically correct and that the end result is displayed meaningfully (Bronzino 1977, 1982).

This last point requires a little elaboration. In the clinical environment a visual display (that is, cathode-ray oscilloscope or integrative display terminal) is the most effective means of providing patient-related information. The storage and recall facilities of modern computers display trends in the various physiologic measures being monitored and provide this information in a graphic format that can in many cases facilitate the decision-making process. On-line operation of the computer enables this type of information to be displayed in real time (that is, as the events actually occur). In this way a plot of the measurements of interest (such as blood pressure, temperature, and so on) can be presented to the health professionals for their evaluation and interpretation. Thus the visual display is an essential means of communicating changes in a patient's condition to the physician or nurse responsible for that patient's care.

However, monitoring is only a small part of the role played by automated systems. Documentation of clinical observations and logging of events are also of great importance. For example, several kinds of observations and numerous types of events may be entered at variable frequency in the course of a patient's critical illness. It is often necessary that these observations and events be classified adequately and presented chronologically in a readable and didactic form. How often this is done is determined primarily by the type of illness and the condition of the patient. Computerized systems can provide this information efficiently and easily.

The continuous survey for arrhythmias (that is, ventricular fibrillation or premature ventricular contractions) is viable today only because of the availability of computerized systems. A decade ago automated systems for detecting and classifying arrhythmias were still on the drawing board. However, the detection, recording, storage, and alarm in the presence of true arrhythmias are presently available and extensively used (Green 1983). In a typical application, if the patient is suspected of having a cardiac rhythmic disturbance, a patient is admitted to the ICU, is affixed with appropriate sensors, and is monitored via a computer programmed to watch for problems. The computer can be instructed to analyze the patient's ECG, and if premature ventricular beating occurs, the physician can be alerted as to the frequency and site of origin within the heart.

Similarly the blood pressures from the arterial line and the Swan-Ganz catheter can be plotted, and any abnormal trends promptly detected. Currently this type of computer can be programmed to assess all monitored functions; all laboratory data including blood pH, gases, and chemistry; and given information about medication; and show the effect of all therapy without delay. An abnormal situation can be identified promptly, and appropriate action taken.

In essence the concept behind the use of automated equipment in critical care situations is similar to that which enables humans to fly jet aircraft. In both cases the human is an integral part of a feedback system and is supplied with information regarding specific levels of performance. In both cases automated systems handle some of the work by monitoring many parameters, performing the necessary calculations, and providing the human interpreter with information so that specific corrections or modifications can be made. The ability of health professionals to integrate the displayed information and select appropriate courses of action is the key factor in ensuring that patients receive immediate and proper treatment.

Economic Issues

There is little disagreement with the idea that ICUs are necessary.* The economic questions that surround ICUs concern, rather, the degree to which they are needed or should be used. For example, which patients should be put in which intensive care settings and for how long? How many ICUs should a hospital have? Should intensive care be given to patients who are not critically ill simply because such hospital beds are available? Do some hospitals overemphasize the use of ICUs to pad revenues and profits by switching patients into higher-paying diagnosis-related group (DRG) classifications? What types of machines and nursing professionals

*There is, however, one British study of cardiac patients in which patients not receiving intensive care after myocardial infarction fared better than those who were admitted to an ICU (Cochrane 1972). The study randomly allocated patients in the sample to different types of care and therefore controlled for bias in the type of patient put into the different therapies.

are necessary to provide appropriate levels of care? To what degree is the sophistication of a hospital's ICU used as a marketing tool to attract patients in profit-making hospitals? Each of these questions is important and affects the costs associated with the use of ICUs and the nation's total health care bill.

Clearly ICUs are used to provide monitoring and support for patients in need of critical care in circumstances where the status of the patient may change dramatically from moment to moment. The benefits of such care are that small changes in the patient's condition that might portend the onset of a critical or life-threatening condition can be identified and cared for by appropriate health care personnel. Although ICUs as such did not exist until the 1950s (Russell 1979), once the concept was developed, it was adopted quite rapidly by hospitals in the United States. Today virtually all hospitals of any significant size (that is, in excess of 100 beds) have at least one ICU with at least four beds.*

The Cost of Intensive Care

ICUs are expensive to operate relative to other types of hospital wards on a cost per patient-bed basis. Studies carried out by Russell (1979) and Wagner and colleagues (1983) clearly indicate that on average daily patient care in ICUs costs three to four times more than does regular hospital health care, although the actual difference in the cost of care for a specific patient depends on the type of case (Wagner et al. 1983). Surprisingly the main cause of the high cost of intensive care is not the sophisticated monitoring and resuscitation equipment that is required. Most of that equipment is relatively cheap. For example, a state-of-the-art ventilator, a relatively expensive piece of ICU technology, costs between $12,000 and $15,000. It has a working life of several years, and therefore the equipment charge amounts to less than $20 per day, less than 10 percent of the charge for normal ward care in a hospital. *The major sources of increased costs are the higher staffing requirements*

---

*Russell has noted that an ICU requires at least four beds to be "economically efficient." Some very small hospitals have ICUs with fewer than four beds because the ICU is intended only to provide short-term care in emergencies.

associated with ICU monitoring and testing procedures (Wagner et al. 1983). These consist of additional hours of physicians' services, nursing, and specialized services provided by radiation therapists, X-ray specialists, and so forth. If, for example, a four-bed ICU required the services of one additional full-time nurse over the course of a year relative to a four-bed ward, the annual cost of that nurse to the hospital would amount to approximately $50,000 (assuming a yearly salary of $25,000 and allowances for benefits such as health insurance, retirement, and training programs that typically more than double the total cost of the worker). This sort of outlay for personnel dwarfs the expenditures by the hospital on equipment used in the ICU.

Why Are There So Many Intensive Care Units?

If, as is generally the case, it is taken as axiomatic that ICUs are necessary, then the questions outlined at the beginning of this section come into play. The first concerns how many ICUs a hospital requires. This question is determined in part by the size of the hospital and in part by the mix of patient "caseloads" it handles. Other things being equal, the larger the size of the hospital, the more ICUs it requires. In her study of the rate at which hospitals developed ICUs in the 1960s and early 1970s, Russell found that larger hospitals introduced intensive care sooner than did smaller hospitals. Also the number of ICU beds in a larger hospital represented a greater proportion of all beds than in a smaller hospital. There are at least two reasons for this: First, there are economies of scale in operating ICUs. Per bed a four-bed ICU is much more expensive to operate than an eight-bed ICU. Thus the larger the hospital (up to a point), the cheaper ICU beds become. Second, larger hospitals are more likely to be involved in critical care specialties (for example, heart transplantation and coronary bypass procedures) that require ICUs for pre- and postoperative care.

The patient case mix with which a hospital deals also affects the number of ICUs it has. Evidence on this point is also provided by Russell. She found that hospitals with a teaching function had more

ICUs and acquired ICUs more rapidly than hospitals of the same bed size that did not have such a function. A major reason for this is that teaching hospitals deal with a more diverse patient population than do other hospitals because of the diversity of clinical expertise in such institutions.

Recognizing that different hospitals have different functions and therefore different needs with respect to ICUs is one thing. It does not answer the question as to whether, in the aggregate, there are too many or too few ICUs in the U.S. health care system. In 1982 approximately 5 percent of the 1,360,000 hospital beds available for patient care in the United States were located in ICUs. Thus ICUs account for approximately 15 percent of all expenditures on hospital care (Russell 1979, Jennett 1986). In Britain by contrast only about 1 percent of all hospital beds are located in ICUs, accounting for only 3 or 4 percent of hospital care expenditures (Jennett 1986). The proportion of all beds allocated to ICU beds is also lower in most European countries than in the United States (Jennett 1986). The differences do not necessarily imply that there are too many ICUs in the United States. But the data do give one pause for thought.

Some of the explanations for the relatively large investment in critical care in the United States are as follows. First, it is argued that U.S. government health care policy is responsible. As was pointed out in chapter 2, the federal tax structure exempts employer-funded health care benefits from income tax. This in turn increases the level and scope of health insurance for many U.S. citizens, making the price of health care to those users close to zero. Moreover, until the past few years, the Medicare and Medicaid programs also provided coverage for almost any procedure that could be regarded as providing some hope of patient benefits, thereby encouraging heavy use and almost certainly, from an economic welfare perspective, overuse and increased costs. One category of health care that often yields only small increases in life expectancy and quality of life is the ICU (Jennett 1986).

A second argument used to support the view that hospitals have too many ICUs concerns the type of patient placed in them. Patients whose lives otherwise could not be sustained often are kept alive in the ICUs at a relatively high cost to patients' families or the government (and the taxpayer) or other users (through implicit subsidies embedded in hospital charging practices). Venal profit motives are often attributed to hospitals by some people who comment on this issue. However, once a patient is placed in an intensive care environment, the legal position of a hospital is extremely tenuous with respect to the removal of life support systems. The quality of life for the patient may be very low, but society as a whole has not determined any legal criteria for defining what constitutes an unacceptable quality of life. Although this ethical question is discussed in detail in the next section of this chapter, it is worth pointing out here that government policy as defined through court decisions about such issues provides incentives for the extended use of ICUs to maintain patients who would otherwise die.

A third reason put forward to account for the large complement of ICUs in the typical U.S. hospital is simply that American physicians are trained to regard state-of-the-art medicine as the only acceptable form of patient care. Russell found evidence indicating that a larger number of ICUs were to be found in hospitals of the same bed size with larger rather than smaller numbers of physicians on staff. This finding could be interpreted as support for the thesis that physicians are enamored with high-tech medicine, including ICUs. Finally, given that most hospitals are private (rather than government) institutions, investment in ICUs (as in other high-technology therapies) is carried by hospitals to demonstrate to potential customers that the hospitals are able to provide them with state-of-the-art health care. To some extent therefore ICUs represent a marketing tool for institutions (hospitals) that sell their commodities (different types of health care) in the marketplace. This argument is closely related to issues concerning the demand for health care. If in fact patients use only hospitals with up-to-date technologies, then the hospitals are simply meeting the (dollar backed) needs or wants of the patients. In the most basic terms patients in the United States demand large amounts of intensive care. Whether they would demand such large amounts of ICU care without government subsidies to the price of that care is another matter, one we have already discussed, and the answer is No.

Do Hospitals Overuse Intensive Care Units?

Two other economic issues must be addressed before we examine the ethical concerns that surround the use of ICUs. The first is the question of the inappropriate and perhaps even fraudulent use of ICUs. It is often asserted that a patient is classified as requiring ICU monitoring simply because the hospital has an unoccupied ICU bed that is too expensive to be left empty. Thus at the outset of treatment, the patient may deliberately be diagnosed incorrectly by the hospital staff and the patient or the relevant insurance agencies subsequently charged for unnecessary services. Given the current widespread use of DRGs by insurance companies and Medicare and Medicaid agencies, that type of action now would be considered illegal in most circumstances. (It has always been immoral.) Moreover, since the advent of the use of DRGs in the mid-1970s, insurance companies have had a vested interest in checking up on hospital diagnostic practices using their own expert physicians. Consequently the likelihood of such abuses has probably diminished. Any precise assessment of how widespread such practices have been or still are is impossible. That such practices do occur is brought home to the United States at large by television news programs, such as "60 Minutes" and "57th Street," at least once a year. No real evidence exists as to the size of the costs they impose on the patients, insurance companies, and taxpayers who underwrite the costs of the health care system.

A second question concerns what constitutes appropriate technology in the ICU. For example, the major issues regarding the capabilities of new ventilators available for use in hospitals have been identified by Spearman and Saunders (1987). They state that the current generation of ventilators is capable of sophisticated monitoring that provides huge amounts of data. Much of the data, however, is never used and, outside of a research setting, is of little help to the health care professionals monitoring the patient. In addition many state-of-the-art ventilators have the capability of delivering ventilation at a rate far in excess of what would ever be needed for therapy.

The complex nature of the ventilators appears to create two problems for the hospital. First, the ventilators are more sophisticated than the hospital requires, and the sophistication comes with a high equipment price. In other words hospitals may find themselves buying more expensive "Cadillac" ventilators instead of cheaper "Volkswagen" ventilators that would serve them just as well. Second, the more sophisticated ventilators require more extensively trained personnel for their operation. Thus the more sophisticated technologies do not necessarily reduce staffing requirements. The upshot may be that in this case technological innovation results in higher-cost but poorer-quality health care. The potential for adverse effects on the quality of care arises from the increased risk of human error in the interpretation of more complicated data from the more sophisticated machine.

Who if anyone is accountable for this phenomenon? One set of candidates is the manufacturers of the equipment. In a market of limited size served by a relatively small number of suppliers (an oligopoly), one means by which the suppliers implicitly collude to exploit potential market power to raise the effective price of their product is by unnecessarily increasing the quality of the product. A well-known example of this phenomenon was the change in the quality (and price) of the average Japanese automobile imported to the United States after the imposition of voluntary import quotas in 1982. A second group may be physicians who have become overly attracted to high-tech machinery. By and large, final purchase decisions regarding medical equipment are made by clinicians. Finally, it may well be that by obtaining expensive state-of-the-art equipment, hospitals are simply responding to the demands of their clients, the patients. Patients may want to receive the best possible health care, especially given that most of them do not have to pay any significant part of the price of the health care they receive but are free to select among competing hospitals for the delivery of that care.

The extensive use of ICUs in U.S. hospitals has been criticized because ICUs are three to four times more expensive to maintain and operate than nonintensive care units. Moreover the health care services of other countries, whose populations appear no less healthy, allocate a much smaller proportion of their total hospital beds to ICUs. Consequently some observers have argued that ICU beds should be rationed on a nonprice basis in the U.S. health care system.

Rationing, or even banning, the use of health technologies has often been proposed as a means of limiting health care expenditures. In fact both types of actions have been implemented in the United States and abroad. In the United States, for example, drugs cannot be sold until they have been approved for use by the FDA—clearly a ban (albeit temporary). In Britain dialysis for end-stage renal disease is only provided by the National Health Service to patients who are both under 65 and in otherwise good health (unless there are truly exceptional circumstances)—clearly a policy of rationing. Although both types of programs may be considered appropriate, even optimal, social policies, they generate costs as well as benefits. Although the FDA's procedures have prevented many drugs with dangerous adverse effects from being improperly used, they are also often accused of causing unnecessary pain, suffering, and even death by delaying the advent of new treatments in the United States. Similarly Britain's dialysis rationing policy has helped to keep health care costs under control, but it has also resulted in many patients dying much sooner than they might have if given dialysis.

Limiting the availability of ICU beds through rationing in the health care system would probably reduce health care costs, but such a policy would also have some adverse effects on the quality of health care. It would certainly create a number of difficult problems. First, if the number of ICU beds has to be limited, the question arises as to what constitutes the "right" number of beds. Currently 5 percent of all hospital beds in the United States are located in ICUs, whereas in other countries with sophisticated health care systems the figure is as low as 1 percent. Should four out of five ICU beds be converted into ward beds, or only two out of five? And who is to decide which hospitals should give up how many ICUs? A natural response is that an appropriate regulatory agency could make these decisions. Unfortunately the creation of such a regulatory agency would represent an extension of the power and activities of government. Right or wrong, many people look with great disfavor on any increase in government activity, if only because they view expansion of government decision making as restricting individual freedom. Even if the responsibility is taken up by an existing agency, it would still represent a broadening of powers and be subject to similar criticism.

A second problem with imposing rationing for any commodity is that it leads to wasted resources because of what is called "rent seeking." Hospital administrators must believe they (or the people they represent) benefit from owning the ICUs they operate: otherwise they would not continue to operate them. Reducing the number of ICU beds, especially if it involves removing all ICUs from some hospitals, therefore involves removing assets that are of value to the owners and managers of hospitals. Many hospital administrators would be willing to spend resources that, in the limit, are as large as the net benefits that are derived from owning the ICUs to prevent their loss. Typically, if government decision makers are perceived to be serious about their nonprice rationing efforts, these resources will be expended on lobbying to ensure that others are forced to lose their ICUs. Of course competitors will also lobby for the same purpose, resulting in a merry-go-round of lobbying or rent seeking, which can be very costly to both the rent seekers and society.

A third problem with rationing is that it is often inefficient. Consider a rationing scheme that is based on permits that are allocated by a regulatory agency. Under this scheme many hospitals would lose some of their ICU beds or even complete units. There is no guarantee, however, that the ICU beds that are lost will be those of least value to the health care system. In fact if the allocation of permits is made on a once-and-for-all basis, it is bound to become inefficient over time, even if it did not start out that way. The reason can perhaps be seen most clearly through a simple example. Suppose hospitals A and B each serve populations of

30,000 and are each allocated 20 ICU beds. Over time the population served by hospital A grows, whereas hospital B draws patients from a declining population (as people near B move to the area served by A). Clearly the ICU beds in A will become more valuable as the demand for their use will increase (probably in proportion to the increase in population). The opposite will be the case for hospital B. Thus even if the initial allocation of ICU beds was efficient in that beds in both hospitals yielded equal net benefits, over time it would become inefficient because ICUs in hospital A would become more valuable than ICUs in hospital B. The obvious solution, from a social welfare perspective, is to reallocate ICU beds from hospital B to hospital A. However, that is easier said than done. Although hospital A may enthusiastically support such a switch (after all it stands to gain additional revenues and profits from the adjustment), hospital B will resist (because the consequence will be lost revenues and perhaps lost profits). Moreover the community served by hospital B will also oppose the reallocation if the consequence is perceived to be a cutback in the quality of health care services available to its members.

One way of circumventing this inefficiency problem is to use "marketable permits." Instead of allowing a hospital to have a fixed number of ICU beds, the regulatory agency could allocate a quota right for the same number of ICU facilities to the hospital. The hospital would then be free either to use the ICU permits to operate the ICU beds itself or to sell some or all of them to other hospitals. The total number of marketable permits issued to all hospitals would equal the number of ICU beds the regulatory authorities believed would be optimal for the system. The permits could then be traded freely across hospitals. Those hospitals that value ICU permits highly (because of demand pressures) could then acquire them from other hospitals that find the ICU permits less valuable.

The price at which the ICU permits trade would be determined by market forces, in just the same way that prices for New York taxi medallions and some types of air pollution rights are determined. Moreover the permit market would allocate ICU beds to those hospitals in which they yield the highest net benefits. When an increase in the number of ICU beds is perceived to be justified, the regulatory authority could simply create an additional supply of new permits. These permits could either be given away, probably to existing hospitals on the basis of existing patient loads or ICU beds, or alternatively offered for sale to the highest bidder in the existing permit market. The latter approach has the attractive feature (for the government and taxpayers) of raising revenues for the regulatory authority. The former approach would simply provide hospitals with a valuable asset at no cost; that is, it would redistribute wealth toward the owners of health care facilities.

A caveat must be added here. If a new issue of ICU permits resulted in a reduction in permit prices (which it well might), then hospitals that had acquired permits in the past would now hold an asset whose market value has depreciated. They could legitimately claim to have been damaged by the new issue. However, if the government were to announce at the outset of the program that it would sell new permits—say, on an annual basis in amounts systematically linked to population growth, changes in the age distribution of the population, and changes in per capita real incomes—then the effects of these sales on permit prices would be negligible. The reason is quite simple. If buyers of existing permits knew that sales of new permits would take place in the future, they would adjust the prices they were willing to offer for existing permits before those new permit issues accordingly. In other words the market would take account of the effects of the new issues on permit prices before they occurred.

In this discussion of rationing ICUs on a quantity basis, we have assumed that reducing the number of beds located in ICUs in the United States is a valid objective. But what is the right number of beds? From an economic welfare perspective, the "right" number of ICUs may be determined as follows: The last ICU bed should yield social benefits (including the direct benefits for the patient) that are just equal to the social costs of creating and operating the facility. There are at least two a priori grounds for believing that, along with every other medical service, there are too many ICUs.

First, the price of health care is subsidized for third-party group insurees by the tax system (as discussed in chapter 2). Second, it is subsidized completely, or almost completely, for Medicare and Medicaid patients (again, see chapter 2). Thus at the margin a large number of users pay less for health care than it costs to produce. The private benefits derived from such care are often less than the social costs of that care. If it is true that the elderly use a large proportion of ICU care and that that care is (for the most part) funded under Medicaid, this problem is likely to be relatively severe in the case of ICUs. In other words it is quite possible that there are too many ICUs. (Of course if society believes that the elderly should receive extensive subsidies for health care in their old age and that social benefits from such care exceed the value placed on that care by the recipients in terms of their willingness to pay for it, then this conclusion would not hold.) But what applies to ICUs also applies to many other types of health care, including coronary bypass operations, prenatal nutrition programs, and heart transplantations. If indeed the health care system is overused, it might be better to get rid of the subsidies that lead to its overuse than to solve the problem by centralized or decentralized rationing schemes. However, if rationing schemes are to be used, it is almost certain that they will be less costly if they are decentralized and based on marketable permit arrangements.

Ethical Issues

The technologies used in critical care units include a wide variety of supportive and resuscitative devices. These devices have provided health care professionals with the ability to sustain life in circumstances where, until recently, death would have been imminent. However, it is essential to ask if this recently acquired ability to sustain life should always be exercised. That is, should health care providers act to forestall death as long as possible in *all* patients? The answer to this question can be Yes only if the assumption is made that life is always valuable, that there are no circumstances where continued life is not something of value in its own right. To pose this question is to raise a longstanding and extremely difficult issue in medical ethics, namely, the moral justifiability of euthanasia.

An act is an instance of euthanasia if and only if it hastens an person's death for that person's own good. The clause *for that person's own good* is crucial. Only when the act of hastening a person's death benefits him does it qualify as an instance of euthanasia. It is this aspect of euthanasia (whether it is ever justifiable or not) that distinguishes it from other acts that hasten a person's death (such as murder). If life is always valuable, then to forestall death is, in every case, to act in a manner that benefits the person whose life is extended. Given the longstanding medical ethical duty of beneficence—the obligation of physicians always to choose actions that benefit the patient—if life is always valuable, then health care professionals are morally obligated to use their power to sustain life whenever possible. If they did otherwise, they would fail to be beneficent.

Furthermore if life is always valuable, the very notion that a person can be benefited by having his death hastened is nonsense. To hold that life is always valuable is to regard death always as a harm, probably the greatest of all harms. On this view nothing can be of value to a person unless that person is alive. Accordingly on this view no deprivation can be worse than death, thus a person can suffer no greater loss than the loss of life. Because an act that hastens a person's death is an instance of euthanasia if and only if it benefits him, then, according to this view, there can be no genuine instances of euthanasia. You may believe that you are acting to benefit a person by hastening her death, but that belief is necessarily incorrect if life is always valuable.

But surely this position is mistaken. Being alive is not always better than being dead. The notion that individuals may find themselves in circumstances where, although alive, they have been rendered incapable of valuing anything and find continued life itself an unbearable burden is neither absurd nor incoherent. In circumstances of this sort death indeed could be a blessing, a welcomed relief from suffering. Consider being terminally ill and in great but unrelievable pain, or the victim of an incurable degenerative disease that progressively destroys the capacities you most treasure. For some people life might have value even when they are in such dire straits. The value they place on their lives might then justify the use of life-sustaining technologies. For others life in such circumstances would be devoid of value, in fact probably a continuing misery. In this sort of case use of the life sustaining capacities provided by modern contemporary health care technology would violate the second longstanding norm of medical ethics—the duty of nonmaleficence, the physician's obligation to avoid harming the patient.

What then is the value of life? As John Harris (1985) has noted,

Like the question of the meaning of life, that of the value of life, when put in such broad terms, seems unanswerable. Not because there is no answer, but rather because there are so many answers! There are likely to be, and perhaps are, as many accounts of what makes life valuable as there are valuable lives. Even if we felt confident that we could give a very general account of what makes life valuable for human beings, perhaps by singling out the most important or most frequently occurring features from the lists of what they value of a large cross-section of people, we would have no reason to suppose that we had arrived at a satisfactory account.

Harris's point is that the value of a person's life is an irreducibly subjective matter. Every person who values life does so for his own reasons, and there are no grounds for holding that one person's reasons for valuing his life is more or less valid than another's. Although this may mean that there is not one exclusive answer or set of answers to the question of what the value of life is, as Harris has noted, we can nonetheless know which lives are valuable (Harris 1985):

. . . the point of the question lies not so much in our arriving at a satisfactory account of the value of life, but rather in our discovering a way of knowing when we are confronted with valuable lives. Our interest is in knowing which other beings have valuable lives, and there may be good reasons for being much surer of this than of the value of any of the features that are supposed to make life valuable! If there are, as there may well be, as many accounts of what it is that makes life valuable as there are valuable lives, these accounts in a sense cancel each other out. What matters is not the content of each account, but rather that the individual in question has the capacity to give such an account.

The point is this: if we allow that the value of life for each individual consists simply in those reasons, whatever they are, that each person has for finding their own life valuable and for wanting to go on living, then we do not need to know what the reasons are. All we need to know is that particular individuals have their own reasons.

The value of life for any individual consists simply of those reasons he has for valuing his life, whatever those reasons happen to be. This entails that an individual's life can have value only if he is capable of having reasons for valuing it. From this argument derives: A life is valuable only if it is the life of an entity that has the capacity to value its life. Only an entity that can value life can have a valuable life.

The Concepts of "Person" and "Human Being"

But what sorts of entities can value life and thus can have valuable lives? The answer to this question depends initially on the long-standing philosophical distinction between the notion of a human being and the notion a person. Although the terms *person* and *human being* often are used as synonyms in ordinary language, many philosophers have found it useful to give the terms separate and distinct meanings. A person is an entity that is self-conscious, aware of its own awareness, and is sufficiently rational to "consider itself the same thinking thing in different times and places" (Harris 1985). In other words a person is a self-conscious entity aware of itself as the same entity (or being) in different times and places. Given that a human being is simply any entity that belongs to the species *Homo sapiens*, then, according to the philosophical definition of a person, it is possible to be a human being without also being person. An example would be a human being in a brain-dead or chronic vegetative state, a human being totally and permanently devoid of self-consciousness. Of course a human being can also be a person. All human beings who are self-conscious and aware of their existence as persisting through changes in time and place* are necessarily persons according to the philosophical definition of personhood.

But how does the philosophical distinction between person and human being affect the question of how valuable lives are to be recognized as such? The capacities which a being must have in order to be able to value its life, and thus to have a valuable life, are also the capacities that constitute personhood (Harris 1985):

In order to value its own life, a being would have to be aware that it has a life to value. This would at the very least require something like Locke's conception of self-consciousness, which involves a person's being able to "consider itself as itself in different times and places." Self-consciousness is not simply awareness, rather it is awareness of awareness. To value its own life, a being would have to be aware of itself as an independent center of consciousness, existing over time with a future that it was capable of envisaging and wishing to experience. Only if it could envisage the future could a being want life to go on, and so value its continued existence. The capacity to value existence in this sense is a fairly low-level capacity; it does not require rationality in any very sophisticated sense of that term, merely the ability to want to experience the future, or to want not to experience it, and the awareness of those wants.

Therefore to recognize which lives have value is to recognize which entities are persons, that is, which individuals are self-conscious in the relevant sense.

*Harris quite correctly notes that this notion of personhood "is value- and species-neutral. It does not imply that any particular kind of being or any particular mode of existence is more valuable than any other, so long as the individual in question can value its mode of existence" (Harris 1985). In principle what sorts of beings can be persons besides human beings is an open question.

Definitions of Death

Irreparable termination of heartbeat, respiration, and blood pressure have traditionally been the conditions jointly taken to indicate death. Before today's impressive supportive and resuscitative devices were developed, any individual who had suffered extensive brain damage would also have suffered all these conditions. Today, because heartbeat, respiration, and blood pressure can all be mechanically maintained for an indefinite amount of time, the traditional criterion of death is no longer adequate. Indeed many states have already made total and irreparable loss of brain function their legal standard for death.

Arriving at a new legal definition of death does not settle the moral issue. Legally sound practices are not necessarily morally sound. Therefore it is appropriate to ask whether a brain-dead individual whose respiratory and circulatory functions are being maintained mechanically is morally indistinguishable from a corpse. A brain-dead individual has no capacity for self-consciousness of any sort whatsoever and therefore cannot be a person. The kind of life possessed by a brain-dead individual then is a life without value and, as such, a life without any moral standing. From a moral point of view, hastening the termination of such a life would be, at worst, a matter of moral indifference. (Given that keeping a brain-dead individual on life support consumes resources that might be used to benefit living human beings who are persons, hastening the demise of such an individual might be a morally sound course on essentially utilitarian grounds.) Failing to keep a brain-dead individual on life support would not then rank as an instance of euthanasia. Because such an individual is morally indistinguishable from a corpse, from a moral point of view its loss of life is neither a benefit nor a harm to it.

Though many people would not have difficulty accepting this conclusion, they might feel somewhat differently about an individual in a chronic vegetative state, a condition where there is sufficient brain function to make mechanical maintenance of respiration and circulation unnecessary. Patients in a persistent vegetative state exhibit no evidence of self-awareness and no purposeful response to external stimuli. Their eyes open periodically. They may exhibit sleep-wake cycles. Some even yawn, appear to chew, or may swallow spontaneously. A variety of simple and complex reflex responses can be elicited from individuals in a persistent vegetative state, responses that cannot be obtained from brain-dead people. But the chance that consciousness will ever be regained is extremely small. The problems created for society by this condition were exemplified in the Karen Ann Quinlan case.

James Rachels (1986) described the situation created by Quinlan's condition:

In April 1975 this young woman ceased breathing for at least two fifteen-minute periods, for reasons that were never made clear. As a result, she suffered severe brain damage, and, in the words of the attending physicians, was reduced to a "chronic vegetative state" in which she "no longer had any cognitive function." Accepting the doctors' judgment that there was no hope of recovery, her parents sought permission from the courts to disconnect the respirator that was keeping her alive in the intensive care unit of a New Jersey hospital.

The trial court, and then the Supreme Court of New Jersey, agreed that Karen's respirator could be removed. So it was disconnected. However, the nurse in charge of her care in the Catholic hospital opposed this decision and, anticipating it, had begun to wean her from the respirator so that by the time it was disconnected she could remain alive without it. So Karen did not die. . . . Karen remained alive for ten additional years. In June 1985 she finally died of acute pneumonia. Antibiotics, which would have fought the pneumonia, were not given.

An individual in a chronic vegetative state is not brain dead, but is her condition different from brain death in any morally significant respect? Again the crux of the matter is personhood. Chronically vegetative individuals are not capable of consciousness and are extremely unlikely to regain that capacity. Such individuals are not persons, and their lives cannot have value. Chronically vegetative life is therefore life that has no moral standing, and acting to hasten the termination of a neocortically dead individual does not constitute an act of euthanasia. As in the case of brain death, there is no person benefited or harmed by such an act. (Here too there is no moral basis for objecting to according an individual in a chronic vegetative state the legal status of a corpse.)

Harris (1985) aptly put the general point as follows:

The reason it is wrong to kill a person is that to do so robs that individual of something they value, and of the very thing that makes possible valuing anything at all. To kill a person not only frustrates their wishes for their own futures, but frustrates every wish a person has. Creatures that cannot value their own existence cannot be wronged in this way, for their death deprives them of nothing that they can value.

Brain-dead individuals and chronically vegetative individuals are creatures of precisely this kind.

For the hastening of an individual's death to constitute an act of euthanasia, it must be done for the sake of that individual. The purpose of the act must be to benefit the individual to whom it is done. Thus for the hastening of an individual's death to be an act of euthanasia, that individual must be more than merely "alive." Because the only sorts of individuals who can possibly benefit from having their deaths hastened are persons, the question that must be answered is, When is it morally permissible to end a person's existence with the aim of thereby benefiting him? To answer this question is to determine the justifiability of euthanasia.

One of the reasons that euthanasia is so morally worrisome is the very real problem of not being able always to distinguish success from failure. Though the intention behind an act of euthanasia intrinsically may be laudable, hastening the death of an innocent person who in fact is not benefited by that act is to end a life that has value. This would be a very serious wrong indeed (perhaps one of the gravest wrongs that can be done to a person). To avoid wrongful acts of this kind, euthanasia must be limited to only a certain class of persons—persons who can really be made better off by such acts. If we cannot distinguish such persons from others, we cannot know when the intention behind euthanasia ought to be acted on. And we run the very serious risk of hastening death when we ought not to do so.

How then are persons who are appropriate subjects of euthanasia to be identified? Suppose an individual has a degenerative illness, say, ALS (Lou Gehrig's disease). Suppose also that the prospect of living with fewer and fewer capacities, of watching these capacities diminish, is one he views as sufficiently horrible to make him place a negative value on life. The fact that he finds continued life to be unacceptable can be known because he can communicate that information and can request that his life not be extended.

It is important to note that such a request should not automatically be regarded as proof that a person disvalues his life. Only when the request is made by a person who is competent enough to know what he values, what makes life worthwhile for him, should it be taken as proof that his life is devoid of value. Such a request should not be acceded to if it is made under extreme but transient duress or on the basis of false beliefs or under the influence of medication that impairs judgment. In all likelihood there are numerous conditions, some relatively transitory and others relatively permanent, that can undermine a person's competence to judge the value of life, and all of them should be taken into account when evaluating such a judgment. As is the case with many matters of ethical significance, there are likely to be instances where no confident determination can be made. In such instances what should be done depends on what is believed to constitute the greater evil, hastening the death of a person whose life has value or prolonging the life of a person who finds that life to be a fate worse than death.

Suppose, though, that a clearly competent person requests that efforts not be made to prolong her life. Is euthanasia morally justified in this type of case? If the traditional medical ethical norm of beneficence is applied, the answer to this question is clearly Yes. For in such a case euthanasia would be beneficent (and furthermore allowing it in such a case has the virtue of respecting individual autonomy).

Notice though that what has been justified here is *voluntary euthanasia*, euthanasia freely requested by a competent person. There are other forms of euthanasia that must be considered in their own right, namely, *involuntary euthanasia* and *nonvoluntary euthanasia*. If carried out against the subject's will, an act of euthanasia is involuntary, whereas if carried out in circumstances where the subject's will cannot be determined, it is nonvoluntary. But before considering these forms of euthanasia, we should answer this question: In a case of voluntary euthanasia, is the manner by which the person's death is hastened morally significant? Does allowing a person to die differ morally from actively inducing death when the aim in either case is to benefit the person whose death occurs?

Active versus Passive Euthanasia

In discussions of the moral justifiability of euthanasia, often a distinction is drawn between active euthanasia and passive euthanasia. The distinction rests on the difference between killing a person and letting a person die, which in turn rests on the difference between an act of commission and an act of omission. If a person's death is hastened by not taking steps that could effectively forestall it, it occurs as the result of an act of omission, and we have an instance of letting a person die. If a person's death is caused by doing something to bring death about (for example, giving a lethal injection), then death is hastened by an act of commission, and a person has been killed. Active euthanasia has taken place.

Does the difference between passive and active euthanasia, a difference in how death is caused, have any moral significance? The American Medical Association seems to believe that it does. According to a statement adopted by the House of Delegates of the American Medical Association on December 4, 1973 (Rachels 1978),

The intentional termination of the life of one human being by another—mercy killing—is contrary to that for which the medical profession stands and is contrary to the policy of the American Medical Association.

The cessation of extraordinary means to prolong the life of the body where there is irrefutable evidence that biological death is imminent is the decision of the patient and immediate family. The advice of the physician should be freely available to the patient and immediate family.

In response to this position Rachels (1978) wrote,

The AMA policy statement isolates the crucial issue very well, the crucial issue is "intentional termination of the life of one human being* by another." But after identifying this issue and forbidding "mercy killing," the statement goes on to deny that the cessation of treatment is the intentional termination of a life. This is where the mistake comes in, for what is the cessation of treatment in these circumstances [where the intention is to release the patient from a fate worse than death], if it is not "the intentional termination of the life of one human being by another?"

As Rachels correctly argues, when steps that could prolong a person's life are omitted for the person's own good, such an omission is just as much the intentional termination of one person's life as when active steps are taken to end a person's life for his own good. For example, failing to place a patient on a respirator or turning the respirator off to end his suffering would be just as much an act intended to terminate life as the administration of a lethal injection. Indeed in many cases the main difference between the two actions is that the latter releases the individual from suffering a life that is disvalued much sooner than does the former. Dying can take a long time and involve considerable suffering even when no measures are undertaken to prolong life. Active killing, by means of lethal injection, for example, can be carried out so that death is caused instantaneously and quite painlessly. This difference certainly does not render killing, in this context, morally worse than letting a person die. Indeed, insofar as the motive is merciful (as it must be in euthanasia), because the person killed is released more

*Rachels uses the term *human being* here to mean what we mean by *person*.

quickly from a life that is disvalued than would otherwise be the case, the difference between killing the individual and letting him die provides support for the active killing (if the difference has any moral significance at all).

According to Rachels (1978), the most common response to this argument is the following:

The important difference between active and passive euthanasia is that in passive euthanasia, the doctor does not do anything to bring about the patient's death. The doctor does nothing, and the patient dies of whatever ills already afflict him. In active euthanasia, however, the doctor does something to bring about the patient's death: he kills him. The doctor who gives the patient with cancer a lethal injection has himself caused his patient's death; whereas if he merely ceases treatment, the cancer is the cause of death.

This response alleges that although in active euthanasia someone must do something to bring about the patient's death, in passive euthanasia the cause of death is the patient's illness rather than someone's action. But this is surely incorrect. Suppose a physician deliberately decides not to treat one of his patients who has a routinely curable ailment and the patient dies. Further suppose the physician were to try to show himself to be blameless by saying, "I did nothing. The patient's illness caused his death, not me." Under current legal and ethical norms, such a response would not be credible. For as Rachels notes, ". . . it would be no defense at all for him to insist that he didn't 'do anything.' He would have done something very serious indeed, for he let his patient die." The physician would be blameworthy for the patient's death just as surely as if he had actively killed him. If causing someone to die is justifiable under a given set of circumstances, whether it is done by allowing death to occur or by actively bringing about death is morally irrelevant. Similarly if causing someone to die is not justifiable in a given set of circumstances, whether it is done by allowing death to occur or by actively bringing death about also is morally irrelevant. So if voluntary passive euthanasia is morally justifiable in light of the duty of beneficence, so is voluntary active euthanasia. Indeed given that the benefit to be achieved by euthanasia often can be generated more quickly by means of active euthanasia, there may be cases where it is morally preferable to passive euthanasia.

Involuntary and Nonvoluntary Euthanasia

What about involuntary and nonvoluntary euthanasia? An act of involuntary euthanasia is one that hastens an individual's death for his own good but against his wishes. To take such a course of action would be to destroy a life that has value; that is, a life that is valued by its possessor. Therefore it is no different in any morally relevant ways from unjustifiable homicide. There are only two legitimate justifications for hastening the death of an innocent person, self-defense and saving the lives of a larger number of other innocent persons. Involuntary euthanasia would not fall into either of these categories for it, by definition, is done for the sake of the person who is euthanized and not for reasons of defense of self or saving others. So no act that counts as an instance of involuntary euthanasia is morally justifiable.

An act of hastening a person's death is an instance of nonvoluntary euthanasia, if it is done for the person's own good and the person to whom it is done is not capable of expressing assent or dissent. Suppose that it is clear that a particular individual is sufficiently self-conscious to be regarded a person, but cannot make his wishes known. Suppose further that he is suffering from the kind of ailment that in fact causes many persons to disvalue their lives. Would hastening his death be permissible? The answer is Yes if there is substantial evidence that he has given prior consent. The individual might have told relatives and friends that should certain circumstances come about, he does not want efforts made to prolong his life, or even that he wants active efforts to be made to bring about his death. Or he might have recorded his wishes in the form of a Living Will or on audio or video tape. Where substantial evidence of prior consent exists, the decision to hasten death would be morally justified. For such a case in fact would be virtually a case of voluntary euthanasia.

The Living Will

TO MY FAMILY, MY PHYSICIAN, MY CLERGYMAN, MY LAWYER—If the time comes when I can no longer take part in decisions for my own future, let this statement stand as testament of my wishes: If there is no reasonable expectation of my recovery from physical or mental disability, I, _____, request that I be allowed to die and not be kept alive by artificial means or heroic measures. Death is as much a reality as birth, growth, maturity, and old age—it is the one certainty. I do not fear death as much as I fear the indignity of deterioration, dependence, and hopeless pain. I ask that drugs be mercifully administered to me for the terminal suffering even if they hasten the moment of death.

This request is made after careful consideration. Although this document is not legally binding, you who care for me will, I hope, feel morally bound to follow its mandate. I recognize that it places a heavy burden of responsibility upon you, and it is with the intention of sharing that responsibility and of mitigating any feelings of guilt that this statement is made.

Signed_____

Date_____

Witnessed by

_____

_____

Living Will statutes have been passed in at least 35 states and the District of Columbia. For a Living Will to be a legally binding document, the person signing it must be of sound mind at the time the will is made and shown not to have altered his opinion in the interim between the signing and his illness. The witnesses must be people not able to benefit from the individual's death.

But what about an instance where such evidence is lacking and cannot be obtained? Suppose the individual at issue is clearly a person but one who has never had the capacity to competently consent to or dissent from decisions concerning his life. It simply cannot be known whether life is valuable to this sort of person. Here, too, what should be done is a matter of what is taken to be the greater evil—ending the life of an innocent person who values life or making a person endure a life that he disvalues.

Should Voluntary Euthanasia Be Legalized?
Suppose it is conceded that euthanasia is morally permissible in cases where it is voluntary. Some would argue that social policy should prohibit it nonetheless because of likely adverse consequences. According to this point of view, the trouble with euthanasia lies not in its impact on the person whose death is hastened but rather in its impact on society as a whole. In other words the overall disutility of allowing voluntary euthanasia outweighs the good it could do for its beneficiaries. The central claim is that legalization of euthanasia would very likely erode respect for life and eventually become a policy under which persons judged to be "socially undesirable" would not be rescued when the power for rescue exists or even into a policy where such persons are actively put to death. Nazi Germany is often cited as evidence for the truth of this claim. It is said that what began in Nazi Germany as a policy of euthanasia soon became a policy of killing persons deemed non-Aryan or otherwise undesirable. The argument goes on to claim that this could happen in our society as well. Once social policy encompasses efforts to hasten the death of persons, respect for human life in general is eroded and the door is open to all sorts of abuses.

Though it may not be possible to absolutely guarantee that such events would not occur, there is no reason to believe that they are inevitable or even remotely likely. For the medical ethical duty of beneficence justifies only voluntary euthanasia. It validates hastening a person's death only for his own good and only with his consent. The actions of killing or refusing to rescue persons judged socially undesirable are quite different and violate the medical ethical duty of nonmaleficence. As long as only voluntary euthanasia is legalized, and it is clear that involuntary (and perhaps nonvoluntary) euthanasia is not and should never be legalized, no degeneration of the policy need take place. And degeneration is not likely to take place if the beneficent nature of voluntary euthanasia is clearly distinguished from the maleficent nature of involuntary euthanasia. Euthanasia decisions must be scrutinized to ensure that the cases that actually occur are only voluntary, and severe penalties must be established to deter abuses.

Summary

In this chapter we have traced the development of and described the technologies used in contemporary intensive care units. Because cardiovascular disorders often require intensive monitoring and care, particular attention has been given in this chapter to pulmonary physiology and failures of the cardiovascular system that respiratory therapies are designed to ameliorate. Technologies for respiratory therapies have been described, including negative-pressure, positive-pressure, and high-frequency jet ventilators and extracorporeal membrane oxygenation. In addition the development of patient monitoring systems has been examined, and current monitoring practices have been described.

The recent innovations in respiratory therapy *and* monitoring have been shown to owe a great deal to developments in other scientific fields. The interface between chemistry and textiles research, for example, resulted in the development of the silicone membranes that make extracorporeal membrane oxygenation feasible. Independent developments in computer science have resulted in more reliable computer-based monitoring systems. Computer-based monitoring innovations in fact represent a classic example of how technology can improve lives. Use of these innovations relieves health care professionals of often tedious and exhaustingly repetitive tasks, tasks which nonetheless can be critical to patient well-being. Because modern monitoring systems perform these tasks more reliably and release health care professionals for other types of therapy, their adoption has improved both the quality of monitoring (and therefore the quality of patient care) and also, to some extent, has reduced the costs of that care for the individual patient.

The capabilities of technologies used in ICUs and the relatively high costs associated with operating them have created some difficult economic and ethical problems for our society as a whole. Many of the economic concerns arise not from the technologies themselves but form the extent to which ICUs have been integrated into the U.S. health care system. Public policy and private self-interest seem to have resulted in excessive use of ICU beds as a means of patient care. Efforts to control excessive use and reduce health care costs could lead to rationing the availability of ICUs (a technique to control health care costs for many different types of care in other countries). It has been shown that if rationing is to be used to limit access to this and other types of health care, then a system of marketable permits is likely to be one way to ensure an efficient allocation of that care across regions and between health care facilities.

Some of the fundamental ethical problems raised by ICU-based technologies, the issue of voluntary and involuntary euthanasia and active versus passive euthanasia, result from the fact that health care professionals can now sustain human life even when the patient has lost all capacity to function as a person. The final section of this chapter has dealt with considerations for and against euthanasia, the value of life, the nature of personhood, and the tragic dilemmas that individuals such as Karen Ann Quinlan's parents and terminally ill patients have to face. In considering the debate over active versus passive euthanasia, it was noted that if relief from suffering is taken to be the aim that justifies passive euthanasia (withdrawal of life support and thus allowing a patient to die), then in some circumstances even greater relief can be provided by active euthanasia (taking active steps to hasten the patient's death). Allowing the process of dying to take its course can mean unrelievable suffering for the patient in some instances and hence may be an inferior option to active euthanasia. This does not entail, of course, that where euthanasia is justified, its active form is superior to its passive form. This depends on the particulars of the individual case and the suffering involved. Obviously, if euthanasia is morally unjustifiable, then neither form can be accepted.

References

Alderson, S. H., and Warren, R. H. 1984. Ventilatory management of muscular dystrophy patients following spinal fusion. *Respiratory Care* 29:829-832.

Bartlett, R. H., Gazzaniga, A. B., Fong, S. W., Jeffries, M. R., Roohk, H. V., and Haiduc, N. 1977. Extracorporeal membrane oxygenator support for cardiopulmonary failure: Experience in 28 cases. *Journal of Thoracic and Cardiovascular Surgery* 73(3):375-386.

Bronzino, J. D. 1977. *Technology For Patient Care.* C. V. Mosby.

Bronzino, J. D. 1982. *Computer Applications for Patient Care.* Addison-Wesley.

Browdie, D. A., Deane, R., Shinozaki, T., Morgan, J., DeMeules, J. E., Coffin, L. H., and Davis, J. H. 1977. Adult respiratory distress syndrome (ARDS), sepsis, and extracorporeal membrane oxygenization (ECMO). *Journal of Trauma* 17(8):579-586.

Burford, J. G., and George, R. B. 1984. Some recent advances in respiratory therapy. *Respiratory Therapy* 14(3):17-28.

Callahan, J. A., and Bahn, R. C. 1974. Cardiac care unit. In Ray, C. D. (ed.): *Medical Engineering.* Year Book Medical Publishers.

Chatburn, R. L. 1984. High frequency ventilation: A report on a state-of-the-art symposium. *Respiratory Care* 29:839-849.

Cochrane, A. L. 1972. *Effectiveness and Efficiency: Random Reflections on Health Sciences.* Nufield Provincial Hospitals Trust, London.

Comroe, J. H., Jr. 1965. *Physiology of Respiration.* Year Book Medical Publishers.

Egan, D. F. 1977. *Fundamentals of Respiratory Therapy.* 3rd ed. C. V. Mosby.

Feldman, C. L., Singer, P. J., and Hubelbank, M. 1972 (Sept). An on-line eight patient heart rhythm monitor. Presented at the International Congress of Cybernetics and Systems.

Friedman, R. B., and Gustafson, D. H. 1977. Computers in clinical medicine, a critical review. *Computers in Biomedical Research* 10:199-204

Geddes, L. A., and Baker, L. E. 1968. *Principles of Applied Biomedical Instrumentation.* John Wiley and Sons.

German, J. C., Worcester, C., Gazzaniga, A., Huxtable, R. F., Amlie, R. N., Brahmbhatt, N., and Bartlett, R. H. 1980. Technical aspects in the management of the meconium aspiration syndrome with extracorporeal circulation. *Journal of Pediatric Surgery* 15(4):378-383.

Howard, R. 1986. Pulmonary assist and monitoring devices. In Bronzino, J. D. (ed.): *Biomedical Engineering: Basic Concepts and Instrumentation.* PWS, pp. 106-135.

Hardesty, R. L., Griffith, B. P., Debski, R. F., Jeffries, M. R., and Borovetz, H. S. 1981. Extracorporeal membrane oxygenization: Successful treatment of persistent fetal circulation following repair of congenital diaphragmatic hernia. *Journal of Thoracic and Cardiovascular Surgery* 81(4):556-563.

Harris, J. 1985. *The Value of Life.* Rutledge and Kegan Paul.

Jennett, B. 1986. *High Technology Medicine: Benefits and Burdens.* 2nd ed. Oxford University Press.

Kirkpatrick, B. V., Krummel, T. M., Mueller, D. G., Ormazabal, M. A., Greenfield, L. J., and Salzberg, A. M. 1983. Use of extracorporeal membrane oxygenization for respiratory failure in term infants. *Pediatrics* 72(6):872-876.

Leggitt, M. S. 1984. Facility report: Respiratory therapy department at Hartford Hospital. *Respiratory Therapy* 14(3):43-46.

Lindsay, J., Jr., and Bruckner, N. 1975. Conventional coronary care unit monitoring. *JAMA* 232:51-53.

Manning, W. G., Newhouse, J. P., Duan, N., Keeler, E. B., et al. 1987. Health insurance and the demand for health care. *American Economic Review* 77:251-278.

McPherson, S. P. 1977. *Respiratory Therapy Equipment.* C. V. Mosby

Rachels, J. 1978. Active and passive euthanasia. In Rachels, J. (ed.): *Moral Problems.* 3rd ed. Harper and Row.

Rachels, J. 1986. *The End of Life: Euthanasia and Morality.* Oxford University Press.

Ray, C. D. 1974. Instrumentation for pulmonary function. In Ray, C. D. (ed.): *Medical Engineering.* Year Book Medical Publishers.

Romhilt, D. W., Bloomfield, S. S., and Chou, T. 1973. Unreliability of conventional electrocardiographic monitoring of arrhythmic detection in coronary care unit. *American Journal of Cardiology* 31:457-467.

Russell, L. B. 1979. Technology in hospitals: Medical advances and their diffusion. In: *Studies in Social Economics*. The Brookings Institute.

Sheppard, L. C. 1979. The computer in the care of critically ill. *Proceedings IEEE* 67:1300-1307.

Sondak, V., Schwartz, H., and Sondak, N. 1979. *Computers and Medicine*. Artech House.

Spearman, C. B., and Saunders, H. G. 1987 (June). The new generation of mechanical ventilators. *Respiratory Care* 32(6).

Taykor, J. P. 1978. *Manual of Respiratory Therapy*. 2nd ed. C. V. Mosby.

Terdiman, J. 1974. Physiological monitoring systems. In Collen, M. F. (ed.): *Hospital Computer Systems*. John Wiley and Sons, pp. 241-273.

Wagner, D. P., Wineland, T. D., and Kraus, W. A. 1983 (Fall). The hidden costs of treating severely ill patients: Charges and resource consumption in an intensive care unit. *Health Care Financing Review* 5(1).

West, J. B. 1974. *Respiratory Physiology—the Essentials*. Williams and Wilkins.

Wetmore, N. E., Bartlett, R. H., Gazzaniga, A. B., and Haiduc, N. J. 1979. Extracorporeal membrane oxygenization (ECMO): A team approach in critical care and life support research. *Heart & Lung* 8(2):288-295.

Zapol, W. M., Snider, M. T., Hill, J. D., et al. 1979. Extracorporeal membrane oxygenization in severe acute respiratory failure: A randomized prospective study. *JAMA* 242(20):2193-2196.

Zapol, W. M., Snider, M. T., and Snider, R. C. 1977. Extracorporeal membrane oxygenization for acute respiratory failure. *Anesthesiology* 46(4):272-285a.

6

Computers in Health Care

During the past two decades the expansion of the use of computer technology has been explosive, involving almost every facet of human activity. Today computers are nearly everywhere, and yet these devices are still considered to be in their infancy. This observation is especially true in the use of computer technology in the clinical environment. In recent years the introduction of computerized systems into the medical environment has greatly accelerated, and there is every indication that this will continue. In the future it is almost certain that computer systems will be even more extensively used by health care systems than they are today. Major areas that have been and will continue to be important include patient care, the development of diagnostic support systems (such as electrocardiogram interpretation and medical imaging devices that include computerized tomography, diagnostic ultrasound, and nuclear medicine), and the development of medical expert systems to help clinicians diagnose and treat their patients. As a result these evolving computer technologies will have a major effect on U.S. health care procedures and policies into the twenty-first century.

The digital computer is clearly the lynch pin for modern state-of-the-art information processing systems. This is especially true in modern health care institutions. The techniques actually used to collect, store, retrieve, and analyze data on the status of a patient, however, have been, until quite recently, slow and inefficient, even rudimentary. Because the quality of patient care depends on the rapid processing of accurate information about the patient's status, comprehensive computer-based health care information systems have been increasingly used for these purposes. Despite progress made so far, patient care could be further enhanced in many hospitals and other health care facilities by the adoption of current hardware and software data-processing systems. Moreover in the future the types of information-processing tasks that can be carried out by health care professionals will expand even further as new applications of computer-based technologies are developed. In this section we briefly review the present status of computerized medical information systems—other computer applications in health care are considered in other chapters. For example, intensive care unit (ICU) monitoring techniques are discussed in chapter 5, and medical imaging technologies are presented in detail in chapter 7.

Since the early 1960s health care technologies based on the digital computer have become more and more pervasive. Advances in almost all types of medical research have come to depend heavily on computer-based technologies, either through the acquisition of accurate data from experimentation or the analysis of that data. Efficient financial management of health care facilities has also become increasingly dependent on high-speed data collection and retrieval of patient-related information. Only with such systems is it possible to achieve economic management of hospital inventories of drugs and equipment and provide accurate accounting of patient care costs and billing.

Since the 1970s patient care computing systems have been developed to provide comprehensive information about the health care status of the patient. These systems use computer hardware and software to collect, retrieve, and present clearly a wide variety of patient-related information (Jenkins 1978, Bronzino 1982). The development of such patient care systems has significantly altered and improved the way in which health care professionals use information about the patient's health care status. For example, patient care computing systems

1. provide clinicians with access to sophisticated and often widely dispersed resources,
2. simplify the assembly of patient-related data for review by all types of health care professional, and
3. assist health care professionals in making fundamental decisions about patient health care strategies.

As current systems are more widely adopted, the scope of the services they provide expands, and patient care computing will enhance both clinical performance and patient care. Specific patient care–related computer applications can be (approximately) allocated into the following categories:

- clinical laboratory analysis
- acquisition of patient data and the evolution of automated multiphasic testing systems
- computer-based medical record data collection, storage, and information retrieval systems for patient management
- automated patient monitoring
- diagnostic support systems (such as electrocardiographic interpretive systems)
- medical imaging
- artificial intelligence and expert systems

With the exceptions of medical imaging (discussed in chapter 7) and patient monitoring (discussed in chapter 5), each of these applications is examined in some detail in this chapter. First, the role of computers in the clinical laboratory is examined. Second, computerized methods for acquiring patient data, multiphasic testing, and the consequent development of standardized computer-based patient medical records are explored. Finally, the role of artificial intelligence and expert systems in providing diagnostic support services is considered.

Computers in the Clinical Laboratory

The primary information required for any clinical decision is usually the medical data collected during the care of patients. These data are often processed in clinical laboratories that have historically provided the chemical, hematological, microbiological, and blood banking services for hospitalized patients and outpatients (Williams 1969). The traditional organization of clinical laboratories usually includes the disciplines of hematology, chemistry, and microbiology; the breakdown of activities within these disciplines is as follows (Martinek 1972):

1. *Hematology* is concerned primarily with monitoring the activity of a single physiological process—the formation and development of erythrocytes (red blood cells).
2. *Clinical chemistry* is based on the technical science of chemistry and is oriented primarily to the application of clinical methods without regard for any specific physiological system or anatomic group of organs or diseases. According to the International Federation of Clinical Chemistry, clinical chemistry encompasses the study of the chemical aspects of human life in health and illness and the application of chemical laboratory methods to diagnosis, control of treatment, and prevention of disease.
3. *Clinical microbiology*, which includes bacteriology, serology, immunology, virology, and other related disciplines, is concerned primarily with the detection of a large number of extraneous biological agents (the microbes). In addition it involves blood bank operations limited to the collection and preservation of blood and its characterization for compatibility and therapeutic use.

The activities within clinical laboratories are primarily concerned with providing information regarding the presence of any "pathological" condition. If physiology is defined as the study of the normal function of cells, tissues, and organs, then pathology is the study of alterations in the human body that are induced by disease. The various laboratory procedures (which include chemical tests, microscopic examinations, tissue cultures, and the like conducted within the clinical laboratory) are vital if the appropriate health professionals are to properly diagnose and monitor the course of a particular disease. To be of clinical value, this information must be accurate, easily accessible, and processed as rapidly as possible.

A clinical laboratory today has the primary function of obtaining appropriate medical data about a particular patient for review by the attending physician. To accomplish this task, the laboratory must collect samples, process tests, report the results, and somehow store all the relevant data collected on each patient. In essence all of the steps required to develop modern microcomputer-based instrumentation equipment are found here.

Over the years the performance of these steps by manual methods has become increasingly difficult as the type of determinations have grown in complexity as well as volume. In 1960, for example, clinical laboratories had an armamentarium limited to twelve or so different biochemical determinations (Dickson 1969), whereas today more than 400 different tests are available within the confines of this facility. These tests range from a rather simple physical measurement of specific gravity to such sophisticated techniques as atomic absorption spectrophotometry. The number of such tests conducted each year has also grown at an ever-increasing rate, reaching the 10-billion mark by the mid-1980s (approximately 50 tests per year for each person in the United States). These staggering numbers illustrate the importance of scientifically based information in the treatment of patients.

Over the past 20 years a number of extremely successful systems have evolved to handle this type of information, some of which are available from commercial vendors. Microcomputers have been directly interfaced with a wide variety of laboratory instruments for on-line analysis of data, and automation of rather complex computations has become indispensible in many laboratories (Whitcomb et al. 1978, Lincoln 1978). Presently computer systems are being designed for use in clinical laboratories to store, retrieve, route, sort, and verify the flow of laboratory information and thereby provide and efficient and cost-effective means of handling patient data. As computer technology continues to become less expensive and more flexible, computerization of a clinical laboratory requires that laboratory personnel as well as engineers assist in the conceptual planning of the computerized system being designed for a particular laboratory.

Traditionally attempts to computerize the clinical laboratory have tended to separate the data acquisition functions of the laboratory from those tasks that are essentially clerical (for example, cumulative and patient-, ward-, and physician-oriented reports, status reports of particular tests, and worksheet and label generation). The managerial aspect of potential computerization, however, is by far the more complex and more difficult to implement of the two tasks (Lipkin and Lipkin 1975, Wist et al. 1982).

A somewhat similar, but perhaps more productive, view of the laboratory is to consider the following two levels of potential computerization: (1) the instrumental/process control level (IPC) and (2) the central laboratory computer (CLC). This division reflects a distinction between decentralized autonomy, which increases processing efficiency within a unit, and central regulation, which promotes patient-oriented integration of data from a variety of sources. To achieve a successful system, a balance between these two levels in the laboratory should be obtained.

The Instrument Level

The first level, the IPC concept, consists of a set of computers or simple processors, each characterized by (1) inputs that are specimens and associated identification data and (2) digital outputs that represent results, analytic conditions, and associated identification data. Typically, sequential multiple analyzers and Coulter counters fall into this category, and their high-volume activity makes computerization quite attractive.

The reliability of the analytic results and the capture of all data as they emerge from these devices are still important concerns in selecting computer techniques to automate these devices. More important, however, the IPC concept emphasizes (1) the modularity and pseudoindependence of individual IPC units, which in turn permit (2) uniformity in data protocols, (3) local control of analytic conditions where the computation process is sufficiently understood and methods are accurate, thereby insuring confidence in results, and (4) the addition or modification of laboratory instrumentation without major interruption of overall system operation.

At the individual instrument level computerization can be made cost effective. The program to run a Coulter counter (figure 6.1), for example, can be essentially the same anywhere. The differ-

Figure 6.1
Coulter counter used in the counting of blood components (Courtesy of Saint
Francis Hospital and Medical Center, Hartford, CT)

ences among such instrumental programs are at input (that is, specimen identification data) and output levels (for example, communications protocol peculiar to the local installation). Thus each device may be designed, constructed, and programmed independently to perform its particular function.

Additionally if the data obtained from the clinical laboratory can be combined with information pertaining to each patient, it becomes possible to generate "patient records." With this in mind, several of the presently available automatic analyzers now include small built-in computers that allow computations to be carried out and improve data presentation. One such device is the sequential multiple analysis plus computer (SMAC) system. The SMAC system (figure 6.2) was a direct result of further development of the continuous flow systems in which samples, separated by air bubbles, are distributed to different channels, dialyzed and heated, measured and recorded. In the SMAC system, however, the computer is involved in virtually every phase of system operation. The computer calibrates each channel at preselected intervals, monitors all channels and modules, aids in troubleshooting, identifies samples, calculates data, and reports results. *It is an excellent example of a computerized laboratory instrument.*

The SMAC system is capable of measuring 20 parameters of a patient's specimen at a rate of 150 patients per hour as compared with 12 per hour for its predecessor, the 12-channel continuous type analyzer—the SMA 12/60. It contains its own microcomputer, with a cathode ray tube or video terminal and teletype writer as the primary input/output devices. Its operation depends not only on the apparent hardware components, but on sophisticated software routines that permit those with little computer experience to operate the system.

It operates as follows: Test requests and pertinent patient data are entered at the keyboard of the SMAC video terminal. This may include the patient's name, hospital number and location, physician's name, and so forth. To assist the operator, the SMAC video screen presents a step-by-step format to ensure correct order of data entry. Test requests may be entered by total profile, organ profile, single

Figure 6.2
The sequential multiple analysis plus computer (SMAC) system

test, or any combination of tests, which are displayed on the video screen for the convenience of the operator. The SMAC system stores all the input information and automatically matches the test requests with the sample, significantly reducing data collection and clerical errors (see figure 6.2). The computer associated with the SMAC system also performs function checks on basic system performance during start-up. Functions checked include reagent absorbance referenced to water, proper standardization factors, proper dwell time for each individual test, as well as checks for shift and noise. After completing the check procedure, the computer produces a detailed report that lists values for all of the parameters. This document then becomes part of the daily laboratory quality control procedures.

As the sample enters the flow cell, the computer begins a series of control and monitor routines that check the overall shape and characterizations of the "peak curve," comparing it with an experimentally derived ideal curve for the particular chemical element being evaluated. If the curve is within certain preset limits, absorbance is converted to concentration and outputted when all other parameters for that patient have been determined. If these limits are exceeded, on the other hand, the computer brings the malfunction to the attention of the operator by means of an audible alarm and provides an indication of which module is malfunctioning, the test or tests affected, and the nature of the problem. The operator can then decide whether to correct the problem immediately or to shut down the affected channel and attend to the problem later. It therefore provides valuable information regarding continuous overall performance.

These systems run quickly and efficiently—a virtual assembly line for chemical reactions. There are, however, some concerns about the use of such systems. The older analytic instruments used for their standards a pure substance in an aqueous base. For example, glucose dissolved in water gives a known glucose concentration. Running this through the older analytic system provides a good calibration and check on the accuracy of the devices. However, with the complexities of the SMAC systems, it is not possible to make up a complete standard against which all samples will be compared. That is, it is difficult to know if the standards used to calibrate the systems are in fact accurate in all the compound concentrations contained in the standard. Though the differences found so far are small, they are of concern to those interested in the degree of accuracy of the data provided to the physician—who must make a clinical decision based on it.

The introduction of more and more microprocessors into the clinical laboratories (Westlake 1975, Titus 1977, Williams 1975, Engler et al. 1983, Preston et al. 1984) offers other possibilities— the interfacing of these devices to a central computer for data processing and integration of medical data with other pertinent patient information. Such an approach would also help the laboratory with its managerial functions.

The Management Level
The incorporation of a computerized management system or CLC is possible at two different levels. The first concerns primarily the control and integration of various laboratory instruments to transfer data processed directly form the various analytic devices into a computerized patient record. The second deals with the clerical functions of the laboratory, for example, recording, verifying, and reporting medical data.

In the first case a controlling laboratory computer (figure 6.3) can be used to "call" an instrument by presenting it with an appropriately identified specimen or reference standard. The analytic device performs its tests, computes the results, and outputs a string of data in an appropriate digital format. In this type of modular general system, each determination is performed by a specific instrument that presents its own digital face to the world. Each instrument then is specifically adapted to its own particular task, and contains a digital interface, buffer, and controller. These are, vis a vis the analysis, specifically adapted to the procedure, its calibration, and the anticipated vagaries of analytic conditions. The output, on the other hand, consists of a small series of standard-length digital words that represent the result, identification, and any

Figure 6.3
Block diagram illustrating the type of input obtained from data-acquisition stations in an integrated clinical laboratory/intensive care unit system and subsequent output routes that are available once the information is assembled and processed

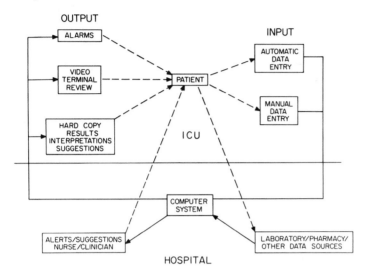

other parameterization data for each determination. Such output is delivered to the CLC for final processing and storage.

For this type of situation the computerization of the clinical laboratory's instruments involves not only the hardware aspects but software complexities required to properly integrate each instrument. Each device must give its data to the main computer before the data are replaced by new data. Some instruments may have more data-items per second than other instruments. The computer is in the middle of the information exchange process, continually integrating these instruments; ". . . all right number 4, tell me what you have; is that all? O.K. now number 5, is that all? O.K. now number 6—everyone wait a minute, I've finally got to talk to number 1 . . ." and so forth. Standardized digital circuits and connector "buses" make easier the task of getting the correct signals to flow. Interfacing the instrumentation in the clinical laboratory therefore results in the creation of one overall computerized system. There are many problems in timing, data transmission, and the like, but such interfacing can be accomplished.

Clinical laboratories offer various services to the hospital, the private physician, the clinic, and comprehensive health testing organizations. The basic job of the laboratory is providing reliable data on biological specimens to support a physician's preliminary diagnosis. It can be used to monitor the course of therapy and in the future, one hopes, identify a disease while there is still time to prevent its further development. These data are obtained through chemical, microbiological, and morphological analyses on blood, urine, spinal and synovial fluids, and tissues. In view of the ever-increasing demands and requirements placed on these laboratories, they can continue to be successful in performing high-quality specimen analysis only by using fully automated equipment. The term *automated* is obviously a broad one that can be used to include modules such as automatic pipettes, specific systems such as automatic cell counters, and general systems such as computers.

Available systems, however, represent only the beginning. Automation, as it exists today, has made few if any inroads into other clinical laboratory sciences, such as microbiology. It is generally agreed that before further real progress can be made in the automation of clinical laboratories, the entire field of laboratory science must be intensely reviewed, and its approaches revised. In general, however, major development can only come through the development of new instruments capable of detecting minute amounts of biological compounds and the incorporation of data-processing equipment and techniques designed to provide auto-mated analysis. There is no need for a large computer in most laboratories. Mini- and microcomputer systems have been devel-oped that can easily assemble and accumulate one day's laboratory data at a cost considerably less than with large machines. The microcomputer, the smallest processor available yet, will become an integral part of many new systems in the future.

This process in underway. As we move toward the end of the twentieth century, the use of computers and microcomputers will continue to change many instruments and the methodologies used in the clinical laboratory. There will be further emphasis on (1) upgrading the data-handling processes within these laboratories, (2) more automatic collation of test results on the same patient from all over the laboratory, (3) more instrument self-monitoring fault-finding and operator alerts, (4) greater availability of ongoing automatic quality control data, and (5) widespread use of effective patient and sample identification systems (Alpert 1979).

In fact it is quite likely that by the turn of the century clinical laboratories will be designed around the computer. Indeed all analytical instruments will have become computer accessories, interconnected with a central computer that will collate results for a patient from several instruments and call for the tests to be performed in logical order. The correct processes used by labora-tories and others not yet even conceived will be carried out by cheaper, faster, smaller, and more powerful computing facilities that can be expected from the postsilicon era of the late 1980s and early 1990s (Kulikowski and Weiss 1979).

The benefits that can be derived from the continuing evolution of automated clinical laboratory systems are quite appealing. For example, automation enables the laboratory to provide more rapid service with a reduction in the time elapsed from the receipt of a biological sample to the delivery of information to the physician from days or hours to minutes. It also enables the physician to arrive at a proper diagnosis more quickly through a computer presentation of all tests performed on each patient on a single printout sheet or the screen of an interactive display terminal. Because of the scarcity and cost of skilled labor, the exploitation of automation in this setting provides for optimal use of professional and technical personnel. Finally, the repetitive nature of most laboratory tests makes them ideal for automation. For example, only 10 to 20 tests account for almost 14 percent of the workload in the average clinical laboratory. With an automated chemistry system two technicians can run more than 40 times the number of tests than a more highly qualified worker can run using the same manual chemistry techniques.

Throughout the United States there are approximately 7,000 clinical laboratories (including those within hospitals) supervised by clinical chemists or clinical pathologists, microbiologists, and the like. However, it is surprising to note that these laboratories account for only 25 percent of the total clinical laboratory work performed in this country each year. The other 75 percent is performed by the small laboratories in general practitioners' and internists' offices and by the small commercial laboratories scattered throughout the country. Thus the magnitude of the economic effect of clinical laboratory services has been considerable and will only increase as the activities of even these smaller laboratories are tied into regional systems.

In recent years the automation of many clinical laboratories has to a large extent not only been successful in technical terms but also has gained wide acceptance within the medical community. It is generally agreed that these automated hospital facilities represent a cornerstone of a patient management system that can be expanded to include automated multiphasic health testing and computerized patient records.

Patient Data Base Acquisition, Multiphasic Testing, and the Development of the Patient Medical Record

Information obtained from clinical laboratories represents only a small portion of the patient data base. The patient care process also involves the gathering and processing of information pertaining to the patient's history and to physiological data. Patient data acquisition therefore consists of conventional medical history information, subjective impressions, physical findings, and various physiological measurements. It may involve consultant reports and other data contributed by various support personnel operating essentially as physician extenders. Multiphasic health testing represents one method to obtain a comprehensive medical profile of each patient. Computers have been used to automate this process, assemble the resultant data, and provide the final results for clinical review.

In the United States this development is a direct consequence of the increasing demand by the public for low-cost periodic health examinations and the increasing availability of automated equipment and computerized data-processing systems (Collen et al. 1974). The advent of automation and the introduction of computers have increased the speed, accuracy, and efficiency of multiphasic health testing techniques and offers the promise of improvements in cost, service, and quality of health care.

In the past the computer was used primarily to expedite or automate an existing health-care procedure. When used in the area of multiphasic health testing, however, it significantly affected the procedure itself. Advances in computer technology have made it possible to use digital processing techniques in gathering and preparing patient data for physician review, thereby facilitating developments that were inconceivable a decade ago. In many instances the availability of this type of computerized system has made feasible the acquisition and processing of large quantities of data for patients—even before they leave the laboratory. Standard laboratory tests and measurements can now be made available for review by the physician before the patient-physician interview. With such a comprehensive medical profile, decisions concerning

diagnosis and treatment can be based on more complete information, thereby ensuring better treatment for the patient (figure 6.4).
Although in the past the primary use of *automated multiphasic
health testing* (AMHT) has been screening for disease, programs
have been expanded so that this approach is increasingly being used
for (1) admission and preoperative examinations for hospital patients, (2) periodic health examinations for ambulatory patients, (3)
evaluation of the general health of a particular population for
special purposes (for example, military, industrial, insurance), and
(4) the conduction of diagnostic surveys to evaluate the quality of
care one receives from the health professionals within an institution. In addition there are those (Garfield 1970, 1974, Collen et al.
1974) who view AMHT as the "point of entry" into the health care
system. Consequently AMHT can be effectively used to direct and
refer the consumer to appropriate services and thereby affect the
nature of the entire health care delivery system. In this regard the
availability of computerized health testing makes it possible to
introduce modern technology into primary care areas and, via
telecommunications, to move this technology out of specialized
central facilities into small satellite health centers. Entering the
automated health testing laboratory requires insertion of patient's
identification card. Once this identification is validated, all data
can be referred under the patient's unique record number.

Clinical medicine is inherently an information-processing activity that depends on the quality of clinical data and the speed of
placing this data in and retrieving it from the patient's medical
record. As a result the medical record and the communication of the
medical information it contains are critical elements in the practice
of high-quality medical care.

Historically the principal use of the medical record has been as
a repository of information concerning the patient. In its most
general form, it is composed of a basic sheet with patient identification and chronological data, including date of admission and date
of discharge; an initial medical history and record of physical
examinations in narrative form by one or more attending physicians; clinical observations (such as body temperature, height,

Figure 6.4
Entering the automated health testing laboratory requires insertion of patient's identification card. Once this identification is validated, all data can be referred under the patient's unique record number.

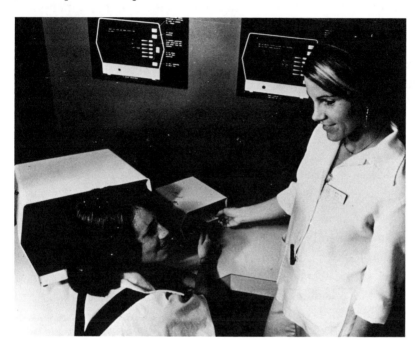

weight, pulse, arterial and venous blood pressure, and patient attitude) recorded throughout the hospital stay by physicians, nurses, therapists, and social workers; precise therapeutic instructions in chronological order; consultation reports, including judgments and therapeutic recommendations; operation reports in narrative form; clinical pathology laboratory determination; results of functional tests, including pulmonary function studies, electrocardiograms, electroencephalograms, and electromyograms; special reports by departments that provide selective services, such as radiation therapy, physical therapy, radiology, nuclear medicine, and cardiac catheterization laboratory; and a discharge summary that contains a synopsis of the patient's history, examination, course of treatment, final diagnosis, and outcome.

The structure of the medical record has essentially remained unchanged since the early part of the twentieth century. Recording practices still rely almost completely on the traditional manual record folder where physicians' notes are handwritten and merged with laboratory data and the other patient material. It is interesting to note that only 40 to 50 years ago the family physician—able to recall the family background, past medical history, and significant medical encounters, such as previous surgeries, pregnancies, and so on—was the guardian of the patient's past medical record. Today, in a mobile society with specialization in medicine, fragments of the patient's record are scattered, kept in hospitals, clinics, X-ray departments, and even by insurance companies. Consequently by default *each patient has become the carrier of his or her own medical history.* As a result expensive and time-consuming reconstruction of the patient's the past medical history has to take place every time the patient changes hospitals or visits a different doctor. More important perhaps is the fact that over time the quality of the medical history given by the typical patient often becomes quite unsatisfactory because he or she cannot remember the names of drugs used, the exact diagnosis, and so forth, in previous treatments.

There is no question that medical and legal needs require the maintenance of patients files or records that store the results of laboratory work and the health profiles generated at patient data acquisition stations or centers. In the past these requirements have been met either by inserting laboratory slips into the chart or by rather laborious transcription procedures, that is, transcribing the values from each laboratory result into the patient's master record. Unfortunately all such methods result in delays and additional expense, whereas the records themselves are often unreliable and unpredictable. Additionally the communication of patient-related information between health care professionals and other relevant agencies is made even more difficult because most medical records lack uniformity or standardization with respect to content, form, and so forth. As a matter of fact it is probably safe to say that, until recently, no two hospitals have kept exactly the same medical record, in part because of the lack of availability of a common language for use in medical records.

These factors have made it extremely difficult for medical information to be easily located, transmitted between, and understood by different medical institutions. Today a major goal is the establishment of a computerized medical information system that will

1. Enable the health professional providing medical care (doctors, nurses) to communicate individual patient data into the patient's own computer medical record. Once this record is established it could then be communicated to other professionals (for example, dietitian or therapist) and hospital services (for example, radiology or clinical laboratory) to inform them of the needs of the patient.
2. Permit patient data obtained from subsystem components (for example, automated multiphasic screening laboratory or intensive care unit) to be inserted into the patient's computer medical record.
3. Provide a means by which clinical services (that is, nursing stations) can communicate with auxiliary services (for example, cardiac or radiology laboratory) regarding a particular patient.
4. Establish procedures and communicate information for the scheduling of patients, for the delivery of specific health care services.
5. Establish a data base for administrative and business functions.

6. Establish a medical data base that can fulfill research and teaching objectives.

Comprehensive computer hospital information systems that provide all these functions do not yet operate anywhere in the world. Large medical centers, however, have some subsystem components, such as AMHT, patient monitoring, admission and bed census, outpatient scheduling, and hospital data base systems.

An Example of a Clinical Information System

The Beth Israel Hospital, a major 449–adult bed teaching hospital of the Harvard Medical School, has a clinical research center, an active emergency ward, an ambulatory care division, and general medical and specialty clinics. Over *20,000 patients* are admitted each year, and there are more than *16,000 outpatient* visits. The hospital is served by a clinical computing system that includes programs for inpatient and outpatient registration, scheduling and admission, medical records, utilization review, nursing, nutrition, surgical pathology, radiology, pharmacy, cardiology, blood bank management, and chemistry, hematology, microbiology, blood gas, and neurophysiology laboratories (Bleich et al. 1985, Safran and Porter 1986).

In the inpatient admitting department the computer collects all demographic information, schedules and preadmits patients, tracks patient transfers, assigns medical record numbers, and prints managerial reports. Clinic, emergency ward, and private ambulatory patients are registered by computer, and all outpatient visits are scheduled on the computer, which in turn keeps track of all appointments. When a patient is admitted to the emergency ward, a priority request for the patient's medical record is automatically generated, and the record is delivered via dumb waiter from the medical records department within minutes. For clinic appointments the computer requests a patient's records the night before a patient's visit. With few exceptions a terminal in any clinic can be used to change a patient's demographic information or schedule or change appointment for any other clinic.

With computerized registration the entire hospital shares a common on-line data base of over 600,000 patients; information entered at any terminal is immediately available (if appropriate) at terminals located throughout the hospital. A patient who registers in the emergency ward need not provide the same information again to obtain ancillary services or to be admitted to the hospital. The computer assigns all unit numbers, and no patient can be seen and no specimen analyzed without first having a record created in the computer.

In utilization review the computer captures working diagnosis, assigns a provisional diagnosis-related group (DRG), and assigns review dates. It prints coordinators' worksheets, generates physician advisory reports, and keeps track of all extensions and terminations and helps monitor the hospital's patient case mix. It produces reports on magnetic tape.

In the radiology department the patient registers at a computer terminal, which then generates film and jacket labels, technologists' work cards, and transportation tickets, and notifies the file room to pull old films. Upon completion of the examination, the radiologist either codes the report directly into the terminal for immediate delivery (55 percent) or dictates it (45 percent) for subsequent transcription into a word processor (on the same computer), which then prints the radiology report. Often a preliminary chest X-ray report will be available in the PATIENT LOOKUP system before a patient is transported back to his or her hospital bed.

In the pharmacy the computer assists with inventory management by allowing surveillance of medications by manufacturer and by functional class. In the outpatient pharmacy all prescriptions are entered into the computer, which then performs the clerical and fiscal tasks and displays the medication profiles.

In the inpatient pharmacy intravenous orders are entered into the computer, which then organizes the preparation and distribution of the medications. When a unit dose order, including discontinuation, is entered at any terminal in the pharmacy, it is immediately reflected in the screen displays, from which medications are picked for distribution to floors, and in the printouts, from which carts are checked. Because all systems are integrated, a patient's renal function can be checked in the chemistry lab, and antibiotic sensitivities can be checked in the microbiology lab before an antibiotic is sent to the patient's floor.

In the hematology, bacteriology, chemistry, and blood gas laboratories, specimens are logged in at computer terminals, which then assign accession numbers and capture the data and the time of transaction. Automated equipment, such as the Coulter counters and SMACs, are interfaced with the main clinical computers via microcomputers. If there is a problem with the central computers, these microcomputers can temporarily store several days' laboratory data. White blood cell differential and reticulocyte counts are transferred to the computer via terminals that beep when the appropriate number of cells have been counted. Unexpected results, such as outlier values or values that have changed markedly since a previous determination, are flagged for approval by supervisory personnel before they are filed. All laboratory results must be verified by a technologist before they enter the patient data base.

In the blood bank all products received from the Red Cross are tracked by light-pen reading of bar codes. Patient transfusion histories are recorded and made available to providers. By using the computer to cross-check blood products and patient identification near the patient's bedside, blood products no longer need to be signed out by doctors or nurses in the blood bank.

In the department of cardiology electrocardiograms, echocardiograms, exercise stress tests, and cardiac catheterizations are reported on the clinical computing system. Although the wave forms and pictures generated by these tests are not kept on-line, when a staff cardiologist approves a final interpretation of the results, the text of the report is immediately available in the PATIENT LOOKUP system.

Results from the hematology, chemistry, blood gas, and bacteriology laboratories, as well as radiology reports, cardiology reports, neurophysiology reports, medication profiles, and inpatient and outpatient admission histories for all patients, are immediately available to providers from 500 terminals throughout the hospital. All systems are integrated, and all share a common data base. All clinical data are kept on-line for approximately one year. When a patient is admitted to the hospital, the first reports include the results from recent outpatient visits.

An important characteristic of these programs is that it is almost impossible to perform a service without simultaneously capturing the appropriate information in computer-readable form. Thus it is the computer that generates the encounter form used by the physician, the workcard used by the laboratory technician, the film label used by the radiology technician, and the prescription label used by the pharmacist. To provide the highest quality data for patient care, medical research, hospital management, and fiscal operations, the data are captured and checked by the computer at the time of the initial transaction, rather than later by manual entry from written forms.

Confidentiality is protected by personal codes, or keys, uniquely assigned to each user. When a terminal is idle for five minutes, information on the screen is either erased or replaced with a test pattern. Each terminal has access only to programs appropriate to its proper user. Some keys are limited to reading, some permit a subset of the data to be edited, and some permit all data in a particular system to be edited and new keys for that system to be assigned. All work must be "signed" with the user's key, and the date, time, system, terminal, and key are recorded for each transaction.

During a typical week the patient registry is accessed over 50,000 times by hospital employees who use the system to perform their work and by physicians, nurses, and medical students who choose to look up a patient's records on the computer rather than in the transactional printout report.

In addition to looking up patient information, personnel throughout the hospital use an electronic mail system. Messages can be sent to individuals or groups of people. The mailbox is frequently used to broadcast changes in hospital or laboratory policy to all persons who need to know about the changes. Physicians are automatically sent an electronic message notifying them when one of their patients comes to the emergency ward or outpatient clinic or is admitted to the hospital for any reason.

Diagnostic Support Systems— Artificial Intelligence in Patient Care Computing

In reviewing the accomplishments of patient care computing, it is convenient to group the applications as follows (Blum 1984a,b):

1. *Data* These are the uninterpreted items that are given to the analyst or problem solver. (The Latin word *datum* means "the given.") Examples are a patient's name, a test result, or a diagnosis. Most mathematical computation operates only on numerical data. The use of a formula to compute the body surface area from the patient's height and weight is a simple example of an application that operates only on data.

2. *Information* This is a collection of data (or information) elements that contain meaning. The processing of data usually ends with a display of information. For this definition, however, we insist that the data (and information) be stored in a permanent, accessible data base. An example of information in this context is a patient medical record that includes demographics, laboratory results, and diagnosis. (In the information theory sense, information reduces uncertainty. Thus it is information rather than its atomic units, data, that we seek. As defined here, the objects stored are sets of information. Compare this with electrocardiographic signal processing, in which data are stored and information is displayed but not retained.)

3. *Knowledge* This is the formalization of the relationships, experience, rules, and so forth by which information is formed from data. The formulation may be descriptive—as in the case of a textbook—or it may be computer processible. In the second case the knowledge may be expressed as a formula (as in the case of the body surface area algorithm) or as a collection of structures, relation ships, and rules. The use of knowledge generally suggests the ability to infer information from that already present.

Application of medical computing has always followed the robustness of the technology. For example, the 1950s were primarily a decade of experimentation with new equipment and concepts. In the 1960s, however, there were robust, reliable systems that routinely satisfied many computational and data-processing tasks. The 1970s saw commercialized multiprogramming, data base technologies, and interactive computing enter the medical environment. Finally, during the 1980s advances in each of these areas have introduced new tools and paradigms for knowledge processing (figure 6.5).

For each new technology there is a natural progression that takes place from concept formulation, to research and development, to prototype development, and finally to a more mature technology with continuous refinement. Presently we have the greatest experience with data processing. Our understanding has significantly advanced from the early electrocardiogram analysis prototypes to the computer-controlled implantable monitors in use today. In the case of information system applications, we are now using a mature technology. Most of the installed products, however, are five to ten years old and do not yet take advantage of our current experience. Nevertheless significant gains can be anticipated when this technology is effectively applied. Finally, knowledge-based applications—the focus of expert systems—are just beginning to demonstrate their effectiveness in both the computer laboratory and real-world clinical settings.

The use of computers to model or mimic human intelligence is, in today's jargon, referred to as *artificial intelligence* (AI). The application of this technology to such tasks as planning, deduction, problem-solving, and understanding language (properties of human thought or intelligence) has recently received a great deal of attention and created considerable excitement. In the process the term *expert system* (Waterman 1986, Myers 1986, Firebaugh 1988) has been given to a class of AI programs that have achieved a level of performance approaching that of human experts, and a number of prototype systems have been developed for medical use (Sandell and Bourne 1985).

Figure 6.5
The evolution of computing systems in medicine, from 1950 through the 1980s

	1950's	1960's	1970's	1980's
Data Processing Applications	Research	Prototype	Mature	Refinement
Information Processing	Concepts	Research	Prototype	Mature
Knowledge Based Systems	Concepts	Concepts	Research	Prototype

These new expert, or knowledge-based, systems are being designed to

• Represent knowledge in a nonprocedural form, that is, not encoded as an algorithm
• Provide an inference mechanism that allows the application of facts (data and information) to the knowledge to produce new facts. (Note that the knowledge base and inference engine are separate.)
• Use symbolic rather than coded data so that information is not lost in the encoding process
• Maintain an audit trail so that the knowledge base may be used to justify or explain the recommendations

Expert systems are capable of manipulating symbolic as well as numerical data and have the ability to interact with the user in something approximating natural language. Expert systems consist of two major components: the knowledge base and the "inference engine" (figure 6.6). The knowledge base consists of all identifiable information relevant to the file in question. This information may include assertions, facts, relationships, and assumptions, and may be expressed in a variety of formats. The inference engine, however, is the unique design feature of expert systems that enables them to infer the solution to a problem from a set of assumptions or rules.

The knowledge incorporated into expert systems is typically of four types: (1) knowledge derived from data analysis; for example, the association between gender and thyroid disease can be established as a medical fact through review (analysis) of cumulative data; (2) subjective knowledge (judgment) derived from experience rather than theory or formal statistical analyses, but a key element in medical decision making; (3) theoretical knowledge, such as information about causal relationships or basic mechanisms of action, for example, the effects of psychotropic drugs; (4) strategic knowledge used in problem solving that uses strategies rather than bits of information; the rule of parsimony in developing a differential diagnosis is a strategy.

Figure 6.6
The architecture of a simple expert system

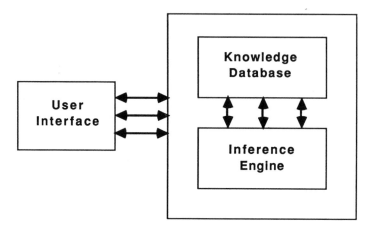

"Intelligent" computer systems do not simulate the detailed structure of the brain but rather mimic strategies followed by humans to reach a decision. The *rule-based approach*, for example, assumes that human experts follow a series of logical rules that can be stated as a series of "if . . . then" statements. *Pattern matching* compares the characteristics of newly encountered patients to profiles of various diseases within the expert system's knowledge base, and when the best match is found, the system constructs and tests an appropriate hypothesis. *Pathophysiologic reasoning* is a more recently developed strategy that requires definitive information about causes not presently available for most psychiatric disorders.

There are *limitations* in the ways in which knowledge is represented and manipulated by expert systems. Schwartz and colleagues (1987) have noted that these systems "capture only the surface behavior of experts, not the reasons they behave as they do." Knowledge derived from data analysis, for example, can be easily incorporated into computer programs, but clinical judgment is more difficult to represent. A rule-based expert system may mimic one type of decision-making process, but human experts often shift from one strategy to another or use multiple techniques in solving a problem. Although many of these human capabilities have not been captured by these systems, some success has been achieved in more simple, well-defined decision-making tasks.

Computer programs to support clinical decision making were first introduced nearly 30 years ago. The initial phase of AI research, however, was based on general problem-solving techniques that were previously applied to problems in mathematics and chess strategies. The computer technology of the sixties emphasized the use of numerical techniques to recognize patterns for modeling physiological processes, the development of algorithms to deal with routine clinical chores, and the application of Bayesian statistics to determine probabilities and occurrence. The goal during this period was to develop good decision-making techniques. In the 1970s the emphasis shifted to developing strategies of knowledge acquisition and representation so these systems could explain their decisions (Shortliffe 1976). Research in the past ten years has led to the development of nonnumerical methods for encoding information that more closely follows the reasoning processes used in complex fields such as medicine (Shortliffe 1986, Rennels and Shortlife 1987). Most prominent among the early applications of expert systems in medicine include MYCIN (Shortliffe et al. 1973, Shortliffe 1976), INTERNIST (Miller et al. 1982), and GLAUCOMA (Weiss et al. 1978). Subsequently more than 50 medical programs were developed (figure 6.7), including EXPERT (Kulikowski and Weiss 1982), NEOMYCIN (Clancey 1983), ONCOCIN (Shortliffe et al. 1981), QMR (Miller et al. 1986), and DXPLain (Barnett et al. 1987).

Although applications of expert system technology to psychiatric problems have been limited, there are several systems that constitute a beginning (table 6.1). HEADMED (Heiser and Brooks 1978, Brooks and Heiser 1979) and BLUEBOX (Mulsant and Servan-Schreiber 1984, Servan-Schreiber 1986), for example, are the most notable and provide assistance in pharmacotherapy decisions. BLUEBOX, for example, provides the clinician with an interactive consultation regarding the suicidial risk of a patient (figure 6.8).

METHUSELAH (Werner 1987), an expert system for geriatric psychiatry, provides guidance in history taking and diagnosis. Beginning with the presenting complaints, it generates a list of candidate disease states that serve as prompts for gathering additional patient data. The system then indicates which of the disorders is most likely.

PSYXPERT (Overby 1987) is a prototype system, designed to assist in the diagnosis of psychotic disorders. It uses a menu-driven user interface to guide the clinician through a series of questions that lead to a diagnosis and recommendations for further evaluation and therapy.

A somewhat different approach is taken by OVERSEER (Bronzino et al. 1989), a prototype expert system designed to perform quality assurance monitoring in the psychiatric hospital. OVERSEER interfaces directly with the hospital's data base and uses treatment

Figure 6.7
Selected expert systems in medicine

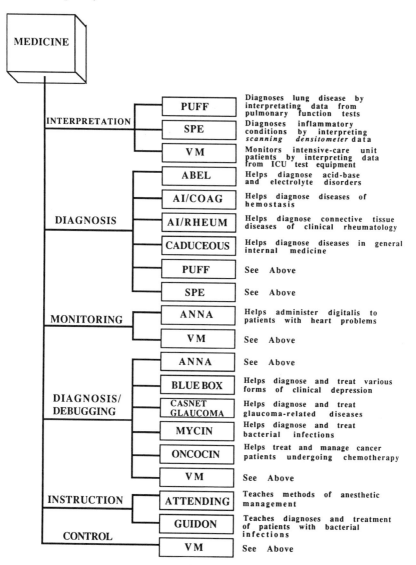

Table 6.1
Research Prototype Expert Systems in Psychiatry

Name (Year)	Developers	Domain	Knowledge Representation (Language)
HEADMED (1979)	Brooks, Heiser	Psychopharmacology adviser	Rule-based (Emycin)
BLUEBOX (1984)	Mulsant, Servan-Schreiber	Depression diagnosis & treatment adviser	Rule-based (Emycin)
METHUSELAH (1987)	Werner	Diagnosis adviser in geriatric psychiatry	Blackboard (PL/I)
PSYXPERT (1987)	Overby	Psychiatric diagnosis adviser	Rule-based (Prolog)
OVERSEER (1989)	Morelli, Goethe, Bronzino	Drug treatment clinical monitor	Rule-based (Prolog)

protocols to monitor drug treatment therapy. It issues alert messages when clinical protocols are not followed or when other apparent problems are detected. It had performed well in tests with actual patient data, and a fully operational system, based on the prototype, is now being designed.

Because physicians are unlikely to accept systems that require lengthy interactions with a computer terminal, as is the case with HEADMED and BLUEBOX, any successful system must pay attention to the data-entry step. This problem can be circumvented by optical scanning devices to input data recorded by clinicians in traditional formats or, more directly, by providing advice as a by-product of a monitoring or data management function, as in HELP (Pryor et al. 1983), ONCOCIN (Shortliffe et al. 1981), and OVERSEER (Bronzino et al. 1989). Other promising developments include advances in user interfaces (Tsuji and Shortliffe 1986) and more efficient knowledge-acquisition techniques (Gallant 1988).

As can be seen, many knowledge-based systems have been developed within the domain of medicine over the past few years. To illustrate how "medical knowledge" can be used by these systems, consider the following rule from MYCIN, a system developed at

Figure 6.8

An interactive consultation with BLUEBOX (Mulsant and Servan-Schreiber 1984). The user's responses to BLUEBOX's questions are given in uppercase. The comments in bold are added to aid understanding. Ellipses indicate where the transcript has been shortened.

(1)   What is the patient's name?
* *   MR BLUE.
(2)   What is the age of Mr BLUE?
* *   60 YEARS OLD
(4)   Please characterize the depressed feelings of Mr. BLUE with any of the following: Sadness, hopelessness, helplessness, anhedonia, anergia, anxiety, guilt, worthlessness, other or none.
* *   ANHEDONIA, SADNESS, ANXIETY ...
(1 2)   Please list (if any) psychotic symptoms of Mr. BLUE:
* *   PARANOID DELUSIONS ...
(1 5)   Is there any history of psychiatric disease or suicidal attempt in Mr. BLUE'S family?
* *   YES.............

(The program has asked basic questions about the patient. It will now try to assess the suicidal risk.)

(2 4)   Is Mr. BLUE suicidal? (If you answer UNKNOWN, I will try to assess the suicidal risk myself.)
* *   UNKNOWN.
(2 5)   Does Mr. BLUE have access to fire arms?
* *   UNKNOWN.
(2 6)   Please grade the quality of family support for Mr. BLUE:
* *   POOR (...)
(3 1)   Has Mr. BLUE accepted to make a No-Suicide decision?
* *   ?

(When the user types a "?" he can get a more explicit question, possibly with some literature references.)

The No-Suicide decision (as defined by R.C. Drye et al., in Am. J. Psychiatry 130:2, February 3) consists of having the patient pronounce the sentence: "No matter what happens, I will not kill myself accidentally or on purpose, at any time."
Expected responses are: YES or NO.
Enter HELP for a list of user options.
* *   UNKNOWN ...

(The evaluation of suicidal risk is over. Using its estimation of the depression type and the suicidal risk, the program has chosen one or more general categories of treatment [electroconvulsive therapy (ECT), tricyclic anti-depres-

*sants, monoamineoxidase inhibitors, neuroleptics, litium, psychotherapy...].
New questions are now asked to discriminate more accurately among these categories.)*

....

(3 3)   You said earlier that Mr. BLUE presents some anxiety. Please characterize the degree of anxiety:
* *   MILD.
(3 4)   Is the treatment of the patient planned in California?
* *   WHY?

(Instead of answering the question, the user can always ask "WHY", i.e., "Why is this particular question asked at this time?" EMYCIN will explain why by displaying the rule that has generated it.)

....

* *   YES.

(The reasoning process is over and the program now presents its conclusions.)

Considering what you have told me about your patient, my recommendation would be: Mr. BLUE needs to be hospitalized. Even though it is difficult to prescribe ECT in California, the severity of the depression being what it is, I would keep ECT in the back of my mind and start with legal proceedures. You could consider treating Mr. BLUE with MAPROTILINE, starting with a dose of 50 mg. t.i.d. Increase the dose by 50 mg every other day until you reach the maximum dose of 300 mg. Remember to warn your patient about the possible side effects. If the side effects are intolerable, consider discontinuation of the treatment. If your patient does not show any improvement after 4 weeks of this treatment, consider switching to another antidepressant; before doing so, you can check the plasma concentration. Since Mr. BLUE presents some psychotic features, you could add HALPERIDOL to the treatment, starting with a daily dose of .5 mg per os t.i.d. Increase the dose by .5 mg every other day until you reach the maximum dose of 3 mg.

Stanford to help physicians determine which antibiotics to use in patients with bacterial infections of the bloodstream. To assist in this process, MYCIN's knowledge base contains 400 to 500 rules, each of which looks like the following:

IF: 1. The site of the culture is blood, and
       2. There is a significant disease associated with this occurrence of the organism, and
       3. The organism's portal of entry is gastrointestinal, and
       4. The patient is a compromised host
THEN: It is definite (1.0) that *Bacteroides* is an organism for which therapy should cover.

Consider another rule from MYCIN:

RULE 280
IF: 1. The infection that requires therapy is meningitis, and
       2. The type of the infection is fungal, and
       3. Organisms were not seen on the stain of the culture, and
       4. The patient is not a compromised host, and
       5. The patient has been to an area that is endemic for coccidiomycoses, and
       6. The race of the patient is one of: Black, Asian, Indian, and
       7. The cryptococcal antigen in the cerebrospinal fluid was not positive
THEN: There is suggestive evidence (.5)* that cryptococcus is not one of the organisms (other than those seen on cultures or smears) that might be causing the infection.

AUTHOR:YU
Justification: Dark skinned races, especially Filipino, Asian, and Black (in that order), have an increased susceptibility to coccidiomycoses meningitis.
Literature: Stevens, D. A., et al. Miconazole in coccidiomycosis. Am. J. Med. 60:191-202, Feb 1976.

*Suggestive evidence meaning 0.5 on a scale of 0 to 1.

Such rules are simple "if . . . then" inferential statements if when we know that certain conditions have been met, we can make a certain conclusion. Each of the "if" assertions will be tested against facts that are true for the case under study, for example, "the infection that requires therapy is meningitis." If all the assertions are valid, then the conclusions in the "then" statement will be added as facts.

The rule is a comprehensible chunk of knowledge. To understand why this is good, think of extracting four lines of code from a random piece of software. There is little chance that the four-line chunk would make sense by itself. But this $1/500$ of the MYCIN knowledge is self-contained and understandable. In evaluating a potential application for MYCINlike technology, it makes sense to ask if this kind of dicing is possible. Successful applications require a domain that can be chopped into a few hundred or a few thousand rule-sized chunks, which are separate, distinct, and comprehensible by themselves (Davis 1984).

Let us turn to the inference engine. We begin by asking for the name of the organism infecting the patient. MYCIN's inference engine reaches into the knowledge base to pull out all the rules that make conclusions about identity. In fact there are many such rules, one of which is rule 27:

RULE 27
IF: 1. The gram stain of the organism is gram-negative, and
      2. The morphology of the organism is rod, and
      3. The aerobicity of the organism is anaerobic
THEN: There is suggestive evidence (.7) that the identity of the organism is *Bacteroides*.

Rule 27 is relevant in this case only if each of its preconditions has been met: the gram stain of the organism is gram-negative, the shape is rod, and it is anaerobic. How is it possible to find out if the gram stain is negative? It is possible by starting all over again, retrieving all of the rules that have anything to say about negative gram stain, and trying each in turn. This is known as backward chaining. The *inference engine* starts with the thing it is trying to accomplish and moves backward through the set of knowledge-base rules. When it finally runs out of rules, it asks the physician a question.

Thus the inference engine searches through the knowledge base using heuristic (that is, rule-of-thumb) techniques to guide and constrain the search. Because the knowledge base is separate from the inference engine, new knowledge can be added without a major programming effort; thus the knowledge base is extensible and cumulative (Blum 1984).

MYCIN has a number of common characteristics found in many expert systems, including

- A performance level equivalent to human experts in the field
- Reasoning in the face of uncertainty, a crucial part of the MYCIN system. The scale ranges from –1 (certainly wrong) to +1 (certainly right).
- Expertise derived from hand-crafted knowledge in a restricted domain, in this case, infectious blood disease
- A special user interface program, TEIRESIAS, developed for providing flexible knowledge acquisition and editing

MYCIN also

- Uses standard production rules rather than inference networks
- Uses backward chaining from the diagnostic hypothesis to the data, rather than forward chaining
- Has been a very influential research program, but to date has not been commercialized

The purpose of MYCIN was to provide physicians with expert diagnostic and therapeutic advice rapidly conveniently. The rationale was that doctors must often make quick judgments based on incomplete or ambiguous knowledge to save the life of a patient and that any assistance in this process would be helpful. The following is a sample of dialog with MYCIN (the human user's entries are in boldface):

PATIENT 1
1. Patient's name: (first-last)
   ** **Dick H.**
2. Sex:
   ** **M**
3. Age
   ** **52**
4. Have you been able to obtain positive microbiological information about a possible infection of Dick H.?
   ** **Y**
5. What is the infection?
   ** **ENDARTERITIS**

[Note that this dialog closely resembles a dialog carried on by a less-specialized physician consulting with expert colleague.]

The first significant organism from this blood culture (CULTURE-1) will be referred to as

ORGANISM-1
9. Enter the identity of ORGANISM-1:
   ** **Unknown**
10. Is ORGANISM-1 a rod or coccus (etc.)?
    ** **ROD**
11. The gram stain of ORGANISM-1:
    ** **GRAMNEG**
12. Have there been POSITIVE cultures yielding organisms about which you will NOT be seeking advice?
    ** **YES**

The physician can respond that the answer to certain questions is unknown. After between thirty to ninety such questions, MYCIN produces a diagnosis. For example:

INFECTION-1 is ENDARTERITIS with BACTEREMIA
   <Item 1> E.COLI [ORGANISM-1]
   <Item 2> SALMONELLA (species unknown) [ORGANISM-1]
   <Item 3> KLEBSIELLA-PNEUMONIAE [ORGANISM-1]

<Item 4> PSEUDOMONAS-AERUGINOSA [ORGANISM-1]
<Item 5> ENTEROBACTER [ORGANISM-1]
<Item 6> PROTEUS-NON-MIRABILIS [ORGANISM-1]

MYCIN, as is often the case with human experts, is ambiguous. The infection may be caused by any of the six organisms listed. Rather than recommend a different medicine for each of the six possible organisms (in which case the cure could be worse than the disease), MYCIN recommends an optimal, mimimum selection of medications for which there is minimal chance for drug interaction:

[Rec 1] My perferred therapy recommendation is as follows:
   In order to cover for Items <1 2 4 5 6>:
     Give: GENTAMICIN
     Dose: 119 mg (1.7 mg/kg) q8h IV [or IM] for 10 days
     Comments: Modify dose in renal failure
   In order to cover for Item <3>:
     Give: CHLORAMPHENICOL
     Dose: 563 mg (7.5 mg/kg) q6h for 14 days
     Comments: Monitor patient's white count
Do you wish to see the next choice therapy?
   ** NO

MYCIN recommends therefore two drugs to cover the six possible organisms. Note that the program only gives advice and that the final responsibility continues to rest with the doctor.

The strengths of diagnostic programs such as MYCIN demonstrate the real advantages of expert systems. Some of the most obvious are (Firebaugh 1988):

• MYCIN does not forget, get tired, or make mistakes as human experts sometimes do.
• Computers can do exhaustive searches, correctly identifying rare diseases that a human expert may never have encountered.
• Computer expert systems can dispense medical advice in the farthest reaches of Alaskan, African, or Australian back country of the quality now available only in major medical centers.

• MYCIN can provide the very latest therapy for use in isolated rural areas where the general practitioner may not have had the opportunity for medical refresher courses in many years.
• MYCIN calculates dosages precisely (based on body weight), which is a refinement that doctors may not have time to do.
• MYCIN can keep track of a broad range of drugs and possible detrimental synergistic interactions between drugs. This information is so complex that human experts have great difficulty coping with it.

Another feature of MYCIN-type expert systems is that their behavior is transparent. A physician can ask why the question the program asks is relevant. The response will be something like

14. Is the patient's illness with ORGANISM-1 a hospital-acquired infection?
     ** Why

[Why is it important to determine whether the infection with ORGANISM-1 was acquired while the patient was hospitalized?]

[2.0] This will aid in determining the identity of ORGANISM-1. It has already been established that
[2.1] the morphology of ORGANISM-1 is rod
[2.2] the gram stain of ORGANISM-1 is gramneg
[2.3] the aerobicity of ORGANISM-1 is facul
Therefore, if
[2.4] the infection with ORGANISM-1 was acquired while the patient was hospitalized then there is weakly suggestive evidence
[2.5] that the identity of ORGANISM-1 is *Pseudomonas*
[RULE050]

MYCIN was using rule 50 and already had certain information about the patient. If the physician adds this fact, then MYCIN will know whether this rule will fire. If it does fire, then MYCIN will have more information about the topic that it is dealing with.

In addition to interrupting a consultation, the physician can wait until the end and ask MYCIN how it reached its conclusion:

** HOW DID YOU DECIDE THAT ORGANISM-1 WAS AN E. COLI?
I used rule 084 to conclude that the identity of ORGANISM-1 is E. coli. This gave a cumulative c.f. of f(.51). I used rule 003 to conclude that the identity of ORGANISM-1 is E. coli. This gave a cumulative c.f. of (.43).

The explanation technology produces an audit trail, reviewing its own logic. By keeping track of the rules that were used, MYCIN can describe how it reached its conclusion.

MYCIN's power lies in its knowledge. In simple tests MYCIN compared favorably with some of the experts in the field, yet the inference engine is trivial. MYCIN achieves its credible record because of the 500 rules stored in the knowledge base. These rules were not found in a manual of infectious disease diagnosis. They were extracted slowly from the experts, which is why it took five or six years to develop the MYCIN system.

In view of all these transparent advantages, why is MYCIN not an integral part of every hospital and doctor's office? The reasons are complex and include the following:

1. One technical problem is that MYCIN is restricted to the narrow domain of infectious blood disease. Human diseases often do not always fall into such nicely defined domains. As the range of infections diagnosed by computer expert systems broadens, their usefulness will increase, and, one would expect, their acceptance will increase along with their capabilities.
2. Its use is not seen as essential by physicians who believe that they have been practicing medicine successfully for years without MYCIN.
3. Physicians take great pride in their intensive training, capability, skill, intuition, and sensitivity to human interactions. If the typical

layperson feels threatened by an "intelligent machine," how much more so will the typical physician when she hears of a machine with equal or superior diagnostic skills? Such antagonism has been clearly demonstrated in comparative studies—one that was open in which the physicians knew a machine was advising, and one that was blind in which the computer's advice was mixed with regular experts' advice. This sociological problem will be the most difficult obstacle to the winning of wide acceptance for medical expert systems.

Economic Issues

The previous section has illustrated the degree to which computers have spread into almost every facet of the health care system. Although in this respect the health care industry is no different from any other sector of the U.S. economy, it has been transformed in several ways that make it a particularly interesting example of how the stream of computer innovations has altered economic and social behavior. In particular, as computers have become more flexible and powerful, they have become more useful and prevalent in the delivery of health care. Recent developments in hardware and software have been especially important because they have resulted in smaller and more convenient devices that have made the accurate implementation of many medical applications much more feasible.

In the process, however, three economic issues concerning the use of computers in medicine have been the subject of particularly extensive debate. First, the question has been raised as to whether the use of computers has reduced or increased health care costs. The question itself involves a confusion between what has happened to the costs of specific health care services and the level of expenditures on those services by health care system users (and their agents such as insurance companies, Medicaid, and Medicare). Once this confusion has been sorted out, it will become clear that most computer-based innovations have led to reductions in the costs of health care procedures, but in many cases may have encouraged increased use of the system and increases in health care expenditures. Second, there has been some debate as to whether the adoption of computers has taken place too slowly or too rapidly. The answer is that both points of view are valid, depending on the criteria used to evaluate the process of adoption. Third, it has been argued that in medicine, as elsewhere, the development of computer technology has resulted in the substitution of machines for people, with the economic consequence of unemployment for certain groups of health care providers and the spiritual consequence of a less humane health care system. None of these issues is unique to medicine. They arise in many other industries in which computers have come to play a significant role. However, they are of great importance and relevance in the health care industry which is increasingly involved with data management problems and the use of sophisticated technology's in the provision of its services.

Has the Introduction of Computers into Medicine Caused Health Care Costs to Increase?

Controversy about this issue has arisen because although computer technologies have been heralded by their advocates as vehicles for cost saving in the provision of health care, the expansion of their use has been associated with increases in health care expenditures in general and even with more extensive use of other types of resources, including those provided by physicians, nurses, and other health care professionals. In fact the controversy centers around a paradox. Computers have certainly helped to reduce the cost of providing existing health care services to specific individuals. They have also improved the quality of many of those services and, in addition, facilitated the development and implementation of new types of health care. On the other hand they also have encouraged heavier use of the health care system with the net result that expenditures by society as a whole on health care services may have increased.

Reductions in the cost of providing existing services or improving the quality of these services are clearly of benefit to both the patient and those who provide the funds to meet the costs of the patient's care. However, reductions in the unit costs of a product to a single user do not necessarily reduce total systemwide expenditures on that product. If health care suppliers are able to provide a particular type of service at lower unit cost, then, irrespective of the degree of competition they face, they will pass on at least some of the cost savings to consumers in the form of lower prices. However, because health care as products obey the law of demand (discussed in chapter 2), the lower prices will increase the quantities of those services demanded by users of the health care system. If the percentage increase in the consumption of the health care services exceeds the percentage decrease in the prices of those services, then total expenditures on those services by all health care users will rise. Suppose, for example, that the cost and price of an electrocardiographic (ECG) evaluation are reduced by 2 percent as a result of introducing computer-assisted diagnostic protocols, and consequently the number of ECG evaluations carried out increases by 4 percent. Total expenditures on ECG evaluations would then rise by approximately 2 percent.

This illustrates the paradox that pervades the issue of the effect of computers on health care costs and expenditures. Computer-based technological innovations may reduce unit costs of providing existing health care services, but the resulting increases in the consumption of those services may cause increases in total expenditures on health care services. However, it is important to recognize that total expenditures may not always increase when the prices to users of health care services drop as a result of computer-based or any other technical innovations. Suppose that the nature of the demand for a particular medical service is such that a given percentage price decrease causes a smaller percentage increase in consumption of the service. In that case total expenditures on the procedure will decline.

In each of these cases the effects of the reductions in unit costs and price on total expenditures were determined by the ratio of the percentage change in quantity demanded to the percentage change in the price of the commodity that caused the consumption adjustment. This ratio is called the *price elasticity of demand for the product.* It is always negative in sign because a price change in one direction (up or down) causes a consumption change in the opposite direction (down or up). If the price elasticity of demand has an absolute value in excess of one, expenditures on the product will rise when its price falls. The opposite also holds true. If the price elasticity of demand is less than one, total expenditures will fall when the price of the product declines.

Economists have carefully constructed statistical estimates of the elasticity of demand for all health care services. They have consistently found that the absolute size of the price elasticity of demand for all existing health care services (relative to other commodities such as food and housing) is much less than one. In fact it appears to be about -0.2 (Manning et al. 1988). Thus to the extent that computer-based technological innovations generally have reduced the unit costs and prices of existing health care services, they have resulted in lower total expenditures on health care than would otherwise have occurred.

Why then have total expenditures on health care services risen during a period in which cost-reducing computer-based technological innovations have taken place? One part of the answer is quite simple and has nothing to do with the efforts of computer technology on medicine. Between 1965 and 1988 the United States as a society enjoyed significant increases in real per capita incomes. As individuals and communities in our society have become better off, they have chosen to use a part of their higher incomes to purchase better health care, and, as a result, to use the health care system more intensively. In fact they would have made that choice even if the price of health care services had remained constant relative to the prices of other goods. Most of the increases in expenditures on health care services during the computer revolution have had little or nothing to do with computer-based medical technological innovations.

In economics, as in other scientific disciplines, diagrams are often useful to present and clarify the issues. Both of the issues concerning the behavior of costs and expenditures on health care can be examined and perhaps better understood using a simple supply-and-demand model of the market for health care services. (The basic concepts used in this type of model are discussed in some detail in chapter 2.) Figure 6.9 illustrates the effects of a cost-reducing technological innovation on expenditures. Before the innovation, the supply curve for health care services is assumed to be the horizontal line SS, whereas the demand for those services is represented by the curve DD. The initial (preinnovation) market equilibrium situation occurs at A, the point at which the initial supply curve (SS) cuts the demand curve (DD). The initial market price is P and the quantity of health care services produced and traded is Q. Initial expenditures on health care (price x quantity) are represented by the rectangular area OQAP.

A cost-reducing technical innovation will shift the supply curve for the health care service downward* to S^1S^1 and create a new market equilibrium at point C. The new (postinnovation) market price is P^1, and the new quantity of health care services produced is Q^1. Total expenditures on health care (the new price multiplied by the new quantity) are now represented by the rectangle OQ^1BP^1. The old and new expenditure rectangles share the common area $OQCP^1$. Whether the new expenditure rectangle is bigger or smaller than the initial expenditure rectangle depends on whether rectangle $QCBQ^1$ is bigger or smaller than area $PACP^1$. This makes common sense as well as geometric sense. Area $QCBQ^1$ represents expenditures on the additional health care services (QQ^1) that are used as a result of the reduction in the price of health care. At the initial price P these services were too expensive. Area $PACP^1$ represents expenditure savings by people who were buying Q units of health care services at the initial price P before the technical innovation and who subsequently have to pay only P^1 for those same services. If this area of expenditures savings by purchasers of the initial quantities exceeds the area representing additional expenditures on new products, then total expenditures after the innovation

*If the industry is competitive and (as implicitly assumed here) faces constant returns to scale and fixed input prices, the vertical downward shift in the supply curve equals the reduction in unit cost association with the innovation.

Figure 6.9
Effects of a cost-reducing computer-based technological innovation on health care expenditures

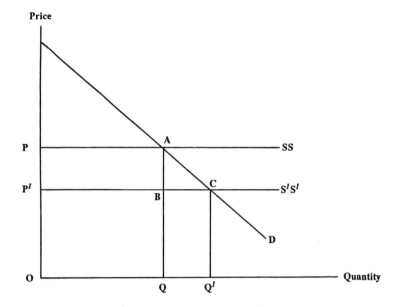

will be smaller than they were before the innovation. The evidence on the demand for health care services suggests that in fact the former effects have exceeded the latter.

Using the same framework, it is relatively easy to see why total expenditures on health care have grown despite the advent of cost-saving computer-based innovations. Over the past thirty years, rising per capita incomes and population growth have caused the demand for health care to increase rapidly. Thus, at any given price, the quantity of health care services demanded in 1988 was much higher than it was in, say, 1960. Figure 6.10 illustrates these effects. In figure 6.10 DD represents the initial (1960) demand curve for health care and $D^I D^I$ the new (1988) demand curve for health care. The initial (preinnovation) supply curve for health care is again SS, and the new (postinnovation) supply curve is $S^I S^I$. The effect of the joint shift in both the demand and the supply curves for health care on expenditures is as follows: The supply shift causes total expenditures to change from OPAQ to $OP^I CQ^I$ (a decline since rectangle $P^I PAC$ is larger than rectangle $QCBQ^I$). However, the demand curve shifts causes use of the health care system to increase further to Q^{II}. Expenditures then rise above the levels associated with just the supply curve shift by rectangle $Q^I CDQ^{II}$ ($Q^I Q^{II}$ x P^I) to $OP^I DQ^{II}$. It is this rectangle, representing the increase in total expenditures associated with the expansion in health care use caused by rising demand that has been the major reason for the rapid increase in total health care expenditures.

One aspect of computer-based innovation in medicine that may have led to increases in health care expenditures is the development of technologies that lead to new types of health care. In the language of Madison Avenue, one consequence of the computer age has been the development of a new set of health care products. To the extent that these new "products" have been used to provide patient care that was previously not available, innovation may have

Figure 6.10
The expenditure effects of a simultaneous increase in the demand for health care
and the introduction of cost-reducing computer-based technologies

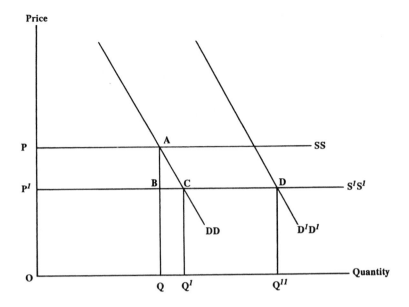

led to some increases in health care expenditures. However, even in the case of new services, computer-based innovation cannot be viewed as having increased the unit costs of those services. Before the advent of the computer technologies, the unit costs of providing such services were so high that no community or individual could afford to obtain them.

Again the use of a simple diagrammatic model of the supply and demand for the new product may clarify the issue involved here. In figure 6.11 the demand curve for the service that would become available with the technological innovation is represented by FD. Notice that this demand curve hits the vertical (price) axis at F, indicating that no units of the service will be purchased by users if the price of the service (and its unit cost) exceeds OF. Before the technical innovation, the price and cost of the service is greater than OF, and therefore it is not provided. (Of course in many cases the price of the service is infinitely high because, quite simply, the therapy has not yet become feasible.) Subsequent to the innovation, the supply curve for the product becomes PS. The price (and approximate production cost) of the product becomes OP, and a market equilibrium with positive use of the service at a level of Q units is now attained. Total expenditures on the service rise from zero to OPAQ; that is, in the case of the new product, the introduction of the new technology has increased health care expenditures and the amount of resources devoted to health care.

This discussion has given some important insights about whether or not computer-based technologies have increased health care costs. Clearly they have not raised the costs of producing health care services. Rather they have reduced production costs. To the extent that computer technologies have enabled health care providers to reduce the cost and prices of existing services, because the elasticity of demand for health care services is small, they have also led to total expenditures on health care services that are lower than would otherwise have been the case. It is true, however, that the creation of new therapies associated with computer technologies has increased expenditures on, but not average (unit) production costs of, some categories of health care (that is, new products).

Figure 6.11
The expenditure effects of introducing a new computer-based technology

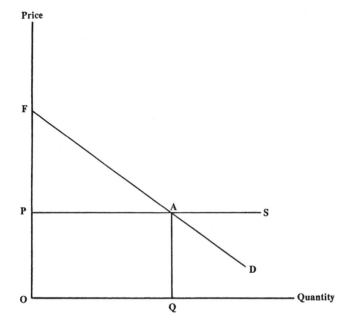

Has the Rate of Adoption of Computer-Based Technologies
in the Health Care Industry Been Too Fast or Too Slow?

This may seem a strange question to ask, but in fact it is one that echoes in the corridors and coffee lounges of every medical facility in the country. As new medical technologies are created, new therapies based on those technologies become available. In the process health care professionals want immediate, or almost immediate, access to them. Their rationale is obvious; they want to improve the quality of care for their patients and their own skills, marketability, and incomes. However, new technologies often represent expensive investments, and the administrators of hospitals and other medical facilities must weigh the potential benefits of those technologies against their costs. These conflicts often are centered on the question of the proper timing of the decision to invest in the new technology rather than on whether it should be acquired at all. For example, should a new microcomputer-based medical instrument that has just become available be acquired immediately, or should the medical facility wait for a few weeks, months, or even a year or two to obtain a perhaps not yet fully developed technology that only promises to provide more services and allow for the use of more effective therapies?

In a world of unlimited resources these concerns would not exist. Everybody could acquire the latest technology as soon as it is available and immediately discard it when something better comes along. However, no institution (for-profit hospital, not-for-profit hospital, private practice, and so on) operates in an environment in which resources are unlimited. So the optimal timing of the acquisition of new equipment that allows access to new technology is an important issue in health care. It is also an extremely complicated economic decision that is surrounded by uncertainty. This is especially true when, as has been the case with computer-based technology, the current rate of technological innovation is very high, and future rates of innovation are likely to be as high or even higher.

What then is the optimal time for a medical institution to acquire new computer-based technology? It is not necessarily the moment at which the technology (perhaps quite unexpectedly) becomes available and the institution's health care professionals become excited about it. For example, the institution may recently have acquired a new hardware-software system that performs the same set of tasks, although perhaps with slightly more person-power and at a slightly slower rate. The resources that would have to be spent to update that system (acquire the new state-of-the-art technology) may well yield greater benefits to the institutions and the communities it serves if allocated to other areas of operation. On the other hand in another institution the data processing/medical record system may be almost completely out of date, consisting primarily of typed and handwritten information on scraps of paper stuffed into manila folders and crammed into files. The medical institution knows it can reduce costs substantially and improve its provision of health care by installing a computer-based medical record system immediately. However, it also recognizes that it may be able to reduce costs by a substantially larger amount if it continues to use its existing system for another six months and adopt a new system based on the next generation of microcomputers.

In each of the examples, what should the rule for technology acquisition be? The answer in both cases is quite simple and rests on the same principle. In the first example the institution should keep the recently installed system if the net benefits the institution expects to obtain from operating it exceed the expected net benefits associated with acquiring and operating the new system. Note that net benefits reflect the difference between the values of the systems to the institution and the costs of operating them. In the second case again the institution should discard its current filing system for the currently available computer system if the net benefits from that decision exceed the net benefits from the decision to endure the current highly inefficient system a little longer and have the advantages of the more sophisticated system that is forecasted to be available in the future.

The principle used to make a decision in both examples is the same. Decision makers should compare the expected net benefits of action A with those of action B (and actions C, D, E, and so forth if more than one alternative is available) and select the action that yields the highest net benefits. If waiting to acquire a technology is the option that yields the higher expected net benefits, then that should be the option chosen. If acquiring the new technology is the option that yields the higher expected net benefits, than that option should be chosen.

What makes decisions about the adoption of computer-based technologies so difficult, however, is that it is extremely difficult to estimate the expected benefits and costs associated with those technologies. Moreover, it is extremely difficult to forecast what new options will become available even in the very near future. Therefore, of necessity, adoption decisions have to be made with incomplete information. In such a decision environment, once a specific choice about the timing of an investment in new technology has been made, that choice may soon appear to have been misguided. In an uncertain world, however, the administrators of medical institutions and the professionals who advise them have no option but to make some mistakes in the selection of technologies. Hence the quarrels in hospital corridors and coffee lounges about the foolishness or wisdom of the institutions' policies with respect to the timing of decisions to invest or not invest in new computer-based medical systems are likely to continue for some time.

Up to this point we have considered what the optimal adoption decision should be for an individual medical institution. There is a broader question that concerns the socially optimal adoption rate for computer-based medical therapies. The social rule for optimal adoption differs from the private rule only in that society should pick the adoption rates that yield the greatest net social benefits. These include benefits and costs that individual medical facilities often cannot afford to consider; for example, equitable access to new therapies irrespective of personal income, geographic location, race, and effects on public health programs. Clearly the federal government has decided that computer-based medical technology has the potential to improve the quality of health care. For example, through the National Institutes of Health it has subsidized an extensive array of research projects in related areas. The adoption of computer-based technologies in medicine has also been facilitated by the increased demand for health care services that has resulted from the direct (Medicare and Medicaid) and indirect (tax-relief for private third-party insurance and direct health expenditures) subsidy programs operated by the federal government (see chapter 2 for a detailed discussion of these programs). Whether the result of the federal research and development and subsidies and demand-enhancement programs has led to socially ideal rates of adoption for computer-based medical technologies is a more open question. It is in fact a question that can only be answered using normative judgments, and because we each have our own value system, each of us must arrive at our own conclusions on this matter.

Has the Adoption of Computer-Based Technologies Caused Unemployment in the Health Care Industry?

The fear that technological innovation will cause large-scale unemployment is not new. Ludditism, as it is called, goes back a long way. The name itself derives from a movement against the introduction of machines in woolen mills in Northern England during the Napoleonic Wars. Luddites, named after a crofter called John Ludd, broke into the woolen mills and smashed new weaving machinery that appeared to be taking away their jobs. In some respects the fears about their jobs that motivated the Luddites were real. The machines did provide services that made some workers' skills redundant. But the new technology enabled the Yorkshire woolen industry to reduce unit production costs, cut prices, attract new customers in existing markets and open up entirely new markets. The medium- to long-term results were a substantial expansion in the size of the industry and a significant increase in the overall level of employment by the industry.

What has all of this to do with the introduction of computer-based technology in medicine? Essentially the types of employment effects associated with these innovations are identical to those that occurred in the Yorkshire woolen industry as a result of a different technological innovation two hundred years ago. Some types of workers in the health care industry have been displaced (a polite way of saying they lost their jobs), but the demand for the services of other categories of workers has increased.

For example, consider the fate of file clerks working in medical records departments before the advent of computers. Most large hospitals had medical records departments that employed dozens of clerks to maintain patient files and make them available "on demand." The functions performed by such people almost completely disappeared with the advent of computer systems, and many lost their jobs or had to be retrained to perform other tasks. Basically those people provided services to their institutions that were very close substitutes for the services provided by the machines. At the same time, however, employment opportunities were created for people whose skills were complementary to the functions of the new equipment; for example, computer programmers, data-entry personnel, and systems analysts.

In addition the introduction of such systems reduced the costs of administering medical records systems and therefore, at least to some extent, the cost of providing health care services. Thus to the degree that cost reductions lead to price reductions, the net result has been an increase in the demand for health care services in general. The consequent increase in the output of health care services also has meant increases in the employment of workers such as physicians, nurses, and other ancillary staff who provide services that neither directly complement nor substitute for the inputs provided by computer-based medical record systems.

What is true for computer-based medical record systems also holds in other areas in which the computer become prominent in the health care industry. Computerization of clinical chemistry and hematology resulted in initial declines in the demand for laboratory assistants. However, reduced costs have led to reduced prices for testing services, increases in the demand for those services, and a recovery in the level of employment of medical laboratory technicians.

These three types of effects on the demand for different types of labor skills associated with technical innovation have been examined extensively by economists for different industries, and formal categories for the effects have been identified. The first type of effect is called the substitution effect. When one type of input, say, a new microcomputer, enables a hospital or health maintenance organization to perform certain tasks more cheaply than previously possible, essentially the price of the services of that input falls. Other inputs that are substitutes for the microcomputer then become less attractive; the relative price of the services they provide rises even though the absolute price of those services remains constant. The result is that the health care institution switches away from the relatively more expensive source of services to the relatively cheaper source of services. The adjustment may be partial or complete depending on the size of the relative price decrease associated with the advent of the new technology and the degree to which it can substitute for the input.

If the goal of the health care system were the provision of a fixed quantity of health care services, then the end result of the innovation would be an unambiguous fall in the use of the substitute input and a decline in the employment of the people who provide that input. There is, however, an important caveat of this conclusion: A technical innovation of the type described lowers the cost of producing health care services, generating a downward shift in the supply function for those services. We have already examined the consequences of such a shift. In figure 6.9 it was shown what results when there is an increase in the production of health care services. The expansion of the scale of the industry will increase the industry's demand for all inputs, including the input whose services are now less attractive because the new technology has created an alternative source of supply. Thus the net effect of the introduction of, say, the microcomputer on the level of employment of the resources for which it is a substitute is ambiguous. The output effect we have just discussed may or may not offset the substitution effect.

A simple numerical example may clarify this issue. Suppose that, before the advent of the new computer, a hospital was producing 500 units of health care services per week (perhaps each unit represents one major operation or 15 outpatient visits). One nurse is required for every 5 units of services, and so the hospital employs a total of 100 nurses. Now suppose that the advent of a new computer-based monitoring system provides services that substitute for some aspect of health care (for example, checking the vital signs of patients in postoperative care every twenty minutes). As a result, subsequent to the innovation, only one nurse is required for every 6.25 units of nursing care. If the hospital's scale of operations remained fixed at 500 units of care per week, the number of nurses employed would fall to 80. However, now the hospital is providing its services more cheaply. Suppose the cost savings are passed through to users (as they would be, at least in part, even if the hospital faces no competition from alternative health care providers), and that consequently the hospital's scale of operations increases by 25 percent. Such an increase in demand would mean that the hospital would be providing 625 units of care each week, and its nursing staff would have to grow back to the original level of 100. Of course a smaller-scale or larger-"output" effect would have meant a net reduction in the employment of nurses, but a larger-scale or larger-output effect would have led to an increase in the employment of people with those skills.

This example highlights the fact that innovations that save labor on a per-unit-of-output basis can still increase the employment of people with the skills for which the new technology provides (partial) substitutes. The second category of inputs for which the new technology has employment implications are complementary or joint inputs. In the case of computer innovations such inputs include programming and data processing, both of which are provided by people. The innovation itself creates a demand for such complementary inputs. Scale effects associated with falling costs and prices lead to even higher levels of demand for such skills and the people who provide them.

The third category of inputs include people whose skills are unrelated to the services provided by the new technology. In the case of ward-based monitoring equipment, surgeons provide such skills. However, the employment effects associated with technical innovation are not zero for such people. The scale or output effects associated with falling costs and prices typically lead to an increase in the demand for services of the providers of those inputs.

The bottom line here is that computer-based medical technological innovations, like almost all technological changes, benefit some groups of workers, but, at least in the short to medium term, harm the employment prospects of others. On balance, however, cost-reducing technological innovations yield net benefits to society as a whole, and it is probably worth the while of society as a whole to ensure that the adoption of such technologies takes place smoothly. An important aspect of public policy that facilitates or inhibits this process is, therefore, the mix of unemployment compensation and retraining programs available to ease the transition of displaced workers into new occupations. In addition the health care industry itself has some incentive to retrain displaced workers for other positions. These workers already have some specialized knowledge about how health care institutions work that may make their training easier and less costly.

Ethical Issues

The previous sections of this chapter have focused on some of the many ways computers are being used in contemporary medicine and some of the economic issues these uses raise. In this section some of the legal and ethical issues that are generated by medical uses of this form of modern technology are examined. Presently computer programs are involved in many aspects of medical care, including patient monitoring; operation of medical instruments and machinery; compilation, storage, and dissemination of medical records; and medical diagnosis. Along with this variety of applications comes a comparable variety of ways medical software can result in harm to patients. Among other things defective programs can result in malfunction and failure of critical medical instruments, misdiagnosis of a patient's condition, or failure to properly monitor a patient. Even when a program is not defective or does not malfunction, it may be used incorrectly or inappropriately. When something of this sort occurs, and a patient is harmed as a result, the question of who should be held liable must be posed. Consider, for instance, the sort of scenario imagined by philosopher Dan Lloyd (1985):

[S]uppose a hospital uses an expert system to make diagnoses, and in a particular case a diagnosis turns out to be fatally incorrect. . . . Consider everyone involved: The expert systems programmer, the experts who were consulted to write the program, the company marketing the program, the hospital which instituted the policy of reliance on the system's diagnoses, the hospital staffer who endorsed the diagnosis and initiated treatment.

As Lloyd correctly notes, significant moral and legal issues will be posed in any effort to determine who, among all the parties involved in a case of this sort, should be held liable for the harm done.

This section also examines the widespread fear that computers pose a threat to medical confidentiality and the privacy this confidentiality is meant to protect. Much of the information that goes into the medical record of any given patient is information that he does not wish to have made public. Individuals enter into a patient-physician relationship with the expectation that such information will be kept confidential. With the increasing use of computers to store, retrieve, and disseminate information about patients, many worry that this expectation may become nearly impossible to meet.

Tort Liability
Living in society necessarily means interacting with and living in proximity to others. Indeed, in good measure, social life is preferable to a life of individual isolation precisely because of this interaction and proximity. Throughout human history social life has made benefits possible that are unavailable to an isolated individual. The many benefits of modern medical care constitute an apt example. Although an individual living apart from others can attain some minuscule portion of these benefits through self-care, he will never be able to provide anything like the enormous level of benefit

that is available to him when his medical care is a social enterprise. Yet along with the benefits of interacting with and living in proximity come costs. Perhaps the most important cost is that each of us is exposed to harms that none of us would face as isolated individuals, that is, harms caused by the behavior of others. One of the ways our society attempts to mitigate this cost of social life is by imposing public standards of conduct on its members in certain circumstances and establishing mechanisms by which to enforce those standards. Tort liability is an example of this type of mechanism.* A particular action constitutes a tort if that action both violates some public standard of conduct and someone suffers harm or injury as a result of that violation. The victim of a tort is legally entitled to initiate litigation against the perpetrator of his injury and thereby to seek compensation for the losses he has suffered. Ideally, this compensation annuls any loss suffered by the victim as well as any gain achieved by the perpetrator. An individual can be said to incur tort liability whenever he is legally bound to pay such compensation.

Adequately confronting this issue in the area of health care computing means answering two questions: when does harm to a patient caused by medical software constitute a tort, that is, an action for which the patient is legally entitled to seek compensation; and, who, among all those involved in the chain of events leading to such harm, may be legitimately subject to tort liability? Answering these questions requires some discussion of the two most important forms of tort liability, liability for negligence (also referred to as fault liability) and strict liability (also referred to as liability without fault or absolute liability).†

---

*The discussion that follows is concerned with liability for harm caused by medical software in cases where no criminal conduct is involved. Therefore issues pertaining to criminal liability are not examined.

†A third form of tort liability is liablity for intentional wrongful conduct. Because it is very unlikely that anyone would intentionally create or use medical software to harm a patient, the discussion of tort liability in this chapter will ignore liability for intentional wrongful conduct. The concern motivating the discussion in this chapter is with who is subject to what form of liability when a medical computer program is an unintentional source of harm.

Liability for Negligence

One of the standards of conduct that our society imposes on its members in certain circumstances is the duty to exercise due care, the duty to "conform [one's behavior] to that level of care that a reasonably prudent person would exercise under similar circumstances so as to avoid injury to another" (Miller et al. 1985). To fail to exercise due care is to act negligently, hence this duty is also referred to as the duty to refrain from negligent conduct. According to Heldrich (1986),

In order to find out whether the particular conduct of a particular person can be called negligent, his conduct must be compared with the supposed behavior of a purely fictitious person who meets the requirements of an ideal citizen. This mythical person, who is always a model of prudence and care, can be found in most legal systems. He is extremely versatile: if the conduct of the driver of an automobile is to be considered, the reasonable man will be the motorist of ordinary sense using ordinary care and skill; if the fatal outcome of a surgical operation gave rise to the issue, he will be the prudent surgeon of the skill and knowledge commonly possessed by members of the surgical profession in good standing; and if the person who caused injury to another lacks one of the physical attributes of ordinary people, his behavior will be compared with that of an ideal man who is equally physically disabled.

As this passage notes, the level of care that must be exercised for one's conduct not to be considered negligent depends on the sort of activity in which one is engaged. Activities that require special skills, training, education, or experience will be assessed against the conduct of a reasonable person with such skills, training, education, or experience.

That is, one's conduct in performing such an activity will be deemed negligent or not by comparison with the care that would be taken by a reasonable person with the appropriate special skills, training, education, or experience. To determine what level of care such a person would exercise, the courts typically use the testimony of individuals who are recognized experts in the activity in question or who have some special expertise relevant to evaluation of the activity in question. If the activity at issue does not presuppose any of these special features, one's conduct will be deemed negligent or not by comparison with the care that would be taken by an ordinary reasonable person.

Mere failure to meet the relevant standard of care is not alone sufficient to render one liable for negligent conduct. An individual cannot successfully sue another for damages on the basis of negligence unless he shows that he was owed a duty of due care by the person being sued. To accomplish this, he must convince the court that some relationship existed between himself and that person. In some cases, the case of a patient and her physician, for example, it will be obvious that a relationship exists. In others, matters will be less clear. In any case the relationship need not be very substantial. For example, any motorist will be regarded by the courts as having the requisite relationship to any pedestrian who is harmed by the motorist's careless driving. Indeed the courts seem to be willing to regard the relevant relationship as existing whenever the conduct of one person poses a risk of reasonably foreseeable and avoidable harm to another.

As well as showing that the defendant owed him a duty of due care and failed to discharge that duty, the plaintiff who hopes to recover damages on the basis of negligence must also show that the defendant's negligence was the proximate cause of the harm suffered (Nycum 1980):

The existence of merely some causal relation or connection between a negligent act and an injury is not sufficient to satisfy the legal requirements for negligence liability. Negligence, regardless of its form, will not give rise to a right of action unless it is the proximate cause of the injury in question. Delineating the contours of proximate cause is a difficult task. What is said to be perhaps the best, as well as the most widely quoted, definition is that the proximate cause of an injury is that cause which, in natural and continuous sequence, unbroken by any efficient, intervening cause, produces the injury, and without which the result would not have occurred.

In other words an individual being sued for negligence can successfully defend himself by showing that the harm suffered would have occurred even without his negligence, or that his negligence resulted in the harm suffered only by virtue of some intervening event such as the negligence of a third party or even the negligence of the victim himself.

What is the moral justification for the practice of holding persons liable in certain circumstances for the harm caused by their negligence? In general two sorts of considerations can be cited: First, appeal can be made to the utility of the practice. Making a perpetrator of harm due to negligence liable to pay damages to his victim will discourage him from failing to exercise due care in the future. And this will in turn motivate others in similar circumstances to refrain from negligent conduct. The net result then of this practice is minimization of harm due to negligence.

Second, an appeal can be made to the dignity of the victim of negligent harm. Because persons are more than mere things, our conduct toward them must be respectful of that status. One way of accomplishing this is by exercising due care. An individual who injures another through negligence has failed to respect the personhood of his victim. Making him liable to pay compensation is society's way of allowing the victim to obtain redress for not being treated with proper respect and of not letting the perpetrator gain by his disrespect. Both the perpetrators and the victims of harm due to failure to exercise due care are thereby made vividly aware of society's commitment to upholding the dignity of its members.

Strict Liability

The second form of tort liability, strict liability, is relatively recent in origin. When an individual causes harm to another, under fault liability he is legally liable to pay damages to his victim only if the harm was caused by his negligence. Liability for negligence presumes some fault on the part of the perpetrator, that the perpetrator reasonably could have acted so as to avoid harming another person, hence the reason it is sometimes referred to as fault liability. Strict liability dispenses with this requirement of fault, and hence the reason it is sometimes referred to as liability without fault or absolute liability. When a person incurs strict liability, she is made liable to pay damages to the victim of a harm that she has caused even if no amount of care would have averted that harm. That is, under the doctrine of strict liability, the victim of a harm need not establish that the perpetrator was at fault, that she reasonably could and should have done otherwise, only that the perpetrator's conduct in fact did cause the harm at issue. Thus in making its members subject to strict liability in certain circumstances, society is placing them, in those circumstances, under a public standard of behavior that is much more stringent than the standard they are placed under its imposition of liability for negligence. Under the latter circumstances society is requiring its members to refrain only from carelessly harming others and so excuses them when harm ensues in spite of their due care. Under the former circumstances it is requiring that they refrain from harming others altogether and therefore disallows the excuse of due care under such circumstances.

The courts typically apply the doctrine of strict liability in cases where the harm provoking legal action was caused by a defective product. Accordingly any manufacturer or vendor of products faces the risk of having to pay damages to an injured person on the basis of this doctrine.* An individual seeking to obtain damages from the manufacturer or vendor of some product must establish more than that he was harmed by that manufacturer's or vendor's product. In addition he must establish that the product was unreasonably defective.† A defect can be either a design or production defect ("Product Liability" in *The Guide to American Law*):

*Two other groups of individuals also face the risk of strict liability: individuals who are engaged in activities that are highly dangerous that society does not see fit to ban altogether, and individuals who own or possess animals. The latter group is clearly irrelevant to discussion of liability for harm caused by medical software. It is possible though that the courts may judge the manufacture, sale, or use of some types of medical software to be highly dangerous activities and so impose strict liability for that reason. This possibility seems highly remote and so is not pursued here.

†The defectiveness of a product can be unreasonable on this doctrine even if no amount of care would have been sufficient to prevent the defect.

Design defects exist when a whole class of products is inadequately planned in such a manner as to pose unreasonable hazards to consumers. In that case, products manufactured in conformity with the intended design, which is faulty in and of itself, would be defective. A production design arises when a product is improperly assembled.

And he must establish that the defective product was the proximate cause of his injury, that is, that the product "reached" him in defective condition.

Can the practice of holding persons strictly liable under certain circumstances for the harm they cause be morally justified? This question is important because it might appear that this practice is patently unfair. After all a person can incur strict liability even when he has done all that he can to avoid harm to others resulting from his conduct. He can incur strict liability although the harm resulting from his conduct is not his fault. How can it be fair to hold individuals liable for harms that are not their fault? Critics of strict liability point out that holding someone responsible for the outcome of his action presupposes that he ought to have prevented that outcome. This in turn presupposes that prevention of this outcome was within his power. But how can it have been possible for him to have prevented a harmful result of his conduct, if no amount of care would have been sufficient? Thus, on this line of reasoning, one could say that although the harm done was an unfortunate and unavoidable outcome of something he did, he is not blameworthy for it, and so should not be held liable for it.

The usual response to this line of reasoning, however, cites the utility of holding individuals liable without fault under certain circumstances. This line of reasoning is apparent in the justifications offered for holding manufacturers and vendors of products strictly liable for the harm caused by those products. In discussing the justification of imposing strict liability on the manufacturers and vendors of products, a legal commentator, Jim Prince, notes that the courts have tended to emphasize three factors, which have been usefully summarized by ethicist Deborah Johnson (1985):

Each [factor] has to do with the role or the position of the vendor. First, to fall within the scope of strict liability, a thing has to be placed in the stream of commerce. This leads to a justification of strict liability on two counts. Since the vendor has placed the thing in the stream of commerce in order to earn profit, he or she should be the one who bears the risk of loss or injury. And, since the vendor has invited the public to use the product, there is implicit in the invitation an assurance that the product is safe. Thus strict liability puts liability where it belongs.

The second consideration has to do with the position of the vendor. The vendor is in a better position than anyone else to anticipate and control the risks in using the program. Holding vendors strictly liable can be effective in giving them an incentive to be careful about what they put into the stream of commerce. In other words, since vendors have some control over what they put into the stream of commerce, holding them strictly liable may encourage them to be very careful about what they do in fact put into the stream of commerce.

Finally, the vendor is in the best position to spread the cost of injury over all buyers by building it into the cost of the product. If some products have inherent risks, then the cost of these can be distributed by building it into the cost of the product. In effect, we make the vendor pay for the injury, knowing that he will recover this cost from the sale of his product. This way, instead of the burden being borne by those who are injured, it is spread to all who buy the product. When you pay for a product, in effect you pay for insurance. If you are harmed by the product, the vendor will pay for it.

By emphasizing these three considerations, the courts are providing an essentially utilitarian justification for imposing strict liability on the manufacturers and vendors of goods. The courts are, in essence, seeking to "place liability where it can be most effective in yielding a result, thereby minimizing dangerous, faulty products" (Johnson 1985).

By itself this line of reasoning does not rebut the criticism that the practice of holding manufacturers and vendors of products strictly liable for the harm their products cause is unfair. A practice may have great utility and be unfair nonetheless. The charge of unfairness laid against the practice of imposing strict liability is based on the claim that it holds a person responsible for untoward consequences of their action even when they are not blameworthy for those consequences, that is, even when those consequences were not avoidable by exercise of even the utmost care. This charge is not refuted by showing the practice to be a source of great utility. A more effective way of responding to the alleged unfairness of strict liability is to note that there is an important sense in which it is not true that manufacturers and vendors are being held responsible for harms they could not avoid causing. Once the practice of imposing strict liability on manufacturers and vendors has been established, they know that they face the risk of being forced to pay damages to persons who are injured by their defective products. With this foreknowledge individuals can choose whether to run this risk in light of their own preferences. Those who feel that the benefits to be gotten by the manufacture or sale of products are worth the risk are free to take that risk. Those who feel otherwise are free to avoid it by not engaging in the manufacture or sale of products.

Accordingly any manufacturer or vendor of products could have avoided the harm caused by one of his defective products simply by not engaging in the manufacture or sale of such products. Just as one cannot justifiably complain about losing money at poker when one knows this to be a risk inherent in gambling for money, so can one not justifiably complain about being held strictly liable for certain harms caused by certain kinds of activity when one knows this to be a risk inherent in those kinds of activity.

Having discussed tort liability in general, we now can address the question of when harm caused by medical software constitutes a tort. Such harm constitutes grounds for a patient to sue for damages if it can be traced either to the negligence of someone who owes a duty of due care to the patient or to a design or production defect. In the first case the patient can sue on the basis of liability for negligence, whereas in the second case the suit could be based on the strict liability of the manufacturer or vendor of the offending software.* Consider the former case first. Who, among the parties involved in the chain of events leading to a patient being harmed by a medical computer program, faces the risk of incurring liability for negligence?

Harm Caused by Medical Software: Liability for Negligence

Answering this question requires a brief sketch of the chain of events that leads to a patient being harmed by a medical computer program (Brannigan and Dayhoff 1981). Once the medical function that the program is to perform has been specified, for example, diagnosis, automatic dispensing of medication, operation of medical machinery or instruments, patient monitoring, and the like, two sorts of computer professionals come into play—the systems analyst and the programmer. The systems analyst has the role of designing the computer program. This entails analyzing into specific tasks the function that the program is to perform and determining "the appropriate order and control mechanisms for each task" (Brannigan and Dayhoff 1981). The programmer's role then is to "create precise computer instructions to accomplish each task and [to] transform those instructions into computer readable form" (Brannigan and Dayhoff 1981). Both the systems analyst and the programmer have the additional task of testing the program and making any alterations necessary to ensure its proper performance.

Once the program has been designed, written, and tested, it is ready to be brought to market. This is accomplished either by the manufacturer, if she is also the retailer of the program, or by a vendor who purchases the program from its manufacturer. Typically once the manufacturer determines that there is a market for the program and so decides to invest in it, she employs the relevant experts to create it, and she, or some vendor who purchases it from her, induces hospitals and physicians to buy and use it. The actual decision to use the software lies with the purchasing hospital or physician. The purchasing hospital can choose to establish a policy

*This assumes that providing a medical computer program is a case of making or selling a product as opposed to a case of performing a service. Strict liability will be incurred only if this assumption holds true.

of using the program. In some cases, such as with diagnostic programs, whether the purchaser of the program is a hospital or a physician, which patients, if any, actually are affected by the program's use is determined by a physician's judgment.

No one standard of care will be appropriate for determining whether due care has been exercised by all the parties involved in this chain of events. Because the design, writing, manufacture, sale, and use of medical software all require special skills and knowledge, the relevant standard of care is determined by asking what level of care would be exercised by the reasonable person possessed of the skill and knowledge required by each of these activities. For the systems analyst the relevant question would be, What level of care would be exercised by the reasonable person with the skills and knowledge required by the activity of designing software? An analogous question would be posed for each of the activities leading to the use of medical computer program, and the answer to each would be based on the testimony of relevant experts.

Suppose that the appropriate standard of care has been determined for each activity, and that in a particular instance a patient has suffered harm because one such activity has not been performed with due care. Given this, the patient will be successful in suing for damages from the negligent party only if he can establish that a duty of due care was owed to him by that party. The moral case in favor of the courts imposing a duty of care to the patients affected by medical software on all parties involved in the process leading to the use of a medical computer program is strong. Negligence at any point in this process, whether by the systems analyst, the programmer, the manufacturer, the vendor, the relevant hospital administrator, or a physician, can lead to a patient's needless suffering. Imposing a duty to meet the relevant standard of care on each of these parties would be morally sound for two compelling reasons: First, there is the utility that would be gained. If each of the various professionals involved in the process that terminates in the use of medical software was faced with the risk of paying damages for harm caused by his negligence, each would have some incentive to exercise due care. This, in all likelihood, would mean that fewer patients would suffer injury because of such negligence. Second, there is society's regard for the dignity of its members. By imposing the risk of liability for negligence on each of the professionals involved in the process leading to use of medical software, society gives them some incentive to take the well-being of the persons who are affected by their professional conduct into account. It thereby acts to reduce the chances that any such professional will forget that her conduct affects persons, not mere things, and that her conduct should be consistent with this fact.* Also, by making it possible for the patient who has suffered because of careless professional conduct to sue for damages, society conveys to him its commitment to respect his dignity as a person.

The courts have already concluded that physicians owe a duty of due care to their patients. So physicians presently face the risk of paying damages to patients suffering injury because of their negligence. In an instance where a physician's judgment leads a patient to being harmed by the use of a medical computer program—for example, basing the patient's therapy on a mistaken computer diagnosis—determination of whether the physician acted negligently would proceed as in any other instance where a determination concerning negligence would be appropriate. That is, the court would use expert testimony to determine if the physician was duly careful in his professional conduct. Typically the point of this testimony is to determine if the physician's conduct conformed to the prevailing custom in his profession, and courts usually do not find a physician guilty of negligent conduct if his conduct did indeed conform to the prevailing custom in his profession.† Hence even if a patient is harmed because his physician based therapy on an incorrect computer diagnosis, that physician will not be deemed liable for negligence if expert testimony proves that his conduct

*This is especially important in the context of medical care. Patients are persons suffering from illness. As such they are dependent persons and so are particularly vulnerable to disrespectful conduct.

†"In medical malpractice cases it has long been widely accepted that the standard of due care that must be met by the physician is conclusively evidenced by custom. 'Where malpractice claims are concerned, custom is the measure of due care'" (Averill 1984).

conformed to the prevailing custom in his profession. The physician's conduct is not negligent simply because it resulted in harm to a patient. Due care does not guarantee that harm will not occur.

Although the courts have most often determined what constitutes due care in diagnosis and therapy by looking to the custom prevailing in the medical profession, it is possible that a physician's conduct would be judged to conform with that custom yet be negligent nonetheless. In certain cases* the courts have ruled that the custom prevailing in a profession's conduct lags behind where it ought to be. In those instances professionals were judged to be negligent for not taking steps that could easily have been taken to significantly reduce the risk to others and would not have been very costly relative to the amount of protection they provided. This raises an interesting possibility concerning negligence and the use of diagnostic programs by physicians. If reliable and relatively inexpensive diagnostic programs become readily available, the courts may someday conclude that failure to use such a program constitutes negligence, even if such use is not the prevailing custom in the practice of medicine. Should such rulings be made, a physician may find herself liable to pay damages to an injured patient on the grounds that she negligently diagnosed the patient's condition. And her negligence would consist of failure to consult a diagnostic program, although doing so is not standard diagnostic practice† (Averill 1984, pp. 221–223)§.

*See Averill's discussion of the case of T. J. Hooper and Helling V. Carey (Averill 1984, pp. 221–222).

†Of course there is always the possibility that should diagnostic programs become highly reliable, use of them will become customary in the practice of medicine. In this case a physician's failure to consult a diagnostic program might be deemed negligent precisely because such consultation is the custom in medical practice.

§Another interesting possibility discussed by Averill is that diagnostic programs may someday become reliable enough to be regarded by courts as valuable sources of expert testimony in cases where a physician is accused of negligently harming a patient. Both the plaintiff and the defendant in such a case might turn to such programs for evidence pertinent to evaluation of the physician's conduct.

We offer no speculation concerning the likelihood that the courts will regard any of the other parties to the process resulting in use of a medical computer program as owing a duty of due care to patients, although the moral case for legal recognition of such a duty is very compelling. If such a duty is imposed by the courts, determining what constitutes due care on the part of the systems analyst, the programmer, the manufacturer, or the vendor may be especially difficult to the degree that any of them is regarded as responsible for adequate testing and correcting (debugging) of medical software. As Deborah Johnson (1985) has written,

> . . . while we expect those who make and sell programs to exercise a duty of care, it is difficult to spell out exactly what this means. And, of course, if we can't spell out what it means, we can't determine when someone has failed to exercise the duty of care. What is needed, at a minimum, is a set of standards for testing. We could then say that a program designer was not negligent if he or she had tested the program to the set level, or had been negligent because of failure to test the program to an "appropriate" level before releasing it for marketing. As things stand now, when the courts have to make a determination about negligence, they must seek to find out what the recognized testing procedures are in the field. This, of course, is not always easy to figure out.

Harm Caused by Medical Software: Strict Liability

It has already been noted that the courts impose the risk of incurring strict liability on the manufacturers and vendors of products and that such persons therefore face the risk of having to pay damages for harm caused by defective products they have marketed, even if the utmost care could not have prevented such harm. The justification for this practice has already been discussed and so is not repeated here. The issue of importance here is whether bringing computer programs to market constitutes provision of a service or the manufacture or sale of products. Only the manufacturers and vendors of medical software—and only if the courts regard computer programs as products—face the risk of being held strictly liable for harm to patients caused by such software.

When is something legally considered to be a product? Although one might think that the crucial question would be whether it is a tangible item, this is not the case. Legal commentator Jim Prince argues that, in general, when the courts have been asked to determine whether a manufacturer or vendor has brought a product to market, they have not limited products to tangible items. Instead the courts have looked to the position of the manufacturer or vendor and have sought to determine whether the utility of holding him strictly liable for harm caused by what he brings to market is greater than limiting his liability to liability for negligence. If the determination is affirmative, the courts regard the manufacturer or vendor in question to be a supplier of products. In other words the rationale for considering the bringing of some good to market as a case of manufacturing or selling a product instead of a case of providing a service is the same as the rationale for imposing strict liability on the supplier of some good. Whenever the utility of holding such a supplier strictly liable is sufficient to justify doing so, that supplier will be judged to be a manufacturer or vendor of products rather than services.

Prince argues that whether the courts will consider computer programs, medical or otherwise, to be products (and thereby whether the rationale for holding suppliers of them strictly liable) depends on the manner in which the software is distributed. Earlier in the discussion of the justification of strict liability, it was noted that the courts look to three factors in determining whether the utilitarian rationale for imposition of strict liability applies to a particular manufacturer or vendor. Whether these three factors are present in the manufacture or sale of software is a matter of how the software is distributed. Prince (1980, pp. 852–855) notes three basic modes of distribution. Consider how a suit of clothing reaches the consumer. First, he may have it tailored especially to fit his particular measurements. Second, he may purchase a ready-to-wear suit, which he will use as it is. Third, he may purchase a suit that is already made, but have it altered to fit him. The ways software may reach a consumer are analogous. First, software may be individualized or custom designed. That is, it may be specifically tailored to the needs of a particular customer. Second, software may be "canned" or "packaged." Purchase of this sort of software is analogous to purchase of a ready-to-wear suit. Third, software may be canned but altered to meet the needs of a particular purchaser.

Prince notes that canned or packaged software, which is sold ready to use, is distributed in the same way as any off-the-shelf good. The purpose is not to yield a program for one unique user, but rather to yield software that can be sold to as many users as possible. The features that the courts focus on in deciding whether an item should be treated as a product are present in the case of packaged software. As Johnson (1985) has written,

The vendor does place them [canned programs] in the stream of commerce, inviting use, and hoping for profit. The vendor is in the best position to anticipate dangers or errors. And the vendor is in the best position to distribute the cost of harm done by building this into the price of the program.

Prince therefore concludes that the courts are likely to consider software that is sold ready to use, so-called canned or packaged software, as a product and to subject its vendors and manufacturers to strict liability for harm to its customers and users that results from defects in the software.

None of the relevant features seems to be present in the case of programs that are individualized or custom designed (Johnson 1985):

The [individualized] programs are not placed by the vendor in the stream of commerce. Rather they are designed for an individual client at the client's request. Also, in this case, the vendor is not necessarily in the best place to determine risk. The buyer may know more about the unique situation in which the program will be used than the vendor does. And, of course, if the program is just being sold to one customer, there is no issue of distributing the cost. Prince concludes, therefore, that tailor-made [that is, individualized] software is not appropriate for strict liability and therefore should not be considered a product.

What about software that is sold as ready to use, but is then altered for the customer's purposes? Prince argues that software of this sort is similar to products that fit the so-called sales/service transaction, where the transaction consists of two parts: the sale of a product and the performance of a service. The distribution of partially modified software falls into this category because the providing of the generic software would the be sale of a product and the modification would be the performance of a service. If a defect is found in the product, the generic software, then strict liability should be imposed for any harm caused thereby. If the defect occurred in the modification of the generic software, then only liability for harm due to negligence should be imposed.

Given Prince's analysis, whether the manufacturer or vendor of medical software should be held strictly liable for harm to patients caused by defects in that software is a matter of whether the software can be classified as (1) individualized, custom-designed, tailor-made software, (2) packaged, canned, ready-to-use, generic software, or (3) partially modified software.

Even if the software at issue does fall into the category of packaged software, the vendor will be held strictly liable for harm only if that harm is due to a defect in the software. An injured patient suing for damages under the doctrine of strict liability must convince the court not merely that the program harmed him but that it harmed him because it was defective. Commenting on liability for harm caused by defective software, Brannigan and Dayhoff (1981) wrote,

Liability is imposed, not simply because someone has been injured, but because the product that caused the injury is defective. Although many definitions of defect have been proposed, none has achieved universal acceptance. The most widely accepted definition, described as the "consumer expectation" test, is set forth in the Second Restatement of Torts: "The article sold must be dangerous to an extent beyond that which would be contemplated by the ordinary consumer who purchases it, with ordinary knowledge common to the community as to its [the software's] characteristics."*

Thus a defective computer program can be defined as one that, considering the environment in which the program is used, performs in a way not contemplated by a reasonable manufacturer or seller of the program and injures a patient.

In addition to convincing the court that the medical program that harmed him did so because it was defective, a patient seeking damages under the doctrine of strict liability must also show that the defective program was the proximate cause of his injury, that the program "reached" him in defective condition. Whether a defective medical computer program will be considered by the courts to have reached the patient in defective condition depends, according to Brannigan (1984), on which of three categories of medical computer programs it fits:

First, there is the sort of software that "directly interacts" with the patient. This category includes programs that are used in the operation of medical machinery and instruments, patient education, and patient monitoring. "In these cases courts would be inclined to find that the product interacts directly with the patient and thus 'reaches' him. Such cases are most similar to current product liability lawsuits." Second, there are

[p]rograms which are ordinarily relied on by medical personnel who do not question the correctness of the output. This type of program is typified by a medical records program. The users of a medical record do not normally challenge or discuss the truth of the record. While the question is less clear, courts have tended to expand strict liability when no independent human intervention is expected.

Third, there are

*For medical software the "ordinary" user would be a hospital or physician, and the relevant "ordinary knowledge" of the characteristics of a medical computer program is the knowledge common to the community of such users.

[p]hysician support programs, which supplement the medical library or act in place of a consultant.* In either case they expand the physician's medical knowledge. It is unlikely that courts would consider these programs as "reaching" the patient. The physician is expected to use his support materials in a critical way, and does not rely on them in the same way as a nurse in an ICU relies on a warning buzzer. He has discretion to ignore them and is expected to evaluate their accuracy. In this case liability to the patient would be based on negligence [of the physician].

Summary of Software Liability

In discussion of liability for harm to patients caused by medical software, we have emphasized three points: First, harm to a patient by medical software constitutes a tort and thereby the basis on which to sue for damages when that harm can be traced to negligence on the part of someone who owes a duty of due care to the patient or when the harm results from a defect in the kind of software that the courts will treat as a product. A suit of the former sort is based on liability for negligence, whereas a suit of the latter sort is based on strict liability, so-called liability without fault. Second, the case for legal imposition of duties of due care to the patients on the systems analysts and programmers who design, write, and test medical software; the manufacturers and vendors who bring it to market and induce purchase of it; and the hospitals and physicians who use it is very persuasive. Although the courts clearly regard physicians as legally subject to such a duty, the question of whether the courts will similarly regard the others involved in the chain of events leading to the use of medical software is open. Third, manufacturers and vendors of medical software can expect to face liability for harm caused by defects in such software when it is of the generic, ready-to-use type and either directly interacts with the patient or is the sort of software on which medical personnel are likely to rely unquestioningly.

---

*Programs for medical diagnosis fit this category.

The Concept of Privacy

Although the concept of privacy is complex and therefore difficult to explain fully, two dimensions of the concept are clearly fundamental. First, "privacy is freedom in one's immediate situation from unwanted intrusions of other people" (Greenwalt 1978). In this sense privacy can be diminished in a number of ways: It is diminished when one wishes to be left alone yet is subjected to the unwanted presence of others, when one is disturbed by unwanted noise created by others, when one's body suffers unwanted physical penetration by others, and when one is subjected to unwanted observation by others (Greenwalt 1978). Although these examples probably do not exhaust the ways in which privacy as freedom from unwanted intrusion can be diminished, they should help to provide an intuitive grasp of what is privacy in this sense.

The second dimension of privacy, and the one most pertinent to the following discussion, concerns control over information about oneself. One has this informational privacy to the extent that one controls what is known about oneself and by whom it is known (Greenwalt 1978):

Informational privacy can be lost when information [about himself] is obtained from a person against his will, either because he is coerced into providing it himself or because it is forcibly taken from his person or an area over which he has control. Informational privacy can also be lost when another person divulges it to a broader audience or has it taken by another audience. This can happen when a friend betrays a confidence, when a reporter publishes an embarrassing story, when a criminal conspirator is given immunity on the condition that he testify against his fellows, or when a physician's files are copied without his awareness.

The Moral Significance of Privacy

The moral importance of privacy, both as freedom from unwanted intrusion and informational privacy, resides primarily in its connection to individual autonomy. An individual has autonomy to the degree that his life is determined by his own uncoerced choices, that is, on the basis of his own preferences and values. Clearly control over what is known about oneself and by whom it is known are crucial to one's autonomy. The kind of job a person is able to obtain and keep, the kinds of relationships she can maintain with others, her ability to obtain a loan, and her ability to avoid being treated unfairly on the basis of erroneous or prejudicial information are but a few of the many important aspects of a person's life that are affected by informational privacy. Each of these aspects of her life and, consequently, her ability to live the life she wishes can be profoundly affected by her ability to control what is known about her and by whom it is known. A society that values individual autonomy must also value privacy. Indeed this appreciation of individual autonomy is part and parcel of why such societies in fact tend to regard privacy as a fundamental right of all persons.

Privacy and Confidentiality
Many of the benefits that are possible only through interaction with others can be obtained only if one provides them with personal information. The benefits of medical care are among the most important of such benefits. Various kinds of information about a person's physical condition, behavior, living habits, most intimate relationships and activities, and the like must be available to providers of medical care if they are to do their best on behalf of the patient's health. Something of a dilemma is posed by this fact. How can an individual obtain benefits of this kind, yet also maintain informational privacy?

The practice of keeping information confidential is an important means of addressing this dilemma. Indeed the medical profession's recognition of the importance of confidentiality in the relation between patient and physician, for example, stretches back to the time of Hippocrates. The oath of Hippocrates, written some time between the sixth century B.C. and the first century A.D., includes the following statement as part of pledge to be taken by physicians:

What I may see or hear in the course of the treatment or even outside of the treatment in regard to the life of men, which on no account must be spread abroad, I will keep to myself holding such things to be shameful to be spoken about.

And more recently section 9 of the AMA Code states that a

physician may not reveal the confidences entrusted to him in the course of medical attendance, or the deficiencies he may observe in the character of his patients, unless he is required to do so by law or unless it becomes necessary in order to protect the welfare of the individual or the community.

By keeping information confidential, that is, by refusing to let it go beyond the bounds of the patient-physician relationship without the patient's permission, the physician makes it possible for a patient to maintain his informational privacy while also obtaining the benefits of medical care. If the physician does in fact maintain confidentiality, then giving personal information to him does not entail that the patient loses control over that information. The practice of confidentiality in medical care allows the patient to receive care without sacrificing his autonomy. Of course the patient cannot receive care and be absolutely certain that his privacy will not be lost. Any time he gives personal information to someone else, he faces the risk that it will be made available to others against his wishes. But insofar as the duty of confidentiality is inculcated in physicians and others privy to personal information about patients, the risk is minimized in the context of medical care. And, insofar as individuals are secure in the belief that providers of medical care will practice confidentiality, they will be willing to seek care. This means that persons who otherwise might not seek care will do so, and the utility this has is bound to be considerable.

Computers, Privacy, and Confidentiality

Why are computers so widely viewed as posing a threat to privacy and confidentiality? According to Johnson (1985),

It is interesting to note that computers can be used to do all sorts of things from modeling systems, analyzing statistics, and regulating industrial processes, to word processing, and yet the privacy issue centers on one of the computer's most primitive functions, its capacity to store, organize, and exchange records. What is it about computerized records that gives rise to this concern? People have been keeping records for thousands of years. Can it be the mere fact that information is being stored on the computer that makes a difference? On the one hand, it would seem to make little difference to the individual how information is kept. Why should I care if my medical records are typed on paper and kept in a file cabinet or stored electronically in a computer? There is nothing inherent in electronic records that makes them threatening. Rather the concern arises because of the possibilities that are created when information is stored in this form.

What are those possibilities? First, the computer vastly increases the amount of information that can be stored in a single place. The desire of medical care providers to improve care for individuals presently being treated as well as for future patients, coupled with the fact that personal information about individuals is crucial to this quest, inevitably leads to use of the vast storage capacity of computers. Information that would not have been collected before is collected now because the storage limitations that were once faced are now gone. As a consequence computerized medical record keeping makes confidentiality more important than ever to patients. For now there is more information that can slip from one's control. Second (Johnson 1985),

[a]s the level of information gathering increases, so do the possibilities for the exchange of records. Once information about an individual is recorded, it can be sold, given away, traded, and even stolen. The information can be spread from one company [or hospital or medical office] to another, from one sector to another, and from one country to another.

The computer makes this kind of exchange of information incredibly quick and easy. The exchange can take place with or without the knowledge of the person whom the information is about, and it can take place intentionally as well as unintentionally. There is unintentional exchange when information is stolen, and also when records are given that contain much more information than is requested.

This means that the possibilities for intentional or unintentional breach of medical confidentiality are increased astronomically by computerized medical records. And this in turn means that the risks of loss of privacy when one provides personal information to a medical professional are also increased astronomically. In short computerization of medical records means that patients face the risk of losing control over much larger quantities of information about themselves than ever before and that the number of persons who potentially may gain access to that information against a patient's will is much greater than ever before. Given the importance of privacy to individual autonomy, although computerization of medical records offers many benefits to patients, these are extremely important risks. Therefore the medical community, computer specialists, and the larger society must give very high priority to the task of minimizing this threat. Medical professionals must resist the temptation not to be judicious and selective in the information they elicit from and about patients. Information should not be collected and stored simply on the hope that it may prove medically useful. Methods of making intentional and unintentional release of confidential medical information especially difficult must be found and constantly improved. And, perhaps most important, the duty to respect privacy must be firmly inculcated in all persons who have access to medical information, and highly effective mechanisms for enforcing this duty must be put in place.

Summary

In this chapter we have examined the use of computer-based medical technology. Particular attention has been given to the use of computer technology in clinical laboratory analysis, acquisition of patient data, multiphasic testing systems, establishing and extracting data from medical records, and diagnostic support systems using artificial intelligence and expert systems approaches. The use of computer technology in these areas of medical practice has increased the ability of medical professionals to obtain, analyze, evaluate, store, and retrieve information that is critical to patient care. As a consequence the quality and efficacy of that care has significantly improved and, in some instances, may have become less costly.

We have also examined three specific economic issues concerning the use of computers in medicine:

First, has the use of computers reduced or increased health care costs? In general, computer-based technologies have reduced the costs of providing specific health care services, although in some cases total expenditures on health care may have increased because lower costs of care have encouraged greater use of the health care system.

Second, has the adoption of computers in medical care taken place too slowly or too rapidly? Whether the rate of adoption of computer-based technologies has been too fast or too slow is more difficult to assess. Determining the optimal rate of adoption associated with a specific technology for a specific health care provider requires precise information about the particular circumstances of the provider.

Third, has the use of computers resulted in the replacement of people by machines and therefore greater unemployment for health care providers? The effects of the adoption of computer-based technologies have been different for different categories of workers. It is clear that employment for some categories of workers (for example, file clerks) has declined. However, employment of people with other types of skills (for example, computer programmers and data management specialists) has increased. In addition, because computer-based technologies have in general lowered health care costs, some new job opportunities have been created for traditional health care providers (for example, physicians and nurses). On balance, employment in the health care industry has probably increased as a result of the computer revolution.

Although none of these computer-related issues is unique to medicine, the issues are of particular importance with respect to the health care industry because of its ever-increasing dependence on clinical data to diagnose and treat patients.

Finally, this chapter has presented two major ethical issues related to the use of computers in the health care industry, that is, legal liability for any harm done to patients as a result of defective computer software and the perception that the use of computers to collect, store, and retrieve patient data poses a threat to privacy. Although the ultimate determination of who will be liable for such harm and what form of liability will be incurred is to be made by the courts, a strong case exists to impose some form of legal liability on all parties involved in the process that leads to use of medical software. Imposing such liability would establish a legal obligation of all parties involved in this process to take measures to protect patient well-being. Concerning patient privacy, it has been argued that the central issue is the patient's ability to control who has access to medically related information about himself. The use of computers allows more information to be collected and stored than ever before. Hence insofar as privacy is an important value, especially in the often intimate context of medicine, the use of computers means that efforts to keep such information confidential and thereby available only to authorized parties have become extremely important.

References

Alpert, N. C. 1979. Laboratory instruments in the year 2000: Streamlined—but much like today's. *Med. Lab. Obs.* 11:120-127.

Averill, K. H. 1984. Computers in the courtroom: Using computer diagnosis as expert opinion. *Computer Law Journal* 5:217-231.

Barnett, G. O., Cimino, J. J., and Hupp, J. A. 1987. DXplain: An evolving diagnostic decision support system. *JAMA* 258:67-74.

Bleich, H. L., Berkley, R. F., and Horowitz, G. L. 1985. Clinical computing in a teaching hospital. *New England Journal of Medicine* 312:756-764.

Blum, B. I. 1984a. Why AI? In Cohen, G. S. (ed.): *Proceedings of the 8th Annual Symposium On Computer Application In Medical Care.* IEEE Publication No. 84CH2090-9, pp. 3-9.

Blum, B. I. (ed.) 1984b. *Information Systems for Patient Care.* Springer-Verlag.

Brannigan, V. M., and Dayhoff, R. E. 1981. Liability for personal injuries caused by defective medical computer programs? *American Journal of Law and Medicine* 7:123-144.

Bronzino, J. D. 1982. *Computer Applications for Patient Care.* Addison-Wesley.

Bronzino, J. D., Morelli, R. A., and Goethe, J. W. 1989. Overseer: A prototype expert system for monitoring drug treatment in the psychiatric clinic. *IEEE Transactions on BME* 36(5):533-540.

Brooks, R. E., and Heiser, J. F. 1979. Transferability of a rule-based control structure to a new knowledge domain. In Orthner, F. H. (ed.): *Proceedings of the 3rd Annual Symposium of Computer Applications in Medical Care*, pp. 56-63.

Clancey, W. J. 1983. The epistemology of a rule-based expert system: A framework for explanation. *Artificial Intelligence* 20:215-251.

Collen, M. F., Danis, L. S., Van Brunt, E. E., and Terdiman, J. F. 1974. Functional goals and problems in large-scale management and automated screening. *Federation Proceedings* 33:2376-2379.

Davis, R. 1984. Amplifying expertise with expert systems. In Winston, P. H., and Prendergast, K. A. (eds.): *The AI Business.* MIT Press.

Dickson, J. R. 1969. Automation of clinical laboratories. *Proc. IEEE* 57:1974-1985.

Engler, P. E., Greisler, H. P., and Stahlgren, L. H. 1983. A versatile mobile clinical microcomputer system. In Karanja, L. L. (ed.): *Computers in Clinical and Biomedical Engineering.* Quest, pp. 77-85.

Firebaugh, M. W. 1988. *Artificial Intelligence: A Knowledge-Based Approach.* Boyd and Fraser.

Gallant, S. I. 1988. Connectionist expert systems. *Communications of the ACM* 31(2):152-169.

Garfield, S. R. 1970. The delivery of medical care. *Scientific American* 222:15-23.

Garfield, S. R. 1974. The computer and the new health care system. In Collen, M. F. (ed.): *Hospital Computer Systems.* John Wiley and Sons, pp. 24-31

Greenwalt, K. 1978. Privacy. In: *The Encyclopedia of Bioethics*. Vol. III. The Free Press.

Heiser, J. F., and Brooks, R. E. 1978. Design considerations for a clinical psychopharmacology advisor. In Orthner, F. H. (ed.): *Proceedings of the 2nd Annual Symposium on Computer Applications in Medical Care*, pp. 278-286.

Heldrich, A. 1986. Torts. In: *The New Encyclopedia Britannica: Macropedia*. Vol. 28.

Jenkins, M. A. 1978. Functions of patient care computing. *Med. Instrum.* 12:213-214.

Johnson, D. 1985. *Computer Ethics*. Prentice-Hall.

Kulikowski, C. A., and Weiss, S. M. 1979. Laboratory computers in the year 2000: Call them intelligence amplifying systems. *Med. Lab. Obs.* 11:150-163.

Kulikowski, C. A., and Weiss, S. M. 1982. Representation of expert knowledge for consultation: The CASNET and EXPORT projects. In Szolovits, P. (ed.): *Artificial Intelligence in Medicine*. West View Press, pp. 21-55.

Lincoln, T. L. 1978. Computers in the clinical laboratory: What we have learned. *Med. Instrum.* 12:223-236.

Lipkin, L. E., and Lipkin, B. S. 1975. Computers in the clinical pathologic laboratory: Chemistry and image processing. *Ann. Rev. Biophys. Bioeng.* 4: 529-577.

Lloyd, D. 1985 (Oct). Frankenstein's children: Artificial intelligence and human value. *Metaphilosophy* 16:307-318.

Martinek, R. 1972. Automated analytical systems. *Medical Electronics and Data* 3:33-39.

Miller, P. L. 1983. Critiquing anesthetic management: The ATTENDING computer system. *Anesthesiology* 58:362-369.

Miller, R. A., Paple, H. E., and Myers, J. D. 1982. Internist-1: An evolving computer-based diagnostic consultant for general internal medicine. *New England Journal of Medicine* 307:468-476.

Miller, R. A., Schaffner, K. F., and Meisel, A. 1985. Ethical and legal issues related to the use of computer programs in clinical medicine. *Annals of Internal Medicine* 102:529-536.

Miller, R. A., McNeil, M. A., and Challino, S. M. 1986. The Intermist-1/QUICK Medical Conference Project: Status report. *Western Journal of Medicine* 145:816-822.

Mulsant, B., and Servan-Schreiber, D. 1984. Knowledge engineering: A daily activity on a hospital ward. *Comput. Biomed. Res.* 17:71-91.

Myers, N. 1986. Introduction to expert systems. *IEEE Expert* 1(1):100-109.

Nycum, S. 1980. Liability for malfunction of computer programs. *Rutgers Journal of Computers, Technology and Law* 7:1-22.

Overby, M. 1987. PSYXPERT: An expert system prototype for aiding psychiatrists in the diagnosis of psychotic disorders. *Comput. Biol. Med.* 17(6):383-393.

Preston, K., Fagen, C. M., Huang, K. H., and Pryor, T. A. 1984. Computing in medicine. *Computer* pp. 294-313.

Prince, J. 1980. Negligence: Liability for defective software. *Oklahoma Law Review* 33:848-855.

Pryor, T. A., Gardener, R. M., Clayton, P. D., and Warner, H. R. 1983 (Apr). The help system. *J. Med. Sys.* 7(2):87-102.

Rennels, G. D., and Shortliffe, E. H. 1987. Advanced computing for medicine. *Scientific American* 255(10):54-161.

Safran, C., and Porter, D. 1986. New uses of the large clinical data base at the Beth Israel Hospital in Boston: On-line searching by clinicians. In Orthner, H. F. (ed.): *Proceedings of the 10th Annual Symposium on Computer Applications in Medical Care*. IEEE Computer Society Press, pp. 114-119.

Sandell, H. S. H., and Bourne, J. R. 1985. Expert systems in medicine: A biomedical engineering perspective. *CRC Crit. Rev. Biomed. Eng.* 12(2):95-129.

Schwartz, W. B., Tatil, R. S., and Szolovits, P. 1987. Artificial intelligence in medicine. *New England Journal of Medicine* 316(11):685-687.

Servan-Schreiber, D. 1986. Artificial intelligence and psychiatry. *Journal of Nervous Mental Disorders* 174(4):191-202.

Shortliffe, E. H. 1976. *Computer-Based Medical Consultations: MYCIN.* Elsevier/North Holland.

Shortliffe, E. H. 1986. Medical expert systems: Knowledge tools for physicians. *Western Journal of Medicine* 145:830-839.

Shortliffe, E. H., Axline, S. G., Buchanan, B. G., et al. 1973. An artificial intelligence program to advise physicians regarding antimicrobial therapy. *Comput. Biomed. Res.* 6:544-560.

Shortliffe, E. H., Scott, A. G., Bischoff, M. B., Campbell, A. B., Van Melle, W., and Jacobs, C. D. 1981. ONCOCIN: An expert system for oncology protocol management. *Proc. 7th International Joint Conference in Artificial Intelligence,* pp. 876-881.

Tsuji, S., and Shortliffe, E. H. 1986 (May). Graphical access to the knowledge base of a medical consultation system. In: *Proceedings of AAMSI Congress 83*, San Francisco, pp. 551-555.

Titus, J. A. 1977. The impact of microcomputers on automated instrumentation in medicine. In: *Proceedings of the 5th Annual Symposium on Computer Applications in Medicine.* IEEE Press, pp. 99-103.

Waterman, D. A. 1986. *A Guide to Expert Systems.* Addison-Wesley.

Weiss, S. M. Kulikowski, C. A., and Safir, A. 1978. Glaucoma consultation by computer. *Comput. Biol. Med.* 8:24-40.

Werner, G. 1987. Methuselah: An expert system for diagnosis in geriative psychiatry. *Computers and Biomedical Research* 20:477-488.

Westlake, G. 1975. Microprocessors, programmable calculators and minicomputers in the clinical laboratory. *Assoc. Adv. Med. Instrum. Clin. Eng. News.* 3: 1-3.

Whitcomb, C. C., Vogt, C. P., and Wilbur, N. M. 1978. Clinical laboratory data processing with a central hospital computer. *Comput. Biol. Med.* 8:197-206.

Williams, C. Z. 1969. Automation in clinical laboratories. In Dickson, J. F.,, and Brown, J. V. H. (eds.): *Future Goals of Engineering in Biology and Medicine.* Academic Press.

Williams, C. Z. 1975. Why automation? In Kinney, T. D., and Melville, R. S. (eds.): *Conference: Evaluation of Uses of Automation in the Clinical Laboratory.* Sponsored by the National Institute of General Medical Sciences, DHEW, pp. 5-15.

Wist, A. O., Horowitz, R. E., and Megargle, R. 1982. Computers in the clinical laboratory. In Schwartz, M. D. (ed.): *Applications of Computers in Medicine.* IEEE Press, Catalog No. TH0095-0.

7
Medical Imaging

Development of noninvasive tools for looking into the human body while posing minimal risk for the patient has been given an extremely high priority in the post–nineteenth century scientific era of medical research and practice. Diagnostic and therapeutic techniques that remove the need for invasive (surgical) exploration often both reduce patient risk and lower the costs of health care. Until recently, however, the health care profession's "black bag" has contained only a relatively small number of noninvasive diagnostic tools that provide images of the body. None was available until the end of the nineteenth century.

The development of medical imaging techniques began with Roentgen's discovery of the X ray on November 8, 1895. Roentgen had been carrying out experiments to discover the effects of passing electricity through rarified gases using a small battery-operated induction coil connected to a partially exhausted (empty) "discharge tube." On that day he covered the discharge tube with black paper to exclude fluorescent light and darkened the room to exclude daylight. During the experiment he noticed that whenever he energized the induction coil, there was a glow in the platinum-barium-cyanide that coated a piece of cardboard lying near the discharge tube. Fascinated, he began to investigate the cause of the glow and rapidly became convinced that he was dealing with an entirely new form of energy, which he later termed X rays (Hodges and Moseley 1974). He reported the results of these experiments to the Wurtzberg Physical Medical Society on January 6, 1896, thereby launching an entirely new field—medical imaging. Appropriately Roentgen later received a Nobel Prize for his discoveries.

We now know that X rays are (1) part of the electromagnetic spectrum that have relatively short wavelengths, (2) generated by collision of atomic particles, (3) travel in straight lines in all directions from the point of origin, and (4) are capable of penetrating many forms of matter opaque to ordinary light. Therefore the classical medical imaging technique, radiology, uses high-energy electromagnetic waves or X rays as a tool to "look" into the patient to view certain anatomical structures. During the early part of the twentieth century, additional improvements in radiography involved advances in technology related to particular components of the system, including intensifying screens and rotating anode tubes.

It is interesting to note that after 1930 profound improvements in radiology were achieved by developing procedures for selectively opacifying regions of interest in the body. These developments were achieved through a variety of approaches, such as the injection of dyes, often invasive to the body, to facilitate visual representation of otherwise invisible organs (Macovski 1983).

Since 1960, through the marriage of computers and sophisticated imaging processing techniques, medical imagery has been able to pursue totally new directions. Startlingly clear images of the internal structures of the body are now created, significantly enhancing the ability of the clinician to arrive at an appropriate diagnosis.

Although modern medical imaging takes many forms, systems of most practical interest today include nuclear medicine, diagnostic ultrasound, computerized tomography, and magnetic resonance imaging. These imaging techniques enable doctors to avoid exploratory surgery and still watch vital organs at work, identify blockages and growths, and detect warning signs of possible future disorders. Because these techniques, for the most part, do not involve penetration of skin, they have been classified as noninvasive. However, patients may face other consequences of their use (for example, the possible destruction of body tissue caused by the radioactive materials used in nuclear medicine, or the increase in tissue temperature as a result of high-frequency ultrasound). Thus it may be more appropriate to consider the relative degree of direct surgical intervention that certain devices entail. Therefore we use the term *noninvasive* to refer only to specific techniques developed in radiology, nuclear medicine, and ultrasound.

The purpose of the next four sections of this chapter is to describe the priniciples that underlie the major imaging modalities and to explore likely future developments of those technologies. The fifth section is concerned with the economic issues that arise because of the high capital costs associated with many imaging modalities, reasons why medicine relies on multiple imaging techniques that appear to duplicate one another in terms of their functions, and the general question of whether it is always optimal for society to prevent the incidence of disease (through diagnosis) than to cure it. In the sixth section we explore the ethical issues associated with the use of technological assessment to determine whether imaging modalities (and other technologies) should be developed or used.

Nuclear Medicine

Nuclear medicine has been defined as the application of radioactive materials (tracers) to medical diagnosis and treatment. It is a classic example of a medical discipline that has embraced and utilized concepts developed in the physical sciences. Conceived as a joint venture between the clinician and physical scientist, nuclear medicine has evolved into a science with its own body of knowledge, techniques, and skills that can be systematically studied and improved. In the process the domain of nuclear medicine has grown to include studies pertaining to (1) the creation and proper use of radioactive tracers (or radiopharmaceuticals), (2) the design and application of appropriate nuclear instrumentation to detect and display the activity of these radioactive elements, and (3) the determination of the relation between the activity of the radioactive tracer and specific physiological processes (Wagner 1968, 1975). To appreciate the present status of nuclear technology in clinical medicine, let us explore what is meant by radioactivity, how it is detected, and what instruments are available to monitor or image the activity of radioactive materials.

What Is Radioactivity?
The seeds for nuclear medicine were planted in 1895 when Wilhelm Conrad Roentgen announced the discovery of a new type of penetrating radiation emitted from a gas discharge tube with which he had been working. Shortly after Roentgen's discovery the

French physicist Henri Becquerel began to investigate the possibility that known fluorescent or phosphorescent substances produce a type of radiation similar to the X rays discovered by Roentgen. As a result of these studies, he announced in 1896 that certain uranium salts also radiated, that is, they emitted penetrating radiations, and that these rays were emitted spontaneously and continuously from all uranium compounds whether or not they were fluorescent. These results were startling, presenting the scientific world with a new and entirely unexpected property of matter.

Becquerel encouraged Marie Curie, at that time a promising young scientist at the Ecole Polytechnique, to investigate what caused radiation in uranium. Subsequently, in 1898, Madame Curie and her husband Pierre, who already had a distinguished reputation as a physicist through his work on piezoelectricity, reported their discovery of radium, a substance that has far more radioactivity than uranium has* (Wagner 1975).

Concurrently scientists in England were developing new insights about the structure of matter. Throughout most of the nineteenth century, Dalton's chemical theory of atomic structure, which held that all matter is composed of atoms that are *indivisible*, received widespread acceptance. Thompson's discovery in 1897 of negatively charged electrons with masses much smaller than the lightest atom shook the foundations on which Dalton's theory had been constructed. Rutherford, in 1911, offered what broadly remains the contemporary view of atomic structure by demonstrating that the principal mass of an atom is concentrated in a dense positively charged nucleus surrounded by a cloud of negatively charged electrons. This insight, incorporated two years later by Niels Bohr in his planetary concept of the atom—a nucleus surrounded by very light orbiting electrons—derived from Rutherford's earlier Nobel Prize–winning research,† which led him to conclude that radioactive emissions involve the spontaneous disintegration of atoms.

*Madame Curie coined the term *radioactivity* to describe the emitting properties of radioactive materials.

†Rutherford received a Nobel Prize in chemistry in 1908 for his research on radiation.

The Curies had recognized that radioactive material emits three distinct types of active radiation. Rutherford arbitrarily labelled them *alpha*, *beta*, and *gamma*. They are now known to be (1) alpha particles, which are positively charged and identical to the nucleus of the helium atom; (2) beta particles, which are negatively charged electrons; and (3) gamma rays, which are pure electromagnetic radiation with zero mass and charge.

The insights developed by Roentgen, the Curies, and Rutherford provided the foundation for our understanding of radioactivity and radiation and created new fields of research in medicine and the natural sciences. Subsequent improvements in measuring radioactive emissions and the development of artificial radionuclides that could be applied in clinical settings all rest on their initial, revolutionary research discoveries.

Elementary Particles

The discoveries of radiation and the properties of radioactive materials provided the scientific blocks on which nuclear medicine, almost fifty years later, could build many of its clinical procedures. Research on elementary particles was also important. By the 1930s three elementary particles of the atom had been identified: the electron, the proton, and the neutron, usually considered the building blocks of atoms. Consider the arrangement of these particles as shown in figure 7.1, a common view of the atom. The atom includes a number of particles: (1) one or more electrons, each having a mass of about 9.1×10^{-31} kg and a negative electrical charge of 1.6×10^{-19} coulomb (the unit of charge 1 coulomb is the charge on 6.28×10^{18} electrons); (2) at least one proton with a mass of 1.6×10^{-27} kg, which is approximately 1800 times that of the electron; and perhaps (3) neutrons, which have the same mass as protons, but possess no charge.

The electrical charge carried by the electron and the mass of a particle are fundamental properties of matter. Its electrical charge is the smallest amount of electricity that can exist, usually expressed as a *negative unit charge* (–1), and in the metric system of units (meters, kilograms, and seconds) it has the value 1.6×10^{-19} coulombs. The charges carried by all atomic particles are therefore integral multiples of this value.

Figure 7.1
Planetary view of the structure of the helium atom. The primary mass of the atom
is located in the nucleus, which contains both neutrons and protons. The nucleus,
which has a net positive charge, is surrounded by smaller orbiting electrons. In
stable atoms the net charge of electrons in orbit is equal and opposite to that of
the nucleus.

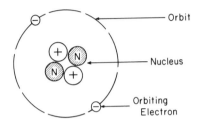

$\ominus$ = Electron: mass = 9.1×10^{-31} kilograms
charge = -1.6×10^{-19} Coulombs

$(+)$ = Proton: mass = 1.6×10^{-27} kilograms
charge = $+1.6 \times 10^{-19}$ Coulombs

(N) = Neutron: mass = 1.6×10^{-27} kilograms
charge = 0

Neutrons and protons that exist together in the nucleus and are called *nucleons*. The total number of nucleons in the nucleus of an element is called the *atomic mass*, or *mass number*, and is represented by the symbol A, whereas the number of protons alone is referred to as the atomic number Z.

Three types of hydrogen atom (hydrogen, deuterium, and tritium) illustrate the various combinations of the neutron and proton that exist in nature. Although hydrogen has a nucleus consisting of a single proton, the combination of one proton and one neutron exists as a single particle called *deutron* and is the nucleus of the atom called *heavy hydrogen*, or *deuterium*. The combination of two protons and two neutrons forms a stable particle—the alpha particle—which exists in nature as the nucleus of the helium atom. Alpha radiation emitted from radioactive substances consists essentially of a stream of such particles. *Tritium*, the third atom of hydrogen, on the other hand, has a nucleus consisting of only one proton and two neutrons.

These three types of hydrogen are examples of atoms whose nuclei have the same number of protons and therefore the same atomic number Z, but a different number of neutrons and thereby a different atomic mass A. Atoms exhibiting this characteristic are called isotopes (from the Greek for "same place"). The term *isotope* has been widely used to refer to any atom, particularly a radioactive one. However, current usage favors the term *nuclide* to refer to a particular combination of neutrons and protons. Thus isotopes are nuclides that have the same atomic number. All elements with an atomic number Z greater than 83 or atomic mass A greater than 209 (or both) are radioactive, that is, they decay spontaneously into other elements, and this decay causes the emission of active particles.

To specifically designate each type of atom, certain symbols have been developed. In the process of reading the literature in this field, it is therefore necessary to become familiar with them. Until recently, in the United States it was the practice to place the atomic number as a subscript before and the atomic mass as a superscript after the chemical symbol of the atom ($_{53}I^{131}$). Because the chemical symbol itself also specifies the atomic number, one often omits it (I^{131}). In Europe on the other hand it was customarily written as a superscript before the chemical symbol (^{131}I). In an effort to achieve international standards, it was agreed in 1964 that the atomic mass should be placed as a superscript preceding the chemical symbol (^{131}I). When superscripts are not used, a more literal form of designation, for example, cobalt 60, is commonly used. Referring to the helium atom illustrated in figure 7.1, then, one simply writes He or $^4_2$He, where 4 is equal to the atomic weight (because of the aggregation of protons and neutrons in the nucleus), and 2 denotes the atomic number of the element.

Radionuclide Production and Decay

In general radionuclides are classified as either natural or artificial (Wagner and Buddemeyer 1974). Naturally occurring radionuclides are those nuclides that emit radiation spontaneously and therefore require no additional energy from external sources. Artificial radionuclides, on the other hand, are essentially synthetic, produced by bombarding so-called stable particles. Both types of radionuclides are important in nuclear medicine.

It has been suggested that there is constant motion of all the particles within the atom in the natural state (Early et al. 1975). In the nucleus this motion results in collisions between nucleons, whereby energy is transferred to other nucleons. If in this process any of the nucleons achieve an energy level greater than the one binding it to the nucleus, the particle is allowed to escape the nucleus. This particle escape is termed a disintegration. The process of disintegration (decay) and the act of particle escape allow the nucleus to reduce the number of protons or neutrons or both to a point of relative equilibrium, in which the binding energy can contain the remainder of the nucleons. In this way, stability is eventually achieved. Stated another way, the method by which a nuclide decays (that is, emits radiation particles) results in the nucleus becoming stable or approaching stability. If the decay process contains only one step, then the resultant new nuclide is stable. On the other hand, if many steps are involved in the decay process, then the nuclide formed after each step is unstable, until eventually the stable form of an element is reached. The ejection of the particle from the nucleus is sometimes accompanied by a release of electromagnetic energy. It is this released energy, or gamma radiation, as well as the ejected particle, that constitutes the phenomenon of radioactivity.

Particulate radiation has two forms: alpha radiation and beta radiation. The alpha particle (two protons and two neutrons) is heavier than the beta particle. Thus it travels much more slowly than a beta particle possessing the same energy. Consequently alpha particles are more likely to interact with matter and to lose energy more rapidly. As a result alpha particles penetrate air only a short distance (centimeters) and can be stopped by a sheet of paper. Beta particles (electrons) on the other hand weigh much less than the alpha particle and are capable of deeper penetration. Beta particles may penetrate centimeters of body tissue, but are easily stopped by a sheet of aluminum. Gamma rays, which are "pure" energy waves, are even less likely to interact with matter and therefore can only be stopped by dense material such as lead. Of the three types of radiation, gamma rays are the most penetrating. Data regarding the relative penetrability of these radiation particles is extremely useful in (1) establishing the required procedure to be followed to properly shield a patient or sensitive equipment from specific types of radiation and (2) selecting and using detector systems that will identify only the radioactivity of interest.

All radioactive materials, whether they occur in nature or are artificially produced, decay by the same types of processes. They emit alpha, beta, or gamma radiations, and the emission of alpha and beta particles involves the disintegration of one element into another. The rate of this decay process is described in terms of disintegrations per unit of time or its half-life (that is, time required for half the original number of atoms in a given radioactive element to disintegrate). Half-lives of radioactive elements range in value from thousandths of a second to millions of years. For example, ^{131}I has a half-life of 8.05 days, whereas ^{60}Co has a half-life of 5.24 years.

The reduction of the number of atoms through disintegration of their nuclei, that is, radioactive decay, is characteristic of all radioactive materials. Unaffected by changes in temperature, pressure, or chemical combination, the rate of the decay process remains constant, with the same number of disintegrations occurring during each interval of time. Furthermore this decay process is a random event; consequently every atom in a radioactive element has the same probability of disintegrating.

As the decay process continues, fewer atoms remain available for disintegration. The fraction of the remaining number of atoms that decay per unit of time is called the decay constant. The half-life and decay constant are obviously related, because the larger the value of the decay constant, the faster the process of decay and consequently the shorter the half-life. In any event the decay constant remains the same throughout the decay process. The critical issue here is that each radionuclide exhibits a distinctive disintegration process because of its inherent properties, that is, its decay constant and half-life. All nuclides decay in the same manner, but not at the same rate.

Consider the information presented in table 7.1, which provides a list of several commonly used radionuclides. The symbol of each radionuclide is given along with its atomic number Z, atomic mass A, and a half-life (Bronzino 1976). Gamma energy refers to the energy contained within the emitted radiation and is a quantitative index of its "penetrating power." In nuclear medicine this energy is expressed in terms of the electron volt (ev), which is defined as the amount of energy acquired by an electron moving through a potential difference of one volt. Because the energy possessed by gamma radiation is usually quite high, it is often expressed in terms of a thousand electron volts (kev) or a million electron volts (mev).

Measuring Radiation—Units

All radioactive substances decay and in the process emit various types of radiation (alpha, beta, or gamma). Because they usually occur in various combinations, however, it is difficult to measure each type of radiation separately. As a result measurement of

Table 7.1
Physical Data for Some Radionuclides Used in Nuclear Medicine

Element	Symbol	Atomic Number	Atomic Mass	Half-life	Prinicipal Gamma Energy(kev)
Cesium	^{137}Cs	55	137	30 y	662
Chromium	^{51}Cr	24	51	27.8 d	322
Cobalt	^{57}Co	27	57	270 d	122
Gallium	^{67}Ga	31	67	78 h	93, 184, 296
Gold	^{198}Au	79	198	2.7 d	412
Indium	^{113m}In	49	113m	104 m	393
Iodine	^{123}I	53	123	13.3 h	160
	^{125}I	53	125	56 d	35
	^{131}I	53	131	8.1 d	364
Iron	^{59}Fe	26	59	45 d	191, 1097, 1289
Mercury	^{203}Hg	80	203	46.9 d	279
Selenium	^{75}Se	34	75	120 d	136, 265, 402
Strontium	^{85}Sr	38	85	65 d	514
	^{87m}Sr	38	87m	2.8 h	391
Technetium	^{99m}Tc	43	99m	6 h	140
Xenon	^{133}Xe	54	133	5.27 d	83
Ytterbium	^{169}Yb	70	169	32 d	63, 93, 177

m = minutes; h = hours; d = days; y = years.

radioactivity is usually accomplished by using one of the following techniques: (1) counting the number of disintegrations that occur per second in a radioactive material, (2) noting how effective this radiation is in producing atoms (ions) with a net positive or negative charge, or (3) measuring the energy absorbed by matter from the radiation penetrating it. As a result three kinds of radiation measurements have been established: (1) the curie (Ci), (2) the roentgen (R), and (3) the radiation-absorbed dose (rad).

The curie is defined as the amount of radioactive material that has the same number of disintegrations per second as 1 gram of radium. Thus the curie is that amount of radioactive material in which there are 3.7 X 10^{10} nuclear transformations per second. Thirty-seven billion disintegrations per second is a large number; therefore in practice the millicurie, which is one-thousandth of a curie, or 37,000 disintegrations per second, is commonly used. Note that the curie defines the number of disintegrations per unit time and not the *nature* of the radiation, which may be alpha, beta, or gamma rays. *The curie is simply a measure of the activity of a radioactive source.*

The roentgen and the rad, on the other hand, are units based on the effect of the radiation on an irradiated object. Thus, whereas the curie defines a source, the roentgen and rad define the effect of the source on an object. One of the major effects of irradiation by X radiation or gamma radiation is the ionization of atoms, that is, the creation of atoms possessing a net positive or net negative charge (ion pair). The roentgen is determined by observing the total number of ion pairs produced by X radiation or gamma radiation in 1 cc of air at standard conditions (at 760 mmHg and 0°C). Because each ion pair has electrical charge, R can be related to electrical effects that can be detected by various instruments. Accordingly one roentgen is defined as the amount of X radiation or gamma radiation that produces enough ion pairs to establish an electrical charge separation of 2.58×10^{-4} coulombs per kilogram of air. Thus the roentgen is a measure of radiation quantity, not intensity. The rad on the other hand is based on the total energy absorbed by the irradiated material. One rad means that 0.01 joule (the unit of energy in the metric system) of energy is absorbed per kilogram of material.

In nuclear medicine human tissue is exposed to various types of radioactive materials. Another unit of measure, the rem (roentgen equivalent human on rem), is often used to specify the biological effect of radiation. Consequently the rem is a unit of human biological dose resulting from exposure of the biological preparation to one or many types of ionizing radiation.

All three measures are used in nuclear medicine. The roentgen, or more commonly its submultiple, the milliroentgen (mr) (one-thousandth of a roentgen) is used as a value for most survey meter readings. The rad describes the amount of exposure received (for example, by an organ of interest on injection of a radiopharmaceutical). The rem values exposure in some personnel-monitoring devices such as film badges (tables 7.2 and 7.3).

Table 7.2
Effects Caused by Different Doses in Humans

Dose (within a factor of 2 or 3)	Effects and Conditions
100,000 r	Spastic seizures; death within seconds; sperm motility stopped
10,000 r	Disruption of central nervous system function; death within minutes or hours
1,000 r	Necrosis of progenitive tissues; 100 percent death in 30 to 60 days
100 r	Mild irritation sickness in part of the organism; no deaths.
10 r	Few or no detectable effects
10 r/day	Debilitation in 3 to 6 weeks; death in 3 to 6 months (projected from animal data)
1 r/day	Debilitation in 3 to 6 weeks; death in 3 to 6 months (projected from animal data)
0.1 r/day	Permissible dose range 1930–1950
0.01 r/day	Permissible dose range 1957
0.001 r/day	Natural radiation

Table 7.3
Recommended Limits on Absorbed Dose Due to External Radiation

Occupational—Critical Organs (includes the whole body, head and trunck, active blood-forming organs, gonads)	Nonoccupational— Individuals in Vicinity of Controlled Areas	General Population— Average Exposure, Exclusive of Background Radiation
3 rems/13 weeks Cumulative for life: 5(age − 18) rems	1.5 rem/year (adult workers occasionally entering radiation areas) 0.5 rem/year (general public, including chidren, residents of controlled areas)	5 rems by age 30

Nuclear Instrumentation—Detection of Radioactivity

Nuclear instrumentation has improved significantly in recent years, especially for radionuclide imaging of organs and tomographic studies, largely because of innovations in computers (Bronzino 1982). The systems for radiation analysis have two major components: the detector and the electronics. The design of the detectors depends on the way in which the interaction of radiation with matter is used. Several devices have been developed to utilize the ionization and excitation of atoms to measure radioactivity.

From the discussion of radiation measurement it should be clear that radiation energy can be measured only indirectly, that is, by measuring some effect caused by the radiation. Included among the various indirect techniques used to measure radioactivity are

1. *Photography* The blackening of film when it is exposed to a specific type of radiation such as X rays. This technique is historically significant because this is how radioactivity was discovered.
2. *Ionization* The passage of radiation through a volume of gas established in the probe of a gas detector produces ion pairs. The function of this type of detector depends on the collection of these ion pairs in such a way that they may be counted. This technique has been most effective in measuring alpha radiation and least effective in measuring gamma radiation.
3. *Luminescence* The emission of light not due to incandescence. Because the flash of light produced by the bombardment of a certain type of material with a penetrating type of radiation can be detected and processed, this technique is extremely useful. As a matter of fact the fluorescent effect produced by ionizing radiation is the basis of the scintillation detector. This type of indirect detection scheme is excellent for observing the presence of all three types of radiation.

Because the majority of modern detector systems in nuclear medicine use probes based on the scintillation principle, we now describe it in greater detail. Scintillation detectors may be used for all types of radiation, depending on the particular type of scintillator used and its configuration. Regardless of the application, however, the general technique is the same for all scintillation probes. Certain materials, for example, zinc sulfide and sodium iodide, emit a flash of light, or scintillation, when struck by ionizing radiation. The amount of light emitted is, over a wide range, proportional to the energy expended by the particle in this material. When the scintillator material is placed next to the sensitive surface of an electronic device called a *photomultiplier*, the light from the scintillator is converted into a series of small electrical pulses whose height is directly proportional to the energy of the incident gamma ray. These electrical pulses can be amplified and processed to provide the operator with information regarding the amount and nature of the radioactivity striking the scintillation detector. Thus scintillators may be used to determine the amount or distribution (or both) of radionuclides in one or more organs of a patient.

Figure 7.2 illustrates a basic scintillation detection system. The system consists of (1) a detector, which usually includes the scintillation crystal, photomultiplier tubes, and preamplifier; (2) signal processing equipment such as the linear amplifier, the single-channel pulse analyzer; and (3) data display units such as the scaler, scanner, and oscilloscope. Once the radioactive event is detected by the crystal and an appropriate pulse is generated by the photomultiplier circuitry, the resulting voltage pulses are still very small. To avoid any serious loss of information caused by distortion from unwanted signals (such as noise) and to provide a signal strong enough to be processed and displayed, the amplifier is used to increase the amplitude of the pulses by a constant factor. This process is called *linear amplification*.

In such a system it should be apparent that because of the wide variation in energies of gamma rays striking the scintillation crystal, the linear amplifier receives pulses that vary widely in "pulse height." This is fine if the goal is to detect all the radiant energy under the probe. If, on the other hand, the operator is interested only in the activity of a specific radionuclide, additional processing is necessary. This is accomplished by using the pulse

Figure 7.2
Basic scintillation detection system

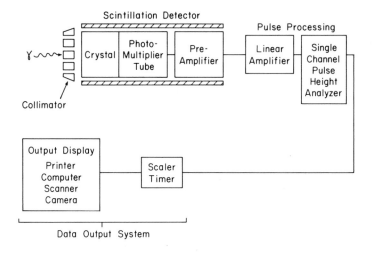

height analyzer, a device that enables the operator to discriminate against all radiant energy other than the one of interest. In this way one can focus on the activity of a specific radionuclide by allowing only those pulses related to it to be processed. Consequently single-channel and multichannel pulse height analyzers are commonly found in nuclear medicine laboratories. The processed data must then be displayed in such a way as to provide information regarding the amount of radioactivity present and its location within the body. This information is quite important in determining the status of the organs under investigation. Studies of the amount of radioactivity present as a function of time enable one to ascertain whether the organ is functioning properly or not, whereas studies of the location of the radionuclide enable one to display the organ or abnormal tissue. Both types of analyses are valuable and require special instrumentation.

For nonimagery applications measuring scintillation events often requires that the events themselves be counted and displayed. The scaler is the electronic device most commonly used in nuclear medicine to accomplish this task. A scaler is used to count the pulses produced by the detector systems and processed by a variety of electronic pieces of equipment. Thus the scaler is an electronic device that accepts signal pulses representing a range of energy levels (energy of the incident radiation) and counts them. The scaler is usually designed so that the operator has a choice between accepting a certain number of counts (preset count) or a predeter-mined period of time over which the counts can be accumulated (preset time). With preset time the scaling device counts the number of events that occur during a set period of time and then shuts off automatically—thus the count is the variable. With preset count the predetermined number of counts are accumulated, after which the scaler is automatically shut off; in this case time is the variable.

Both types of data can be observed directly by the operator from the front panel of the scaler. However, this information can also be supplied to other devices for analysis. Consider the case of operat-ing the scaler in the preset time mode. The information from the scaler can then be presented in a variety of ways: It can be (1) displayed on the front panel of the scaler as counts per unit time (second, minute) for continual observation by the operator, (2) supplied to a digital printer to provide a running tabulation of the counts as a function of time, or (3) directed to a computer that can retain the counts in memory and perform a variety of calculations on the incoming data as they are actually collected (if the process is slow enough) and at the same time provide the operator with a visual display (on the screen of an oscilloscope) in the form of a plot of radioactive decay. Thus once the radiation is detected by the scintillation detector and subsequently measured by the scaler, the operator can be presented with the information in a variety of formats that may be clinically valuable.

Scintillation data may also be processed through imaging devices. In general these devices take the pulse output from a detector and the electronic processing devices described previously and place the pulse in a spatial representation according to its point of origin in the radioactive source. Obviously there is some distortion in this representation, because the point of origin lies in a three-dimensional plane, whereas most of the instruments commonly available are capable of only displaying that point of origin in two dimensions. Despite this obvious shortcoming, this technique has proved to be of tremendous clinical value. The instruments that have become widely accepted by the medical profession are the scanner and the scintillation camera.

Before the 1950s the only method of scanning body organs containing any radioactive material involved manually placing the probe as close to the skin as possible, moving it in small increments over the organ to be studied, and recording the "count" as each new increment of area was surveyed (Early et al. 1975). With this type of manual scan the distribution of radioactivity within an organ could be obtained by plotting areas having the same count and then connecting these points by lines (called *isocount lines*). This technique provided information regarding the distribution of radioactivity as well as an overall representation (anatomical map) of the organ. However, this procedure was too time consuming and required the patience of both operator and patient.

In 1951 Benedict Cassen demonstrated the imaging capabilities of scintillation detectors using the rectilinear scanner. In the prototype the scintillation detector was mounted on a carriage that could automatically be moved over the area to be scanned. Attached to the carriage was a dot-producing mechanism that would respond to the electrical pulses coming from the scintillation counter so that the spatial distribution of radioactive material in an organ could be printed mechanically. Although technological advances have considerably enhanced the versatility, sensitivity, and spatial resolution of this type of device, the basic concepts developed by Cassen still remain the heart of most rectilinear scanning systems.

Thus modern rectilinear scanners move across the organ in one or more planes until the organ or area of interest has been completely surveyed by the scintillation detector. As it moves back and forth across the organ containing radioactive elements, the scanning mechanism produces electrical impulses proportional to the amount of activity present in each area covered by the detector. To produce a picture (image) of the area under survey, a variety of techniques have been developed. One of the most common methods is to display the image on photographic or X-ray film. In this case, as the detector scans the target area in a rectilinear path, one line at a time, its output controls a light beam that, moving synchronously with the detector, "paints" the image onto photographic film (Beck 1975). These images on film are produced not only by the density contrast of the dots but by the overlapping of the dots as well. Consequently variations in concentrations of activity can be displayed as variations in the shades of gray and black, with black representing areas of high concentration of radioactivity, and gray areas representing areas of various lesser concentrations of radioactivity.

Some imaging devices have a color display that may be modulated by the radioactivity under the scanner at each point measured during the scan. One of the first devices used to provide color in nuclear imaging was developed by Gerald Hine in the early 1960s. In this system the imaging was constructed by a colored stylus moving over a sheet of paper. Controlled by a simple electronic linkage, it was possible to contrast areas of isocount values in different colors. More recent devices use computer controls of a color video tube to present isocount values over a wide color spectrum. Various colors can be applied to increase contrast and accentuate different activity levels, areas of interest, or specific physiological functions (Wagner 1975). These displays provide a range of colors from blue—indicative of decreased count—through intermediate colors to red—indicative of a "hot" area. Initially designed this way because of the belief that it is easier to discern colors than shades of gray, it has been the subject of a great deal of controversy. Some users believe that this type of presentation provides little if any additional information and in some cases may even confound the issue, whereas others demand it when ordering new imaging systems.

The other type of scintillation counter is the stationary detector system, which simultaneously detects gamma radiation emitted from all parts of the organ viewed and does not use mechanical scanning. The most commonly used stationary detector system is the scintillation camera, or gamma camera. This device views all parts of the radiation field continuously and is therefore capable of operating almost like a camera—building up an image quickly—in contrast with one built up over long intervals of time as with moving scanners. Initially developed by Hal Anger during the late 1950s, the first gamma, or Anger, camera reflected the convergence of the disciplines of nuclear physics, electronics, optics, and data processing in a clinical setting. Its ultimate acceptance and further development has had a profound effect not only on the practice of clinical nuclear medicine but also on the entire diagnostic process. The initial concepts introduced by Anger, as was the case with Cassen's development of the rectilinear scanner, became fundamental to the art of imaging specific physiological processes.

The Gamma Camera

The operation of the basic Anger, or gamma, camera is illustrated in figure 7.3. The detector of the gamma camera is placed over the

Figure 7.3
Basic elements of the gamma camera

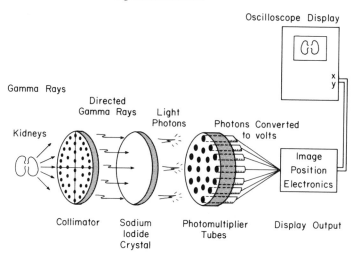

Oscilloscope Display

Gamma Rays

Directed
Gamma Rays   Light
Photons

Kidneys

x
y

Photons Converted
to volts

Image
Position
Electronics

Collimator   Sodium   Photomultiplier   Display Output
Iodide   Tubes
Crystal

organ to be scanned. To localize the radiation from a given point in the organ and send it to an equivalent point on the detector, a collimator is placed over the base of the scintillation crystal. Because gamma rays cannot be bent, another technique must be used to selectively block those gamma rays that, if allowed to continue on their straight-line path, would strike the detector at sites completely unrelated to their points of origin in the patient. This process of selective interference is accomplished by the collimator.

To prevent unwanted off-axis gamma rays from striking the crystal, collimators usually contain a large number of narrow parallel apertures made of heavy-metal absorbers. Consider the multihole collimator illustrated in figure 7.3. It consists of a flat lead plate through which narrow holes are drilled. As can be seen, only a gamma event occurring directly under each hole will penetrate the collimator and be represented at only one location on the face of the crystal. If this gamma event occurred much further away from the collimator, such as in region Y, then it would be represented by more than one location on the face of the crystal. When this occurs, resolution, which may be defined as the ability of the detector to distinguish between two sources at various distances from the collimator, is greatly decreased. For the multihole collimator then the best resolution occurs when the area of interest is close to the collimator. Thus as the patient is moved from point Z toward the detector, the resolution is improves. Obviously, when viewing an organ that lies beneath the surface of the skin, it becomes quite important to closely approximate its distance from the probe relative to the degree of resolution required.

From this brief discussion it can be seen that collimators in conjunction with a scintillation detector essentially focus the radioactivity occurring at a particular point in space within the organ to a particular point on the surface of the crystal. The radiation passing through the collimator then impinges on or interacts with the scintillation crystal.

Today commercially available gamma cameras have a single large circular sodium iodide crystal backed with an array of photomultiplier tubes. This arrangement makes it possible for the image detected by the crystal to be converted to an analogous optical display. Once this has been accomplished, it is then simply a matter of photographing the display on the face of the oscilloscope and recording each event on film. Because each dot on the oscilloscope is a result that corresponds to an event within the detector, recording these events over a period of time results in an image on film that displays the distribution of the radionuclide within an organ modified to some extent by various uncontrollable scatterings and absorptions that occur on the way through the rest of the body. Because it is often desirable to view the distribution of radioactivity as it is being collected, most gamma cameras are equipped with variable persistence, or storage, oscilloscopes. These devices permit the operator to continuously view the image being represented on the screen of the oscilloscope and provide a means for recording a series of images on film or digital tape for dynamic studies. In cases in which there is a constantly changing displacement of the radionuclide in an organ, such as renal function and cardiac studies, the gamma camera provides a distinct advantage over other techniques.

The basic features of the basic gamma camera have been incorporated in almost all modern gamma camera products presently on the market. In these systems each scintillation event that occurs in an organ and enters the detector is displayed as a single dot on a CRT display. The position of this dot precisely corresponds to the position of the originating gamma event in the organ. A photograph can be obtained of this display, thereby recording the dots on film. After recording enough photographic records, a cumulative record or scintigram can be provided that accurately shows the distribution of diagnostic radionuclides in the organ.

In most systems today the data collected are supplied to a computer for further processing and image enhancement. Most gamma cameras are equipped to supply data to a computer for additional processing. The availability of the computer or data processor provides the system with a great deal of flexibility in displaying the collected data in a variety of ways. By using the computer's ability to make rapid calculations on the sequential data stored for each image element, it is possible to program the computer to produce a value for a group of these image elements that may represent a specific physiological function. With this system and the proper photographic equipment, it is then possible to visualize (1) rapid dynamic changes, such as those occurring in cardiac flow studies, and (2) slower changes, such as those occurring in kidney function tests.

The flexibility of such a computerized system also becomes quite apparent when one considers that all the calculated data within the data processor's memory may be stored for display or reprocessing at a later time. The actual counts obtained during each study from injection to completion may also be fully recorded and stored using a high-speed digital data tape recorder. In this way the data can then be replayed through the gamma camera system, just as if it were data coming from a live study. This capability allows a variety of data processing routines to be used on the same data, each ideally providing additional insight into the process under study. Recorded data from relatively rapid dynamic processes (those occurring in cardiac blood flow studies) can be replayed at slower speeds to permit a detailed study of such a dynamic process.

Because the power of an imaging system lies in the capability of the computer to manipulate the data obtained with the gamma camera, future development depends on continual refinement of these computerized packages. The establishment of complete libraries of computer programs that can be "pulled off the shelf" and used to study the function of important physiological processes easily and quickly represents a primary objective for the future.

Advanced Systems

A simple gamma camera system is currently the most important as well as the most basic instrument for diagnostic nuclear medicine (Spencer and Hosain 1986). Interfacing with a digital computer makes the system more versatile. Digital image-processing systems perform data collection, storage, and analyses. Data acquisition requires digitization of the image by an analog-to-digital converter (ADC), which divides a rectangular image area into small elements, or *pixels*, usually a 64 X 64, 128 X 128, or 256 X 256 matrix. One can select a particular region of interest to obtain certain quantitative information. The regional rate of uptake and clearance of radioactive substances, for example, can be obtained from the serial images for any particular region. Interfacing of a computer with a gamma camera is then essential for dynamic studies (such as monitoring cardiac wall motion).

Gamma camera systems are large and heavy and cannot be moved easily. They are regarded as stationary cameras. However, relatively mobile versions are available with computer interface especially for cardiac studies. Gamma cameras often have capabilities for obtaining whole-body images by linearly moving the camera head or the patient bed. Often called *scanning cameras*, they have replaced rectilinear scanner systems (figure 7.4). A major innovation in gamma camera design has been the addition of an extra camera head with rotational capabilities that is used with a computer system known as *single photon emission computed tomography* (SPECT) for carrying optional tomographic work. The technique is known as computer-assisted reconstruction tomography and is similar to X-ray computed tomographic (CT) scanning (discussed in detail later). Tomograms represent the images of isolated cross-sections of the body. They facilitate identification of abnormalities that are otherwise difficult or impossible to identify in a two-dimensional image covering overlying and underlying tissues (Spencer and Hosain 1986).

Clinical Applications
With the discovery of artificial radionuclides, the modern era of nuclear medicine began. The availability of these radioactive elements increasingly encouraged pioneers in this discipline to use radioactive tracer techniques to gain information regarding biological and physiological systems. One of the first physiological systems to be studied both in animals and humans was the metabolism of phosphorus. Using the cyclotron-produced tracer ^{32}P, several investigators observed the efficiency with which the body absorbed inorganic phosphorus and preferentially used it in rapidly multiply-

Figure 7.4
Modern body scanners are (1) versatile, providing several scanning directions, (2) easy to use, and (3) can provide scans in color as well as in black and white. (Courtesy of General Electric Co. Medical Systems Division, Milwaukee, WI)

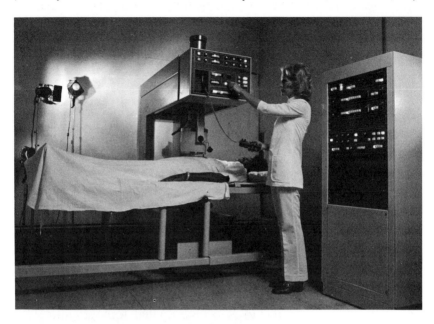

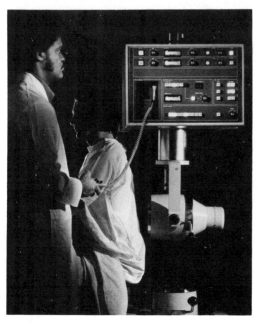

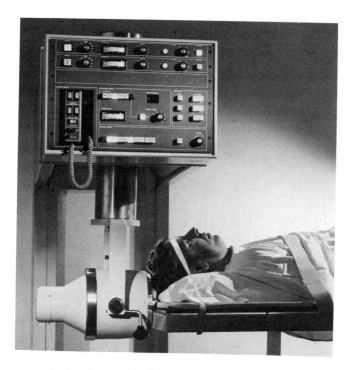

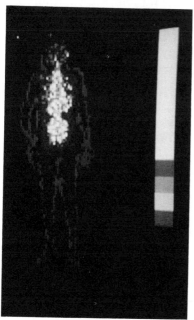

ing tissues, such as those associated with malignancies. In 1936 John Lawrence, seeing the therapeutic implications of these findings, used ^{32}P in the treatment of leukemia and thereby inaugurated the therapeutic use of artificial radionuclides (Wagner 1975).

Although the early uses of radioactive materials in medicine were chiefly in radiation therapy, today most radioactive materials are used to provide useful diagnostic information. These radionuclides can be monitored within the patient (in vivo) or in various bodily fluids removed from the patient (in vitro). For example, it is possible to inject into a patient a radioactive substance that is taken up by a particular organ such as the kidney. By placing a detector over the kidneys, the amount of radioactive material accumulated by the kidney can actually be measured. This uptake test allows the clinician to monitor the activity of specific organs and determine whether they are functioning properly. Another example of an in vivo study involves imaging the distribution of radionuclide within an organ. This can be extremely important, especially because it has been demonstrated that abnormal tissue tends to accumulate more or less of the radionuclide administered than does the normal tissue surrounding it. In this way in vivo measurement can help the clinician delineate the presence of these tissue abnormalities.

The in vitro category includes tests made outside the body. These tests are used to study various chemical, as well as physiological, processes. Today in vitro tests are capable of detecting extremely small amounts of various hormones and chemicals in the blood and have been used to determine (1) insulin dosage in diabetic patients, (2) whether individuals are immune to specific diseases such as hepatitis, and even (3) the proper dosage of digitalis required by cardiac patients. Recent developments indicate that this area of application will continue to grow in importance over the next decade.

The major thrust of the current application in nuclear medicine lies primarily in the use of radioactive tracers in both in vivo and in vitro studies to evaluate various physiological systems within a patient. *The key to success in this application is the specificity of the radioactive material used* (Bronzino 1976). That is, the more closely the radionuclide can be tied to a specific function, the easier it is to examine the dynamic physiological system under study. It is beyond the scope of this text to completely outline of all the possible applications presently used. However, some of the most common clinical applications of nuclear technology in studying major organs are summarized in table 7.4 (Spencer and Hosain 1986). These applications are only representative of the type of information available using nuclear instrumentation techniques.

Although the initial notion of developing radioactive "magic bullets" capable of seeking out and destroying diseased tissue in the human body has not been realized, it is increasingly evident that the techniques embodied in current nuclear medicine clearly allow the acquisition of quantitative information not available by other means. The use of radionuclides makes possible the measurement and visualization of regional function. As a result it is possible to use nuclear technology to determine both the site and activity (rate of change) of essential biological processes, thereby permitting abnormalities, that is, regions of dysfunction, to be detected.

Because most physiological processes within the body are usually in a dynamic state insofar as they exhibit continual change, radioactive tracer techniques have enabled the clinician to accurately monitor the activity of a variety of different physiological systems. These systems range from the pulmonary and cardiovascular system, to renal function, to the microvascular circulation of the eye. Today several of these procedures are widely used in clinical medicine, whereas others are conducted only in research laboratories. As has been the case in other fields of medicine, procedures that are presently viewed as ancillary may become an essential part of the routine medical examination conducted in the future.

Diagnostic Ultrasonography

The application of ultrasound as a noninvasive diagnostic tool is presently widely accepted within the medical community. In fact diagnostic ultrasound studies are often the method of choice

Table 7.4
Clinical Application of Nuclear Technology

Organ	Clinical Use of Scan
Brain	Detection of intracerebral space-occupying disease: primary or secondary neoplasms, abscesses, arteriovenous malformations Evaluation of cerebrovascular disease Detection and localization of intracranial injury after trauma
Lung	Diagnosis and management of pulmonary embolism Evaluation of regional function in emphysema and other forms of chronic obstructive lung disease Evaluation of pulmonary venous hypertension
Thyroid	Evaluation of thyroid size, position, and function Functional assessment of thyroid nodules Diagnosis of functioning metastatic lesions in patients with known carcinoma
Liver	Evaluation of liver size, configuration, and position Preoperative search for metastatic disease when a primary neoplasm exists elsewhere Functional evaluation of cirrhosis and other diffuse parenchymal processes
Spleen	Evaluation of spleen size, shape, and position Evaluation of spleen function Detection of intrasplenic space-occupying disease
Pancreas	Detection carcinoma of the pancreas Detection of pancreatitis Miscellaneous pancreatic problems such as pseudocysts

because they provide information otherwise not available. Because diagnostic ultrasonography is such an important modern medical imaging modality, let us review briefly both its development and the physical basis of continuous and pulsed ultrasonic techniques.

Definitions
The term *ultrasound* is used to describe sound waves (mechanical vibrations) that are beyond the range of human hearing, that is, above 20,000 Hz. Because sound waves of any frequency are essentially vibratory phenomena, their transmission from one point

to another requires the presence of matter. As a result substances best suited for the transmission of sound waves are those with many molecules. Consequently solids transmit sound waves better than do gases, in which there are fewer molecules. In a vacuum, where there are no molecules present, the conduction of sound waves becomes impossible (Goldberg 1973).

Because the composition of each substance varies, ultrasonic or high-frequency sound waves travel through them at different veloxities. Thus it is possible to investigate the internal composition of many materials simply by studying the transmission of ultrasound through them. In addition, just as audible sound under correct conditions produces echoes that can be heard, ultrasound, also under the correct conditions, produces echoes that can be detected with appropriate instruments and then processed and displayed. The application of these basic concepts of *ultrasonic transmission* and *echo-ranging* led to the development of non-destructive testing techniques in industry and eventually to the use of ultrasonic techniques in noninvasive medical testing today. Although the discovery of ultrasound preceded the discovery of X rays by 12 years, X rays found almost immediate clinical application, whereas the use of ultrasound did not become widespread until much later. The primary reason for this delay is that the application of ultrasound had to await the development of appropriate technologies to detect the presence of echoes and display the information in a meaningful manner.

Historical Perspective

During the latter part of the nineteenth century, the Curie brothers discovered that piezoelectric crystals (for example, quartz, Rochelle salts) could act as both generators and receivers for ultrasound waves. That is, not only was it possible to elicit mechanical vibrations when these crystals were subjected to a time-varying or oscillatory electrical potential, it was also possible to produce a detectable change in electrical potential when these crystals were subjected to mechanical stress. These mechanical vibrations, transmitted into the surrounding medium as sound waves, form the basis for today's ultrasonographic techniques.

Initially the major efforts undertaken to apply this acoustic vibration phenomenon were directed toward developing methods for underwater communication and eventually led to the development of sonar (*sound *n*avigation *r*anging), which was extensively used during World War II (Meyer and Gitter 1967). At almost the same time ultrasound became widely accepted in the industrial community as the basis for the development of an internal flaw detector. In 1936 in Leningrad, Sokoloff, while studying the propagation of ultrasonic waves in both solids and liquids, developed a technique for the detection of internal flaws in materials. During World War II these techniques were refined by Floyd A. Firestone with the development of the *reflectoscope*, a device that used echoes to study the reflections of ultrasonic waves coming from any defect present in solid structures. Industry quickly exploited these principles, and "ultrasonic flaw detectors" became widely used.

Using these same techniques, several investigators initiated studies designed to utilize ultrasound in medical diagnoses (Holmes 1974). However, genuine progress depended on the development of efficient ultrasound transducers and the appropriate signal processing equipment. As advances were made in modern electronic technology, ultrasonic instruments designed specifically for medical applications were successfully developed and became widely used.

In the late 1940s and early 1950s, Douglas Howry, working with engineer W. R. Bliss, developed a successful pulse-echo system and lead the effort to improve the quality of the image obtained using ultrasonic techniques. Their results led to a continual parade of ultrasonic imaging instrument systems, including Howry's "Somascope" (Howry et al. 1954, Howry 1957). This device incorporated the principle of compound scanning—the combination of circular, angular, or linear scanning patterns—to produce cross-sectional ultrasonic images. In this compound scanning technique the area to be examined was immersed in water to facilitate the propagation of sound. The transducer carriage was placed on a metal ring, immersed in water, and rotated completely around the patient. During its 360° rotation around the patient, the focused transducer was also moved back and forth to perform a linear scan. The combined linear and circular motion (compound scan) produced a picture that presented a cross section with excellent definition. The ultrasonic images that were initially obtained in this manner actually displayed the internal structure of soft tissue (Howry et al. 1954).

At the same time ultrasound was also found to be effective in indicating the presence and location of tumors. John Wild clearly demonstrated that different echo patterns would be obtained from each layer of tissue present in the specimen. His results provided evidence that ultrasound could be used to detect the presence of any tumor in the exposed preparation. Extending these studies to compare the reflection patterns in brain tissue demonstrated that ultrasonic echo pattern techniques could also be used to assist in the differentiation of brain tissue (French et al. 1950, Wild and Reid 1956). These results represented a significant breakthrough in the clinical use of ultrasound.

Early apparatus, however, was cumbersome and difficult to use because most equipment required immersion of the patient in a fluid bath. A major breakthrough came with the development of contact scanners, devices that could be applied directly to the body to produce images of various internal organs. With the elimination of the sound-transmitting water tank, acoustic coupling was achieved by using an aqueous jelly that allows the probe to make direct contact with the patient's skin. Hand-operated contact scanners subsequently became available and have been incorporated into many of today's ultrasonic systems (Meyer and Gitter 1967).

Basic Physical Concepts

The basic ultrasonic device is relatively simple. It consists of a pulse generator, a transducer, electronic circuitry that amplifies and modifies the signals received, and a display unit. Usually this device is designed to accomplish the following tasks:

1. Convert electrical energy into acoustic pulses
2. Transmit these pulses into the tissue of the patient

3. Detect any reflected energy from within the subject
4. Convert the reflected pulse into an electrical signal
5. Process this signal and present a display that can be evaluated

A block diagram of the essential components of the basic ultrasonic device is shown in figure 7.5. The transmitter consists of an oscillator and a pulse generator. The oscillator establishes the pulse repetition frequency (how often an electrical pulse is allowed to be generated, which is usually in the range of 1000 to 2000 pulses per second), whereas the pulse generator provides an adequate electrical pulse of very short duration that will excite the transducer when passed through a decoupling circuit. The decoupler acts as a connecting link between the transducer and the electronic circuitry and is designed to allow the transducer to be used both as transmitter and receiver.

The transducer converts the electrical pulse into a mechanical vibration, in particular an acoustic wave, and vice versa. The most important component of the transducer is the crystal. Usually fabricated of synthetic ceramic materials that have been processed to have piezoelectric characteristics, the transducer converts the electrical energy supplied to it in the form of an electrical pulse into an acoustic wave having different frequencies, usually in the range of 1 million to 15 million cycles per second (1 MHz–15 MHz). This wave is then propagated at a certain velocity (approximately 1500 meters per second in soft tissue) and in a certain beam pattern toward and into the structure to be examined. When these ultrasound waves strike an acoustic interface, such as might occur at the boundary of an organ, a certain amount of the energy is reflected.

If the reflected energy reaches the transducer, the transducer will convert the reflected acoustic wave back into an electrical signal that is amplified and processed by an audiovisual interface into a form that the operator can evaluate. The audiovisual interface may be the screen of an oscilloscope, a chart recorder, a camera, or a speaker (Brown and Gordon 1967). Because the velocity of sound (velocity = change in distance ÷ change in time) within the body is relatively constant within soft tissue structures, the introduction of

Figure 7.5
Basic diagram of the essential components of a basic ultrasonic detector system

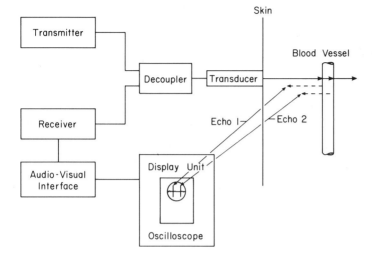

timing circuitry into the electrical processing of the reflected sound wave allows accurate correlation of the signals displayed on an oscilloscope with the depth of the reflecting acoustic interfaces. This is the principle on which pulse-echo diagnostic procedures depend.

Types of Display Modes

Pulse-echo data may be displayed or analyzed in many different ways. The three most common display methods that present the reflected sound on an oscilloscope are designated the A-, B-, and M- (or T-M) modes. In the A-mode, when an electrical pulse is applied to the transducer, it also initiates the trace on an oscilloscope. The echoes are then detected and displayed as vertical lines that represent the relative strength of the echo. Although the echoes visible on the screen of the oscilloscope have undergone considerable electronic processing after being picked up by the transducer, they possess different amplitudes, depending primarily on the strength of the reflection at various acoustic interfaces. Thus it is possible in this way to obtain the reflecting and absorbing (attenuation) properties of the various tissues in the pathway from the transmitter to the medium and then back to the receiver. In addition with this type of display, because the velocity of ultrasound is assumed to be constant in the soft tissue, the time between the start of the sweep of the oscilloscope and the detection of the various echoes can be calibrated to represent the distances to underlying structures. Once calibrated, the location of the vertical deflections along the horizontal axis of the oscilloscope provides a measure of the distance between the acoustic interface represented by the echo and the transducer.

In many cases, however, the amplitude of the echo is not important because the operator is interested primarily in information regarding the actual position of the underlying structure. In this case, the B-mode, or brightness modulation, presentation is used. The arrangement of equipment for this type of display mode is almost the same as that used for the A-mode method, except that the time base that triggers the pulsed generator does not illuminate the screen of the oscilloscope. Instead the echoes are applied to the scope in such a way as to modulate the intensity of the display, thereby representing echoes as dots rather than as vertical deflections. As a result the display consists of a series of dots whose brightness describes the amplitude of the echo and whose position clearly identifies the distances of the reflecting structures below the transducer.

In studying dynamic structures in the body, such as moving heart walls and valves, it is obvious that the position of these acoustic interfaces will change with the physiological events of the cardiac cycle. To record the appropriate pattern of cardiac motion with respect to time, the M-mode is used (Winsberg 1973). The M-mode is simply a B-mode–type display that is allowed to progressively move up the vertical axis of the oscilloscope screen or a photographic paper or film. Therefore the resultant waveforms represent the changing depths of the acoustical interfaces that occur with movements within the body. In cardiology this technique is often used to determine the opening and closing velocity of the mitral valve and to investigate the degree of mitral stenosis.

In the three methods of displaying reflected acoustic energy, the transducer is always held stationary. However, it is possible to produce a two-dimensional cross-sectional representation of anatomy by actually moving the transducer. The resultant B-scan, as it is called, is made with a scanning instrument that moves the transducer through a type of compound scan to view a particular plane or cross-sectional area. The reflection characteristics of various tissue structures require that all tissue interfaces be isolated from as many angles as possible to completely outline their margins. However, a variety of scanning techniques have been worked out (Baker 1974) and have proved successful. In water bath scanners an electrical motor drives a system of levers and gears to make the transducer scan and translate simultaneously. In contact scanners conversely the transducer must be moved similarly by an operator. In this latter procedure the operator must observe the B-mode display as it is being developed to modify or correct the scanning procedure. In both procedures a predetermined set of mathematical equations are solved using the angular position of the transducer to determine the x,y position and direction of the sound beam emanating from the transducer. This information is then finally converted into x and y deflection signals for display on the oscilloscope. The display is once again brightness modulated, but in this case forms a two-dimensional image of the corresponding section across the patient (Wells 1975). These two-dimensional B-mode images, however, are best suited for stationary structures because any movement will cause blurring of images.

However, to interpret these conventional two-dimensional B-scans, the physician must build up, in his own mind, a three-dimensional impression of the scanned anatomy represented by a series of two-dimensional images. This is a difficult task, to say the least. Because the widespread use of diagnostic ultrasound depends on the information being readily obtained and easily understood, it is imperative that these interpretive difficulties be reduced. One approach to this problem is to use the computer to reconstruct the desired image.

One of the latest advances is the accurate display of echo amplitude to form a shaded picture—the so-called gray scale display. With this display technique it is easier to see texture and determine more exactly the limits of certain structures. This is possible because the dynamic range of the image obtained with this technique is sufficiently wide to give some indication of the amplitude of the echoes arising from within different tissues while simultaneously showing the stronger echoes from organ boundaries (Taylor et al. 1973, Puera 1986).

Although the original work on gray scale display was done with photographic recording, a new storage instrument—the scan conversion memory—is capable of comparable results. Initially designed and developed for television applications, it has been incorporated in ultrasonic B-scanning systems so that the stored image may be displayed continuously on a standard television monitor. Scan conversion memories are now available with the diagnostic ultrasound systems of many commercial manufacturers. Although a scan conversion memory is more expensive than a conventional storage system, the convenience of use, the variable dynamic range, and the ability to use cheap television accessory instruments such as monitors frequently justify the cost.

The primary function of these display modes is to provide an image that can be easily and accurately interpreted. The key to future development lies in further use of the computer to provide image enhancement, as well as image reconstruction. The computer can be used to increase the system's resolution and its ability to discriminate between separate targets placed close together along the axis of the ultrasonic beam. For example, by averaging the A-scans obtained over a period of examination, it is possible to reduce some of the uncertainties in the position and amplitude information that this display mode provides. Thus computerization of the appropriate data processing techniques is essential for further refinements in image enhancement.

Doppler Shift Techniques

The ultrasonic devices previously discussed all provide the operator with a visual display of the echoes resulting from the reflection of acoustic waves. These devices are primarily employed for the purpose of imaging specific anatomical structures to detect the presence of any abnormalities. However, there is another application of ultrasound that does not concern imaging, but rather provides information regarding moving acoustic interfaces. This is the Doppler ultrasonic motion detector, the output of which, in its simplest form, is an audible sound rather than a visual display (Goldberg 1973).

Anyone who has ever heard a train pass with its horn blowing or an ambulance speed by with its siren blasting has experienced the phenomenon called the *Doppler frequency shift*. The pitch or perceived frequency of the horn seems higher while the source of the sound is approaching, but suddenly seems lower while the source is moving away. The Doppler frequency shift effect is the change in the apparent frequency of sound caused by a change in the path length between the source and the receiver. This phenomenon is a source of valuable information when used with ultrasonic equipment.

Consider what happens when a transducer is used to introduce a beam of ultrasound through the skin and subcutaneous tissues. A small amount of the transmitted beam is reflected back toward the transducer crystal by the acoustic interfaces that intervene along the beam. If these interfaces are stationary, then the reflected wave has the same frequency as the transmitted wave propagating through the tissue. However, if any of these interfaces are in motion, then the reflected wave will vary in frequency, depending on whether it is moving toward or away from the transducer.

The Doppler ultrasonic instrument detects the reflected wave and then mixes the frequencies of the reflected and transmitted signals acoustically at the transducer and electrically in the receiving amplifier. This change in frequency, which is subsequently presented in the audible range, is then amplified so that the operator can actually hear the reflections coming from a moving interface.

To date the clinical use of this procedure had been primarily for the detection and measurement of blood flow. In this application the sound is actually reflected by the red blood cells in motion within the blood vessel. To illustrate the application of such a system, consider a case in which the target is a stationary blood vessel containing a single clump of moving blood cells embedded in an acoustically homogeneous medium (figure 7.6). In its simplest form a pencil probe that houses coplanar emitting and receiving crystals is used to emit a single frequency f_1 (typically in the range of 2 MHz–10 MHz), which is reflected from the vessel walls and scattered from the blood cells. Because the walls of the blood vessel are stationary, the frequency of the reflected waves coming from these interfaces is exactly equal to f_1. However, this is not the case of the blood cells moving through the vessel at a velocity v. The frequency of the scattered waves coming from these blood cells is the Doppler shifted frequency, which is different from f_1. After subtraction of the received signal from the transmitted signal, the output simply consists of an audio frequency (fn), which represents the velocity component in the direction of the probe. The magnitude of this audio frequency depends on the cosine of the angle and the arrangement of the transmitting and receiving crystals inside the probe.

Figure 7.6

Transcutaneous Doppler frequency shift device. Ultrasound is transmitted from a hand-held pencil probe and backscattered by moving blood cells. The receiving crystal detects the reflected waveform, having a frequency different from the transmitted frequency. The resulting Doppler frequency shift in reflected sound is a function of blood velocity.

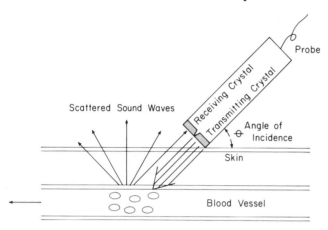

Today Doppler shift–type units are no longer simply concerned with just the magnitude of the change in frequency. Units are now available that provide information regarding the direction of flow as well. These devices enable the observer to note whether flow is toward or away from the probe. With some of these systems it is even possible to resolve simultaneously flow in two opposing directions such as can occur, for example, with an artery and overlying vein.

Range information had now been added to the velocity measurement usually obtained. Pulsed Doppler (that is, burst of high-frequency sound) units have been developed that can be used to detect echoes located only at a specified depth below the skin surface. This is accomplished with "range gating" or pulsing techniques, so that only Doppler signals that originate from preselected range limits (within the range gate) are analyzed at one time. Thus by proper adjustment of this type of range gate, pulsed Doppler systems can be used to obtain information concerning the blood flow velocity and at the same place provide data regarding the dimensions of the blood vessels. Using some of the very high resolution characteristics of pulsed Doppler systems, it has been demonstrated that the velocity profile of flow through a blood vessel (as one moves across the diameter of the blood vessel) can be obtained. These new advances are important steps in enabling the physician to identify volume blood flow in blood vessels under the skin by noninvasive ultrasonographic techniques (Reid 1972).

Clinical Applications of Real-Time Scanners

Real-time ultrasonographic examinations are being used diagnostically to accurately characterize normal and abnormal anatomical structures in many areas of the body (Puera 1986). Ultrasonographic images of the fetus, for example, are especially useful in fetal examinations because there is no electromagnetic radiation danger, and they can be used to guide a needle to harvest ovarian tissue for pathology tests. Today it is estimated that one-half to two-thirds of the pregnant women in the United States have at least one fetal ultrasonographic scan, whereas in Great Britain and Germany fetal examination by ultrasonography is required by law.

Fetal ultrasonographic scans are used to view the fetus and to determine whether the fetus is normal. Fetal anatomical features including the width of the fetal head, size of brain ventricles to monitor possible hydrocephalus, crown-to-rump length to determine the age of the fetus, spinal examination to check for abnormalities, the femur length to correlate with biparietal diameter for assessment of proper physical development of the fetus, and detectable fetal anomalies can be determined.

Real-time ultrasound has been used to perform interventional therapeutic techniques on the fetus in utero (Fleischer and James 1984). Fetal transfusions have been performed using real-time ultrasonographic visualization. Also ultrasonic guidance has been used to drain hydrocephalus, posterior urethral valves, and chylothorax in utero.

The introduction of ultrasonic examinations of the heart has had a great effect on the field of cardiology because the present real-time two-dimensional ultrasonographic images or echocardiograms provide an almost unlimited number of cross-sectional image planes (figure 7.7). Furthermore with real-time scanners additional spatial resolution of the morphology of valves as well as the cardiac chambers is now possible. The recent addition of Doppler flow measurements to cardiac echocardiography imaging also holds great promise for the future because the cardiologist will be able to not only visualize the movement of cardiac structures but also measure the blood flow velocity in the heart chambers and great vessels.

Superficial structures, located less than 5 cm from the skin surface, can also be imaged in real-time using ultrasonographic techniques. Because depth penetration is not a requirement, high-frequency transducers with resultant submillimeter anatomic resolution are used, thereby permitting very small lesions (2 mm–3 mm) to be observed. Superficial anatomical structures that can be imaged include the thyroid, carotid arteries, eye, scrotum, breast, and subcutaneous tissue.

Figure 7.7
Echocardiography enables an operator to monitor heart structure and motion.
(Courtesy of Picker Corp., Cleveland, OH)

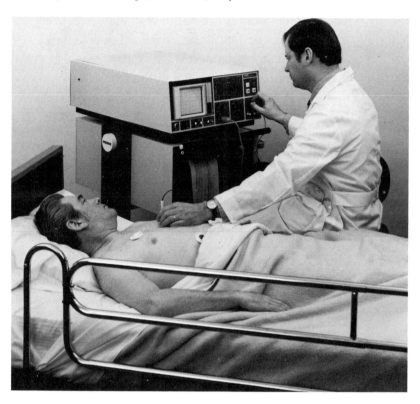

Ultrasonic examinations of the eye are important especially when an ophthalmologist cannot visually examine the eye (in situations in which opaque material such as blood is present or when the patient has cataracts). It would be unfortunate to have a situation in which a cataract was removed only to find that the patient had a retinal detachment or a large melanoma. The major application of ultrasonographic examinations of the eye is in the evaluation of nonmetallic foreign bodies, retinal detachment, and intraocular tumors. The availability of high-frequency transducers makes evaluating the eye for periorbital and intraorbital disorders possible (Fleischer and James 1984). Figure 7.8 shows an ultrasonographic examination of the eye and the resulting images.

The future of medical applications of diagnostic ultrasonography is tied to the progress of electronic and computer technology. As electronic and computer circuits become more powerful for the same hardware volume and cost, more sophisticated ultrasonographic imagers will be developed and new medical applications will be found. Furthermore the future of diagnostic ultrasonographic applications in medicine is contingent on the development of improved digital processing of ultrasonographic data. New image-processing algorithms will handle and quantify the large amounts of data. In contemporary systems, once the data are stored in relatively inexpensive digital memory devices, they can be manipulated to improve the image quality and to make simple calculations of tissue dimensions. Further work remains for improving the image quality, extracting information about the nature of the tissue, and making accurate calculations of tissue dimensions, including volumes, from three-dimensional measurements.

Perhaps one of the most exciting and challenging areas of ultrasonographic imaging deals with tissue characterization. Wells (1977) reviewed the early work in this area. Basically the research relates to whether the difference between the ultrasonic signals (or signature) are different when normal or abnormal tissue is studied. A number of ultrasound parameters have been investigated, including attenuation, scattering, and velocity of propagation. To put it in terms of a physiological example, Is there a way to determine

Figure 7.8
(A) Ultrasonographic exam of the eye, (B) scanning system, (C) retinal detachment with dense vitreous traction and dense vitreous hemorrhage, and (D) reaction. (Adapted from Bronzino, J. 1977. *Technology of Patient Care*. C. V. Mosby)

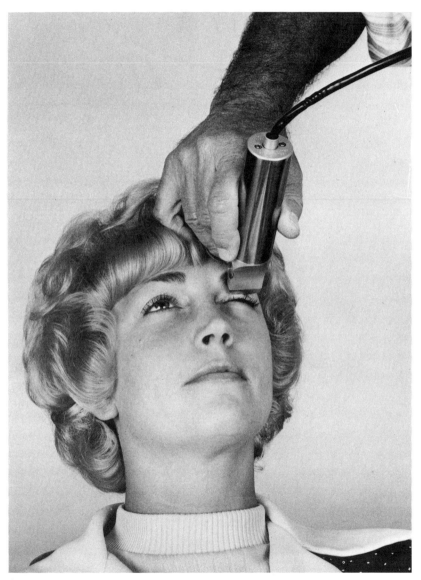

A

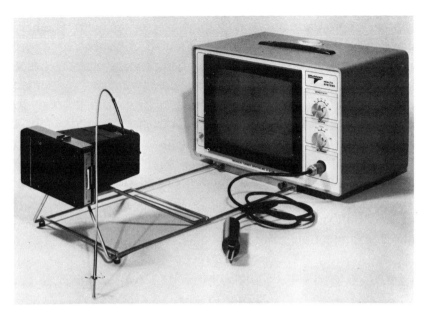

B

C

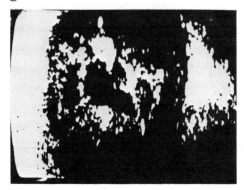

D

from an appropriate ultrasonographic scan of the heart whether a myocardial infarction has taken place and, if so, how severe it was? Similar questions could be posed for benign and cancerous tissue.

Another area of future development deals with the development of specialized transducers that would be placed inside the body for better imaging of a specific organ. For instance, approximately 30 percent of all men aged 50 or over have prostate cancer. It has been proposed that a high-resolution ultrasonic transducer operating at 10 MHz when placed in the rectum could provide high-resolution ultrasonographic images of the prostate cancer. This should give more definitive results than those available from a conventional digital examination for prostate cancer.

Computerized Axial Tomography—The CT Scanner

One of the most significant advances in the realm of medical imaging has been the use of computerized axial tomography (CAT), or simply computed tomography (CT), and the development of the first computer-based medical instrument—the computerized brain and body scanner. This modern piece of X-ray equipment, introduced in 1972, has permitted the visualization of structures and masses in otherwise inaccessible regions of the body. To accomplish this the device utilizes the computer in a most fundamental way. The computer is not used simply to enhance the value of a diagnostic measurement, perform a job faster, or interpret a physiological function more accurately. In CT scanning the computer is much more vital, generating the image and producing the diagnostic features to be interpreted by clinicians. Consequently the resident computer is the most important component—the very heart—of this medical marvel.

Surprisingly CT scanners are relatively simple to use. The patient simply lies down, placing the portion of the body to be imaged in a section of the device that passes an X-ray beam through it. The radiation emerging from the tissue is absorbed by a detector that in turn supplies the resident computer with information regarding the intensity of the radiation striking it. The computer performs the necessary calculations, compiles the results, implements image reconstruction algorithms, and provides an image or visual display for review by the physician. The detail in the resultant image provided by CT scanning is so remarkable that it has permitted different tissues with only slight variations in density to be clearly differentiated. As a result CT scanning has become an extremely valuable tool for diagnostic evaluation. In less than a decade this new technological innovation has revolutionized the field of radiology and has gained wide acclaim throughout the medical world.

The electrification of the medical world caused by CT technology is reminiscent of another era in the history of medicine. In November 1895, when the first Nobel laureate, physicist Wilhelm Conrad Roentgen, viewed the image of the bones in his hand with stunned amazement, a new age of medicine was born. With the discovery of X rays the forces of history converged on Roentgen's laboratory and changed the course of medical science. Less than one hundred years later, the same excitement would follow Allan Macleod Cormack and Godfrey Newbold Hounsfield. Cormack, in the physics department of Tufts University, and Hounsfield, in the research laboratories of the English company EMI Limited, worked independently of each other and, although obviously under the spell of a common dream, accomplished quite different things. Cormack elegantly demonstrated the mathematical rudiments of image reconstruction in a remarkable paper published in 1963. Less than a decade later Hounsfield revealed an incredible engineering achievement—the first commercial instrument capable of obtaining high-resolution images for medical purposes. In recognition of the significant advances made possible by the development of CT, they shared the 1979 Nobel Prize in physiology and medicine. The excitement of their discovery has not yet subsided (Bronzino 1982).

Basic Concepts

The CT scanner is a device that produces a picture of the distribution of X-ray absorption (or transmission) in a cross section of a patient. Its ultimate development depended on the availability of an appropriate X-ray source and detector systems, image reconstruction algorithms, and computer hardware and software packages.

For almost three-quarters of a century, conventional medical X-ray pictures were obtained using a broad X-ray beam and photographic film. With this standard technique X rays are passed through the body, projecting an image of bones, organs, air spaces, and any obstruction, tumors, or foreign bodies onto a sheet of film (figure 7.9). The shadowgraph images are therefore the results of the variation in the intensity of the transmitted X-ray beam after it has passed through tissues and body fluids having different densities. When this procedure is used to project three-dimensional objects onto such a two-dimensional screen, however, difficulties are usually encountered. Because there is often an overlapping of structures represented on the film, it becomes difficult to distinguish between tissues that are similar in density. For this reason conventional X-ray techniques have been unsuccessful in obtaining images of the various sections of the brain, which consists primarily of soft tissue. In an effort to overcome this shortcoming, attempts have been made to obtain X-ray pictures from a number of different angles in which the internal organs appear in different relationships to one another and to introduce a medium (such as air to iodine solution) that is either translucent or opaque to X rays. However, these efforts are usually time consuming, sometimes difficult on the patient, and often are just not accurate enough.

In the early 1920s another X-ray technique for visualizing three-dimensional structures was developed. This technique, known as tomography, permits the imaging of specific planes within the body. With this technique it is possible to obtain a cross-sectional view or slice of almost any internal structure. This has been accomplished in the past by moving the X-ray source in one direction while moving the photographic film (which is placed on the other side of the body and picks up the X rays) simultaneously in the opposite direction (figure 7.9B). As the X rays pass through the body, there is only one plane that does not move relative to the X-ray source and film. It is this plane or cross-sectional view that is displayed quite clearly on film, whereas all the other planes show up only as a blur. Such an approach is clearly better than conventional methods in revealing the position and details of various structures and in visualizing three-dimensional information using such a two-dimensional approach. However, there are limitations in its use. First, it does not really localize a single plane, that is, there is some error in the depth perception that is obtained; and second, large contrasts in radiodensity are usually required to obtain high-quality images that are easy to interpret. This second limitation unfortunately also increases the risk of a patient being exposed to excessive radiation.

CT represents a completely different approach. Consisting primarily of a scanning and detection system, a computer, and a display medium, it uses image reconstruction techniques that combine X-ray absorption measurements in such a way as to facilitate the display of any internal organ in three dimensions. The starting point however, is quite similar to that used in the conventional approaches. A collimated beam of X rays is directed through the body section being scanned to a detector located on the other side of the patient (see figure 7.9C). With a narrowly collimated source and detection system, it is possible to send a narrow beam of photons to a specific detection site. Some of the energy of the X rays is absorbed, whereas the remainder continues to the detector and is measured. In CT the detector system usually consists of a crystal (such as sodium iodide) that has the ability to scintillate, or emit light photons, when bombarded with gamma or X radiation. These photons, or "bundles of energy," are in turn measured by photomultipliers, thereby quantifying the energy absorbed (or transmitted) by the medium penetrated by the X-ray beam.

Because the X-ray source and detection system are usually mounted on a frame, or "scanning gantry," they can be moved together across the object being visualized. In early designs, for example, X-ray absorption measurements were made and recorded at each position laterally traversed by the source and detector system. The result was the generation of an absorption profile for that angular position. To obtain another absorption profile, the

Figure 7.9
Overview of the three X-ray techniques used to obtain medical images. (A) A *conventional X-ray picture* is made by having the X rays diverge from a source, pass through the body, and then fall on a sheet of photographic film. (B) A *tomogram* is made by moving the X-ray source in one direction and the film in the other direction during the exposure. In the projected image only one plane of the body remains stationary with respect to moving film, and all other planes of the body appear blurred. (C) In computed tomography reconstruction from projections is made by mounting the X-ray source and an X-ray detector on a yoke and moving them past the body. The yoke is also rotated through a series of angles around the body. Data recorded by the detector are processed by a special computer algorithm, or program. The computer generates the picture on a CRT screen.

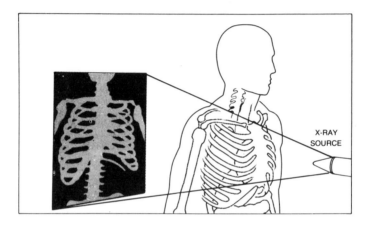

CONVENTIONAL X-RAY PICTURE is made by allowing the X-rays to diverge from a source, pass through the body of the subject and then fall on a sheet of photographic film.

A

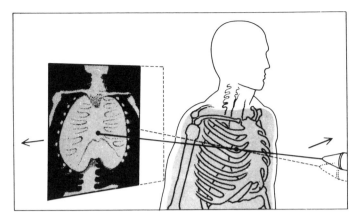

TOMOGRAM is made by having the X-ray source move in one
direction during the exposure and film in the other direction.
In projected image only one plane in body remains stationary
with respect to moving film. In the picture all other planes
in body are blurred.

B

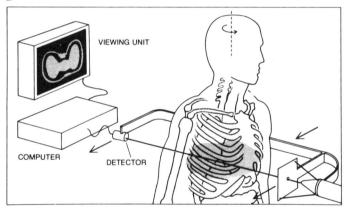

RECONSTRUCTION FROM PROJECTIONS is made by mounting
the X-ray source and an X-ray detector on a yoke and moving
them past the body. The yoke is also rotated through a
series of angles around the body. Data recorded by detector
are processed by a special computer algorithm, or program.
Computer generates picture on a cathode-ray screen.

C

scanning gantry holding the X-ray source and detector was then rotated through a small angle, and an additional set of absorption or transmission measurements was recorded. Each X-ray profile or projection obtained in this fashion is basically two-dimensional, as wide as the body, but only as thick as the cross section.

The exact number of these equally spaced positions depends on the dimensions to be represented by the picture elements that constitute the display. For example, generation of a 160 X 160 picture matrix, which was the format used in the initial design, requires absorption measurements from 160 equally spaced positions. Recall, however, that each one-dimensional array constitutes one X-ray profile or projection. To obtain the next profile, the scanning unit is then rotated a certain number of degrees around the patient, and 160 more linear readings are taken at this new position. This process is repeated again and again until the unit has been rotated a full 180°. When all the projections have been collected, 160 X 180 (or 28,000) individual X-ray intensity measurements are then available for use in forming a reconstruction of a cross section of the patient's head or body (Gordon et al. 1975).

At this point the advantages of the computer are used. Each of the measurements obtained enters the resident computer and is stored in memory. Once all the absorption data have been obtained and subsequently located in the computer's memory, the software packages developed to perform the analysis of the data, as prescribed by image reconstruction algorithms, are called into action. These image reconstruction techniques have been known for some time. However, they have not been used until quite recently because the number of computations required for each reconstruction is virtually impossible without some type of computational aid. Only with the advent of modern computer technology has it become possible to fully exploit these reconstruction techniques.

To develop an image from the stored values of X-ray absorption, the computer initially establishes a grid consisting of a number of small squares for the cross section of interest, depending on the size of the desired display. Because the cross section has thickness, each of these squares represents a volume of tissue, a rectangular solid whose length and width are determined by the size of the matrix and whose depth is equal to the thickness of the cross section. Such a rectangular block of tissue is referred to as a *pixel* (figure 7.10).

During the scanning process each pixel is irradiated by a narrow beam of X rays up to 180 times. Thus the absorption caused by that pixel contributes up to 180 absorption measurements, each part of a different projection. Because each pixel affects a unique set of absorption measurements to which it has contributed, the computer calculates the total absorption due to that pixel. Using the total absorption and the dimension of the pixel, the average absorption coefficient of the tissues in that pixel can be precisely determined.

To illustrate how this may be accomplished, consider the following procedure. X rays can be directed so that they pass through a cross-sectional area of the patient's body in parallel rays. The cross section of interest can be considered to be made up of a set of blocks of material, as illustrated in figure 7.11. Each block has an attenuation effect on the passage of the X-ray energy or photons, absorbing some of the incident energy passing through it. Let us assume that the first block absorbs a fraction A_1 of the incident photons, the second a fraction A_2, and so on so that the nth block absorbs a fraction A_n. The total fraction A absorbed through all the blocks is the product of all the fractions, and the logarithm of this total absorption fraction is defined as the measured absorption. Consider figure 7.11 once again. The set of absorption measurements for an object that comprises only four blocks would be (Pullen 1979)

in position 1: $A(1) = A_1 + A_3$ (1)

in position 2: $A(2) = A_2 + A_4$ (2)

in position 3: $A(3) = A_1 + A_2$ (3)

in position 4: $A(4) = A_3 + A_4$ (4)

Figure 7.10
CT brain section, 160 X 160 matrix

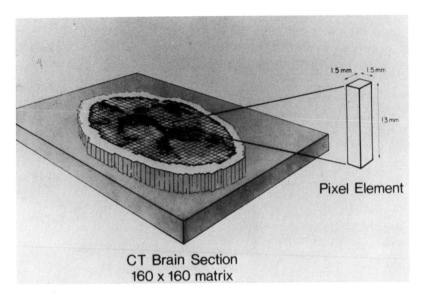

CT Brain Section
160 x 160 matrix

Pixel Element

Figure 7.11
Basics of image reconstruction

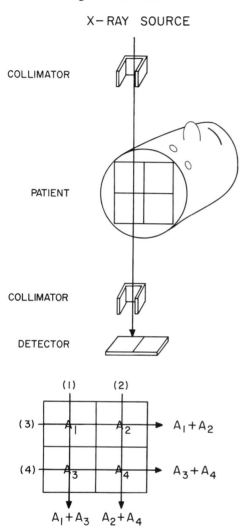

In practice only the measured absorption factors A(1), A(2), A(3), and A(4) would be known. The problem is, therefore, to compute A_1, A_2, A_3, and A_4 from the measured absorption values. The fact that the computation is possible can be seen from equations 1 through 4. There are four simultaneous equations and four unknowns, and a solution can therefore be found. To reconstruct a cross section containing n rows of blocks and n columns, therefore, it is necessary to make at least n individual absorption measurements from a least n directions. For example, a display consisting of 320 X 320 picture elements requires a minimum of 320 X 320 (or 102,400) independent absorption measurements.

These absorption measurements are taken in the form of profiles. Imagine a plane parallel to the X-ray beam passing through the required slice. If the absorption of the emergent beam along a line perpendicular to the X-ray beam is plotted, one has an absorption profile. This profile represents the total absorption along each of the parallel X-ray beams. In general the more profiles that are obtained, the better (in terms of resolution or detail) the resulting picture. In practice it is usually necessary to take at least 180 such profiles at 1° intervals to obtain a diagnostically useful picture.

From these individual measurements a single two-dimensional plane can be reconstructed; and by simply (that is, for the computer) stacking the appropriate sequence of such planes, it is possible to reconstruct a full three-dimensional picture. Thus the technique of image reconstruction of a three-dimensional object is based primarily on a process of obtaining a cross-sectional or two-dimensional picture from many one-dimensional projections (Gordon et al. 1975). The earliest method used to accomplish this was defined simply as "back projection," which means that each of the measurement profiles was projected back over the area from which it was taken. Unfortunately this rather simple approach was not totally successful because of its blurring effect. To overcome this, iterative methods were introduced that successively modified the profile being back-projected until a satisfactory picture was obtained. Iteration is a good method, but it is slow, taking several steps to modify the original profiles into a set of profiles that can be projected back to provide an unblurred picture of the original image. To speed up the process, mathematical techniques involving convolution or filtering were introduced, permitting the original profile to be modified directly into the final one.

Whichever method of reconstruction is used, the final result of the computation is the same. In each case a file is created in the computer memory, usually known as the picture file, which contains an absorption coefficient or density reading for each element of the final picture (for example, some 25,600 for 160 X 160 pictures, or over 100,000 for 320 X 320 pictures). The resultant absorption (or transmission) coefficients for each element of the image calculated in this manner can then be displayed as gray tones or color scales on a visual display. Because most CT systems provide the user with an "image" by projecting the image onto a television screen, using fairly conventional television techniques, and remembering that each element, or pixel, of the picture file has a value that represents the density (or more precisely the relative absorption coefficient) of a point in the cross section of the body being examined, let us illustrate how these absorption values can be displayed.

Consider the scale shown in figure 7.12, showing the values of absorption coefficients using the Hounsfield scale, which ranges from air (−1000) at the bottom of the scale to bone at the top. This scale covers some 1000 levels of absorption on either side of water, which was chosen to be zero at the center. This is done for convenience because the absorption of water is close to that of tissue. In using this scale, the number 1000 is added to the absorption coefficient of each picture element. In the process the absorption value of air becomes zero, whereas that of water becomes 1000 (Hounsfield 1973). Having assigned each pixel to a particular value in this scale, the system is ready to display each element and create the reconstructed image.

In the display provided by any medical imaging device, factors affecting the resolution or accuracy of the image must be recognized and optimized. Because the information provided by a CT scanner essentially reflects the distribution of X-ray absorption in

Figure 7.12
Hounsfield's absorption coefficients

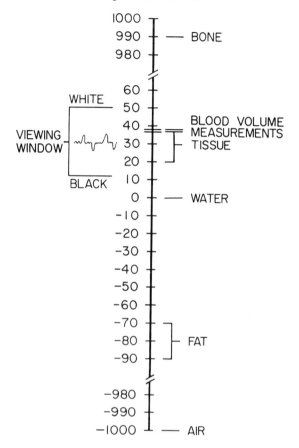

a cross section of a patient, the smallest change in X-ray absorption that can be detected and presented in the image provided by this device is defined as *density resolution.* In addition the smallest distance apart that two objects can be placed and still be seen as separate entities is defined as *spatial resolution.* The density resolution therefore reflects the sensitivity of the method to tissue change, whereas spatial resolution indicates the fineness of detail possible in the image provided by the CT scanner. These two characteristics of resolution are related to each other and to the radiation dose received by the patient being scanned. The clarity of the picture and hence the accuracy to which one can measure absorption values are often impaired by a mottled appearance (or grain), which unfortunately is fundamental to the system. It is caused by the reduction in the number of photons arriving at the detectors after penetrating the body. Fewer available photons, in turn, results in a reduction in the sharpness of the image because the resultant differences between the absorption coefficients of adjacent pixels are not large. Unfortunately it is a situation that must be accepted, as long as low radiation doses are used. In spite of this limitation the reconstructed image can be *enhanced* by the "display," units, which constitute another major component of the CT system.

Display units usually have the capacity of displaying a given number (16, 20, and so on) of different colors or gray at one time. These different colors (or gray levels) can therefore be used to represent a specific range of absorption coefficients. In this way the entire range of absorption values in the picture can be represented by the entire range of available gray tones, colors, or both. To sharpen the resolution or detail of the display, the range of gray tones, that is, between black and white, can be restricted to a very small part of the scale. This process of establishing a display "window" that can be raised or lowered actually enhances the imaging of specific organs. For example, if it is desired to image the tissue of the heart, this window is raised, whereas if the detail within the lung is desired, it is lowered. The overall sensitivity can be further increased by reducing the window width, permitting ever greater differentiation between absorption coefficients obtained from specific organs and the surrounding tissue (Ledley 1976).

Various other techniques are available to make the picture more easily interpreted. For instance, a portion of the picture can be enlarged, or certain absorption values can be highlighted or made brighter. These methods do not add to the original information, but they enhance the image made possible by the data available in the computer's memory. Because a human observer can see an object only if there is sufficient contrast between the object and its surroundings, enhancement modifies the subjective features of an image to increase its impact on the observer, making it easier to locate and precisely measure obscure details (Barrett 1978).

CT scanning has a number of advantages over conventional X-ray techniques. For example, CT provides three-dimensional information concerning the internal structure of the body by presenting it in the form of a series of slices. It is very sensitive and can show differences in soft tissue clearly, which conventional radiographs cannot do. It permits the accurate measurement of the X-ray absorption of various tissues, enabling the nature of these tissues to be studied. And yet the amount of X rays given to the patient using it is often less than the amount emitted by the conventional X-ray technique (Hounsfield 1973).

In comparing the CT approach with conventional tomography, which also images a slice through the body, but by blurring the image of the material on the picture on either side of the slice, it becomes readily apparent that in the conventional approach only a short path of the beam passes through the slice to be viewed, collecting useful information. The other nine-tenths of the beam passes through material on either side of the slice, collecting unwanted, artifact-producing information. On the other hand the X-ray beam in CT passes along the full length of the plane of the slice, thereby permitting measurements to be taken that are 100 percent relevant to that slice and only that slice. These measurements are not affected by the materials lying on either side of the section. As a result the entire information potential of the X-ray beam is used to the fullest, and the image becomes more clearly defined (Hounsfield 1980).

The medical applications of such detailed pictures are considerable. Hard and soft tumors, cysts, blood clots, injured or dead tissue, and other abnormal morphological conditions can be easily observed. The ability to distinguish between various types of soft tissues means that many diagnostic procedures can be simplified, thus eliminating the need for a number of methods that require hospitalization, are costly, and carry a risk to the patient. Because CT can help in accurately locating the areas of the body to be radiated and in monitoring the actual progress of the treatment, it is effective not only in diagnosis but in radiotherapy as well.

The Present Status of Computed Tomography

Since its introduction CT has undergone continuous refinement and development. Significant improvements have been made in

1. X-ray tube performance
2. reduction of scan times
3. reduction of image reconstruction time
4. reduction and, in some cases, elimination of image artifacts
5. reduction of patient dose

Achievements in these areas have focused attention on two key issues: The first is image quality and its relation to diagnostic value. The second is scanner efficiency and its relation to patient throughput and the economic effect it entails. To attain a better understanding of the operation and utility of present CT systems, it is important that some of the evolutionary steps be presented.

Present CT scanners are usually integrated units consisting of three major elements: (1) *the scanning gantry,* which takes the readings in a suitable form and quantity for a picture to be reconstructed, (2) *the data handling unit,* which converts these readings into intelligible picture information, presents this picture information in a visual format, and provides various manipulative aids to enhance the image and thereby assist the doctor in forming a diagnosis, and (3) *a storage facility*, which enables the information to be examined or reexamined at any time after the actual scan (Walmsley 1979).

The objective of the scanning system is to obtain enough information to reconstruct an image of the cross section of interest. All the information obtained must be relevant and accurate, and there must be enough independent readings to reconstruct a picture with sufficient spatial resolution and density discrimination to permit an accurate diagnosis. The operation of the scanning system is therefore extremely important.

During the past decade the CT scanner has undergone several design changes (figure 7.13). The earliest generation of gantries used a system known as *traverse and index*. In these systems the tube and detector were mounted on a frame, and a single beam of X rays traversed the slice linearly, thereby providing absorption measurements along one profile. At the end of the traverse the frame indexed through 1°, and the traverse was repeated. This procedure continued until 180 single traverses were made and 180 profiles were measured. Using the traverse-and-index approach, the entire scanning procedure took approximately four to five minutes. The images it provided had excellent picture quality and hence high diagnostic usefulness. However, this approach had several disadvantages. First of all, it was relatively slow resulting in relatively low patient throughput. Second, streak artifacts were common, and although they normally were not diagnostically confusing, the image quality did suffer. Finally, perhaps the biggest drawback was that because the patient had to remain absolutely still during the scan, its use was restricted solely to brain scans.

In an attempt to solve these problems, a second generation of gantries was developed. The scanning procedures in this approach included a translation step as in the first generation, but incremented the scanning gantry 180° around the patient in 10° steps. In addition, in an effort to obtain more profiles with each traverse, a larger fan beam was used. By using a 10° fan beam, it is possible to take ten profiles, at 1° intervals, with each traverse. By indexing through 10° before taking the next set profiles, it is then clearly

Figure 7.13
Four generations of scanning gantry designs

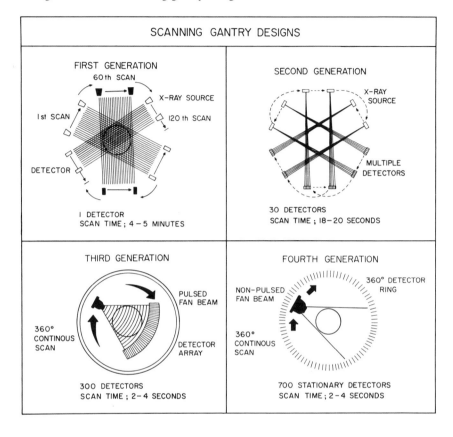

SCANNING GANTRY DESIGNS

FIRST GENERATION
60th SCAN
1st SCAN
120th SCAN
X-RAY SOURCE
DETECTOR
1 DETECTOR
SCAN TIME ; 4 – 5 MINUTES

SECOND GENERATION
X-RAY SOURCE
MULTIPLE DETECTORS
30 DETECTORS
SCAN TIME ; 18 – 20 SECONDS

THIRD GENERATION
PULSED FAN BEAM
360° CONTINOUS SCAN
DETECTOR ARRAY
300 DETECTORS
SCAN TIME ; 2 – 4 SECONDS

FOURTH GENERATION
NON-PULSED FAN BEAM
360° DETECTOR RING
360° CONTINOUS SCAN
700 STATIONARY DETECTORS
SCAN TIME ; 2 – 4 SECONDS

possible to obtain a full set of 180 profiles by making only 18 traverses. At the rate of approximately one second for each traverse, scanning systems of this type usually operate in the range of 18 to 20 seconds. Even in this extremely reduced scan time, this scanning and detection system is able to obtain over 300,000 precise absorption measurements during the complete (that is, over 180°) scan. These values are then used to construct a picture of the slice under investigation. With this new generation of CT scanners, it became possible to obtain cross sections of any part of the body (Ledley 1976). However, it still was not fast enough to eliminate the streak artifacts, nor did it seem possible to further reduce scan time with this approach. In essence the first and second generations had reached their limit. Scanning times could not be reduced further and still maintain an acceptable image quality.

A major redesign ensued, and many of the scanning system components were improved. In the process a third generation of gantries evolved, consisting of a pure *rotational system.* In these systems the source/detector unit, having a pulsing, highly colli-mated wide-angle (typically 20° to 50°) fan beam and a multiple detector array, is rotated 360° around the patient. This single 360°, smoothly rotating movement decreases scan time to approximately three seconds, increases appreciably the reliability of the data because it is taken twice, and increases the quality of the recon-structed image. A by-product of this design is that the pulsing X-ray source can be synchronized with physiological parameters, enabling rhythmically moving structures, such as the heart, to be accurately imaged.

The main advantages of this type of system are its simplicity and its speed, but it has two major disadvantages: First, its fixed geometry system, with a fan beam usually established for the largest patient, is inefficient for smaller objects and in particular for head scanning. Second, it is particularly prone to circular artifacts. These "ring" artifacts are clearly evident in early pictures from this type of machine, although manufacturers now claim that the detector stability problems causing these artifacts have largely been overcome. Unfortunately the magnitude of the difficulty in achieving stability can be appreciated when it is realized that an error of only 1 part in 10,000 can lead to an artifact that is diagnostically confusing (Walmsley 1979). Consequently a fourth generation gantry system, consisting of a continuous ring of fixed detectors (usually numbering between 500 and 700), has been developed. In this system the X-ray source rotates as before, but the detectors remain stationary, arranged in a full circle around the gantry. The fan beam is increased slightly so that the detectors on the leading and trailing edge of the fan can be continuously monitored and can thereby adjust the data in case of shifts in detector performance. As a result the data obtained using the approach are more reliable, and the ring artifact is usually eliminated (Marvilla and Pastel 1978).

The type of CT scanners in current use fall into one of these categories. However, a new gantry design has recently been released. This system has the same basic configuration as the fourth generation scanning system—that is, a rotating tube and fixed detector system—but has eliminated the bulky, noisy electromechanical rotation mechanism. In its place a magnetic induction system is used to rotate the source. This enabled designers to attain a long-sought goal: the 1-second scan.

The whole purpose of the scanning procedure is to take thousands of accurate absorption measurements throughout the body. Taken at all angles through the cross section of interest, these measurements provide an enormous amount of information about the composition of the section of the body being scanned. Because these readings are taken by counting photons, the more photons counted, the better will be the quality of the information. With these factors in mind it is essential that the radiation dose be sufficient to obtain good-quality pictures. Because the total radiation dose is a function of the maximum power of the tube and its operating time, faster scanners have to use a significantly higher-powered tube. These aspects of dose efficiency are very important in tube and detector design.

Currently there is general agreement between manufacturers on the type of tubes that are used. Machines that operate in the region of 20 seconds used fixed-anode, oil-cooled tubes, which are run continuously during the scan time; those that operate in much less than 20 seconds have a rotating-anode, air-cooled tube, which is often used in a pulse mode.

The fixed-anode tubes provide sufficient radiation energy for 20-second scanners. Being oil cooled, they can be run continuously, and the sequence of measurements is obtained by electronic gating of the detectors. Scanners that operate in the region of 5 seconds need tubes of significantly higher power, requiring a rotating-anode tube. There is a tendency to use these in a pulse mode largely for convenience because it simplifies the gating of the detectors. One of the problems encountered with fast machines is the difficulty in providing enough dose within the scan time. For high-quality pictures many of these scanners run at rates much slower than their maximum possible speed. At present the problem of providing sufficient dose in the scan time is one of the major limiting factors in providing faster scanners (Walmsley 1979).

Of equal importance in the accurate detection and measurement of X-ray absorption provided by scanning maneuver is the detector itself. Theoretically detectors should have the highest X-ray photon capture and conversion efficiency possible to minimize the radiation dose to the patient. Clearly a photon that is not detected does not contribute to the generation of an image. Should the detection system of a CT scanner be able to capture and convert a larger number of photons with very little noise, it will produce accurate, high-quality images when aligned with reconstruction algorithms. Consequently the selection of a detector always represents a trade-off between image quality and low radiation dose.

There are three types of detectors in clinical use today: Certain manufacturers use scintillation crystals (such as sodium iodide, calcium fluoride, or bismuth germinate) combined with photomultiplier tubes. Others use gas ionization detectors containing xenon, and still others are beginning to incorporate the latest advances in solid-state detector design, using advanced scintillation crystals with integral photodiodes. The principal factors in judging detector quality are how well the detector captures photons and its subsequent conversion efficiency that results in optimum dose utilization. Those systems using scintillation crystals combined with photomultiplier tubes do not lend themselves to the dense packing necessary for maximum photon capture, even though they offer a high conversion efficiency. The less-than-50-percent dose utilization inherent in these detectors necessitates a higher dose to the patient to offset the resultant inferior image quality. Although gas detectors are relatively inexpensive, compact, and lend themselves to use in large arrays, they have a low conversion efficiency and are subject to drift. Consequently use of xenon detectors necessitates increasing patient dose.

The advantage of scintillation crystals with integral photodiodes extends beyond their 98-percent conversion efficiency because they lend themselves quite readily to a densely packed configuration. The excellent image quality combined with high dose utilization (that is, approximately 80 percent of the applied photon energy) and detector stability ensure that solid-state detectors will continue to gain favor in more modern CT systems.

In conjunction with the operation of the scanning and detection system, the data obtained must be processed rapidly to permit viewing of a scan as quickly as possible. Data-handling systems implementing the latest advances in computer technology have also been developed as changes have occurred in the design of the scanning and detector components of CT scanners. For example, with the evolution of the third generation of scanners, the data-handling unit had to digitize and store increasing amounts of data, consisting of (1) positional information, such as how far the scanning frame is along its scan, (2) reference information from the reference detector that monitors the X-ray output, (3) calibration information, which is obtained at the end of each scan, and (4) the bulk of the readings, the actual absorption information obtained from the detectors. Electronic interfaces are presently available that convert these data into digital form and systematically store them in memory of the resident computer. The next step is the actual process of image reconstruction.

The first stage of this computation process is to analyze all of these raw data and convert them into the set of profiles, normally 180 or more, and then convert these profiles into information that can be displayed as a picture and used for diagnosis. Here we have the heart of the operation of a CT scanner, that element that makes CT totally different from conventional X-ray techniques and most other imaging techniques. The algorithms or computer programs for reconstructing tissue X-ray absorption presently fall into one of the following categories: simple back projection (sometimes called summation), Fourier transforms, integral equations, and series expansions. The choice of an algorithm for a CT scanner depends on its speed and accuracy (Nagel 1988).

In the simple back-projection algorithm, it is imagined that each picture element contributes equally to the projection density. Unfortunately this approach blurs out sharp features in the original (figure 7.14). Because it is difficult to imagine a pattern for which such a blurring effect would not be objectionable, various processing and filtering techniques must be used to achieve an accurate image.

The Fourier method depends on the transformation of the projections into Fourier space in which it is possible to compile all the data pertaining to the image quite effectively. Once the calculations are complete, the image is then reconstructed by inverting this information back into "real space." Due to the nature of the computations required using this method, considerable errors are often introduced. To achieve a high degree of resolution, the number of equations to be solved are often large and can cause considerable computational problems. This approach would benefit considerably from advances in computer hardware and software developments.

The use of integral equations in image reconstruction has demonstrated both accuracy and speed. Using an approach known as the convolution method, it is possible to operate on the various projection densities such that when these projections are back-projected, they form the original image without blurring. One unique advantage of this approach is that it offers the possibility of performing a large percentage of the reconstruction while the data are being accumulated. This is particularly desirable in a clinical environment.

The series expansion approach approaches the problem by assuming that here is a specific linear combination of the properties that may be used to generate the image. The algebraic reconstruction technique (ART) is an example of this method. Unfortunately reconstruction with the ART algorithm must await the acquisition of almost all of the projection data. Hence, in a clinical environment where patient throughput time is usually of considerable importance, this technique may be less than optimal (Payne and McCullough 1976).

To facilitate the use of these computational processes, significant advances had been made in the development of data acquisition and conversion systems by the late 1970s. One of the most important of these developments was the design of the array processor, which has become an integral part of the data acquisition and conversion systems used in the most recent CT systems.

This array processor enables large amounts of data to be handled quickly. To accomplish this, it incorporates the concepts of the *parallelism* and *pipelining*. The significance of parallelism becomes apparent when one realizes that by expanding the number of instructions contained in a single instruction word, many instructions can be carried out in the time it takes to execute only one command. Because the array processor can contain up to ten separate instructions in a single 64-bit instruction word, it is capable of executing one such instruction word during each 167-ns (167×10^{-9} s) machine cycle. Thus parallelism enables the processor to execute at a maximum rate of 60 million instructions per second. In the processing of data for image reconstruction, attention must be paid not only to speed but also to accuracy. Therefore computations, such as addition and multiplication, must be accomplished in floating point arithmetic. In array processors these floating point elements are not only operated in parallel, they are pipelined, which means that once each stage of these elements has data in its input buffers, results are obtained every 167 ns. This enables the machine to deliver a maximum rate of 12 million

Figure 7.14
Images obtained with a modern CT scanner

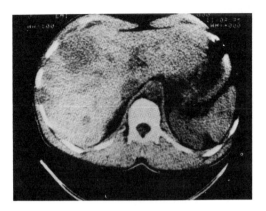

Problem: 33-year old man with metastatic adenocarcinoma in cervical lymph node. Weight loss. Liver scintiscan suggesting a lesion in the right lobe of liver.
CT: Multiple lesion of diminished density in the right lobe of the liver. The lesions vary from 1.5 to several cm in diameter and are consistent with metastasis.
Biopsy: Metastatic adenocarcinoma.

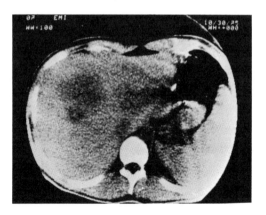

Problem: 21-year old man with abdominal pain, febrile illness.
CT: Lesion of diminished density in the right lobe of liver.
Surgery: Hepatic abscess.

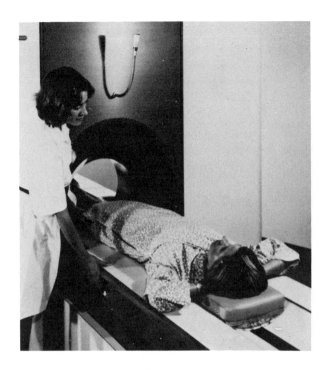

floating point results per second while maintaining a full 38 bits of accuracy. In most array processors, operating in parallel with these floating point units are a number of special components such as a separate 16-bit arithmetic and logic unit, which is used primarily for address calculations, and fast registers, as well as separate memories for often used instruction words, constants, and data. With this new special-purpose, high-speed reconstruction processor, which relies on integrated circuit arithmetic components of advanced design and highly parallel architecture, it has been possible to execute CT algorithms at a rate of hundreds of views per second. As a result data-handling systems associated with modern CT scanners can reconstruct an image in 30 to 60 seconds (Gilbert et al. 1979).

Advances in software design have paralled those in hardware. In the past several years tremendous advances in programming efficiency have enabled the central processing unit (CPU) to perform a number of manipulations on the stored data, including such sophisticated imaging aids as the "region-of-interest" package. The region of interest in the image may be defined by means of an outline that can be either regular, usually a circle or rectangle, or irregular, in which some convenient mechanism, such as a light pen, is used to allow the operator to trace around an organ or other area of interest. Normally calculations are then performed by the computer to improve the statistical quality of this information, and the resultant absorption values strictly within the region of interest are displayed. Another extremely useful manipulation can be provided by combining the information from two or more slices. For example, values from one slice can be subtracted from those in the other slice. If the two slices represent the same section of the body before and after treatment, then useful information regarding the effect of the treatment can be obtained and evaluated.

Once the image is displayed and initially reviewed, it is often important that this information be retained. Consequently present CT systems have available a variety of storage media, often several different types on one machine. The type of storage facility used depends to a large extent on the length of time for which the information is required to be held. Immediate storage is provided by the main memory associated with the computer system, usually a disk that will normally hold a relatively small number of pictures, usually for one or two patients. This storage is used for immediate access, usually while the patient is still being scanned.

When a full set of scans has been completed for a patient, it is normal to transfer the pictures to either a medium- or long-term storage device. The usual medium-term storage device is the floppy disk, which will hold typically six to eight pictures. One or two floppy disks can be appended to the patient's file and kept for the duration of the examination and treatment. Floppy disks are very convenient, but are fairly expensive for very long-term storage. Magnetic tape is the storage medium of choice when storage is required for a number of years. As a result archival information, which some authorities insist be kept for periods of up to seven to ten years, is normally put on magnetic tape, which is relatively inexpensive and extremely reliable for long-term storage.

In addition to the advantages inherent in retaining the data for relatively long periods of time, there are additional advantages to reanalyzing CT scans. Because the three types of storage (main memory, floppy disk, and magnetic tape) are capable of storing complete picture information, modern CT scanning systems are also used to perform a wide variety of image manipulation routines (such as windowing, region of interest, and so forth) at a console away from the testing location. When the data are redisplayed, the clinician is able to view the collected scanning information at more than one specific setting of window width and window level and carefully analyze the data to check a number of different interpretations of the data to verify the diagnosis.

The availability of various storage media therefore provide a means of recording, storing, and viewing patient scans independently of the scanning system's operational viewing unit. Consequently health professionals are provided with around-the-clock access to existing patient scans, irrespective of the in-use status of the basic scanner. With present CT systems patient records can be stored on floppy disks or magnetic tape in less than one minute.

These magnetic storage devices can serve as a means of permanent storage of patients' records or may be erased and then used again.

Obviously, compared with early head and body scanners, present models process information faster, use energy more efficiently, are more sensitive to differences in tissue density, have finer spatial resolution, and are less susceptible to artifacts. Table 7.1 summarizes the CT scanning systems presently available. They may be compared with one another by considering

1. *gantry design,* which affects scan speed, patient throughput, and cost effectiveness

2. *aperture size,* which determines the maximum size of the patient

3. *the type of X-ray source,* which affects the patient radiation dose and the overall life of the scanning device

4. *X-ray for angle and scan field,* which affects resolution

5. *the slice thickness,* as well as the number of pulses and the angular rotation of the source, which are important in determining resolution

6. *the number and types of detectors,* which are critical parameters in image quality

7. *the type of minicomputer* used, which is important in assessing system capability and expandability

8. *the type of data-handling routines* available with the system, which are important user and reliability considerations

9. *the storage capacity* of the system, which is important in ascertaining the accessibility of the stored data

In reviewing the presently available CT systems, it is important to compare them not only in technical terms, but with what one considers an ideal scanner. The ultimate objective of CT is to provide accurate diagnostic information that significantly improves patient care. A CT scanner ideally would provide an unambiguous diagnosis, thereby removing the necessity for further tests. Because financial considerations are also important and CT scanners require a high capital outlay and have considerable maintenance costs, the ideal system would also permit high patient throughput.

Patient throughput is a complex matter, and the parameters affecting it are not always obvious. Studies of several hospitals have found that, in practice, several of the factors that materially affected patient throughput are not confined to the CT scanner itself but to some ancillary operation such as the availability (or lack of it) of hospital ports or the willingness or ability of patients to wait in line for examinations. In considering only the contribution of the scanner operation to throughput, however, several factors, such as patient preparation, scan time, processing time, and interscan recovery times, must be considered. The significance of the contribution of these factors varies from machine to machine, and when considering throughput, it is important to consider the practical case and not simply a theoretical case. Even if the scanner could perform all its activities in zero time, there would still be a finite patient throughput. Operator decisions have to be made, the patient sometimes has to be comforted, and people have to walk around the operational area—all simple human considerations. Any effort to get close to zero time for all scanner activities will result in great expenses of money and of picture quality and ultimately in scanner usability. A good scanner should not significantly influence patient throughput in excess of those times needed to obtain diagnostic information.

A good scanner therefore must satisfy two major criteria: good diagnostic information and high patient throughput (Walmsley 1979). As CT scanning systems continue to improve, these factors must continually be kept in mind.

Magnetic Resonance Imaging

Tissue variations can be measured by methods other than those using X rays. For example, protons, neutrons, high-frequency sound waves, or nuclear magnetic resonance can also be used. Each of these energy-containing sources can be used to generate sets of measurements or projections at different angles across the body, and, in a similar manner, a picture can be reconstructed from them.

In recent years attention has focused on body-scanning devices capable of obtaining chemical as well as structural information about soft tissues. Using magnetic resonance imaging (MRI) techniques instead of X radiation, several systems have produceo images that permit the detection of healthy bone, organs, and tissue, as well as areas of chemical and structural irregularity (Damadian 1977, Lauterbur 1973). This method is noninvasive, penetrates bony structures without attenuation, does not use ionizing radiation, and appears to be without hazard (Marx 1980). As a result MRI has become an essential and well-established technique for analytic, structural, and dynamic investigations of matter in many disciplines, especially chemistry and physics (Andrew 1979), and in recent years has made significant inroads in medicine.

Although MRI is still a relatively new technology—in 1984 only 78 units had been installed worldwide (Steinberg 1986)—it has already changed the definition of optimal diagnostic care in several clinical areas, particularly the brain, spinal cord, and musculoskeletal system. Thus many radiologists, neurosurgeons, neurologists, and orthopedists are incorporating MRI scanning into their practices, either as a single diagnostic test or as a complement to other modalities, such as CT. Numerous other specialists are interested in MRI as well, and reports continue to appear on the efficacy of this modality in a variety of applications (Stone and Taylor 1988a).

MRI Operating Principles

Magnetic resonance imaging is a noninvasive method of obtaining organ images without ionizing radiation. Nuclear magnetic resonance imaging is based on the existence of a magnetic moment in nuclei containing an odd number of protons. The requirement of containing an odd number of protons is met by a large percentage of stable nuclei, radioisotopes, and particularly hydrogen. Virtually every such nucleus has a natural spin, making it in effect a tiny magnet that can be detected. The concept that some nuclei have magnetic properties was first postulated by Pauli in 1924 as a result of his work on optical spectroscopy. This property can be considered as arising from a net positively charged nucleus spinning about its axis, thereby behaving as microscopic bar magnets. The existence of a nuclear magnetic moment was verified that same year by Stern and Gerlack working with molecular beams, but it was not until 1946 that the effort was demonstrated in bulk matter, almost simultaneously by the groups of Block at Stanford University and Purcell at Harvard University. Block and Purcell received the Nobel Prize in physics for their work, and the technique has developed into a major analytical tool (Shaw 1988).

To perform either imaging or material analysis, the nuclei must emit a signal at the precession frequency. This is done by exciting to resonance the precessing nuclei with a magnetic field at the frequency known as the larmor frequency. When the external magnetic field is turned off, the rotating magnetic moment undergoes what is known as free induction decay as the nuclei returns to the ground state. In the process a signal is emitted at the resonant frequency, which is usually detected by the same coils used to produce the external magnetic field. The voltage induced in these coils, in turn, can be related to the spin density (that is, the number of nuclei resonating per unit volume). In most imaging systems repeated excitations are necessary and Fourier spectral analysis techniques are used to realize cross-sectional and three-dimensional images that appear on the system's display device (Stone and Taylor 1988b). The first MRIs were produced by Lauterbur in the early 1970s, using signals received from hydrogen nuclei contained in water molecules. Because water is contained in all body tissues, it is the source of most MRIs currently produced. However, other diagnostically relevant nuclei include ^{13}C, ^{23}N, and ^{31}P.

The value of the magnetic moment of nucleus depends complexly on its composition, that is, on the number of protons and neutrons. Its presence means that these nuclei are essentially arrays of tiny magnets. When these nuclei are placed into an external magnetic field, the nuclear magnetic moments tend to become aligned parallel with the external field. Because the nucleus is spinning, the magnetic moment M responds to the field as a gyroscope precessing around the direction of the field. The precessional or rotating angular frequency is the larmor frequency. Because the precessional frequency depends on both the strength of the external magnetic field and the gyromagnetic ratio, which is constant for each nuclear species, it forms the basis of both analytical and imaging modalities.

From this discussion it is clear that if such an atom is placed in the field of a strong magnet and is irradiated with a pulse of energetic radio waves, the nucleus will respond with a radio signal of its own. In fact each nucleus has its own characteristic signal. Because the chemical environment of a particular nucleus can affect the magnetic field to which it is subjected, the resonance frequency of a given nucleus in one environment can differ from that of the same nucleus in another. Structural information can be gleaned from these differences. It is not surprising that with such properties, MRI has become one of the most powerful tools of analytic chemistry (Lerch 1980).

The nucleus of special interest, as far as medical applications are concerned, is that of the hydrogen atom, which, when bound to oxygen, makes up water, which makes up the largest bulk of our body chemistry. It just so happens that the nucleus of the hydrogen atom, when subjected to an exciting pulse, gives the strongest reply. Because hydrogen protons essentially "wobble" or precess at a definite frequency when placed in a magnetic field, the usual MRI procedure for imaging is to apply a strong magnetic field along the body to be studied. After a short period of time, the nuclei align with their magnetic moments along the field. A radio frequency tuned to the precession frequency of the hydrogen nucleus is then applied at right angles to the main field by means of a set of coils at the side of the body. This causes some of the hydrogen nuclei to precess—all keeping in step. After the radio-frequency field has been switched off, the nuclei continue to precess in phase, generating a similar radio frequency, which can be picked up in receiver coils also placed at the side of the body. These signals detect the water content of the body. Because it takes time for the precessions to die away as the nuclei again realign themselves with the magnetic field, the measurement of this time (which is in the order of tenths of a second) provides important information about the nature of the tissue under investigation.

Thsee procedures refer only to a method of tissue detection. To produce a picture that maps differences in the tissue within the body, it is necessary to independently measure small volumes of material across it. In MRI this is done by applying a small magnetic field gradient across the body in addition to the main uniform field. The nuclei will resonate at different frequencies across the body according to the magnitude of the field gradient. In one method of MRI the frequencies received in the coil can be separated (by Fourier analysis) and the whole spectrum of frequencies will resonate as a series of line integrals across the body, each frequency reflecting the amount of hydrogen nuclei resonating also that particular line. Compared with CT, this is equivalent to one X-ray sweep across the body at a particular angle. A number of sweeps can be repeated at different angles by rotating the gradient field, and sufficient data can be built up to reconstruct a picture similarly to the way a CT picture is constructed.

Using this technique, machines now exist that are capable of imaging entire body cross sections to include proton density. By analyzing a cross section of a human chest for the hydrogen density of its internal tissue, for example, a lung tumor can be detected. The densities of other biologic substances known to differ in tumor tissue, such as the energy carrier adenosine triphosphate (ATP), can also be measured (Damadian 1977).

Such a device offers great promise for the early detection of various tumors without biopsy. It might also be used to detect metastases to help physicians choose cancer drugs on the basis of the tumor's precise chemical composition and to monitor the tumor during treatment. Work on MRI is still in its early stages. Presently CT and MRI should perhaps be viewed not as potential competitors but rather as complementary techniques that can exist side by side. MRI provides information on the chemical composition of tissue, and CT provides a means of visualizing its position and shape.

Comparison of CT Scanners and MRI

MRI systems have a number of advantages over X-ray based CT scanners, one being the use of nonionizing radiation (Stone and Taylor 1988a). Because MRI uses only magnetic fields, multiple scans can be performed without the fear of radiation overexposure or any of the other biological effects of long-term ionizing radiation exposure. Another important advantage of MRI is that because hydrogen nuclei are imaged, dense substances such as bone that normally block X rays become transparent because they contain little water. Soft tissues such as the gray and white matter of the brain are imaged very well as a result of their high water content, and MRI can distinguish between gray and white matter, whereas CT scanning cannot.

Further because MRI provides some indication of tissue biochemistry, it also can differentiate some tumors that CT scanners cannot. Additionally it offers the promise of being able to image other nuclear species besides hydrogen, such as fluorine, iron, phosphorus, and sodium, to provide considerable information about body metabolism. Other advantages are that blood flow can be measured without contrast agents, and that patient motion in MRI only degrades image resolution, whereas in CT scanning it can degrade resolution and cause image artifacts.

MRI systems also have a number of disadvantages. Although MRI does not involve ionizing radiation, the large magnetic fields used can cause problems. For example, pacemaker wearers cannot be examined because of interference with pacemaker operation or programming, and patients whose bodies contain metal surgical clips are also ruled out because the field may cause hazardous clip movement. Further, although extensive testing on the biological effects of large magnetic fields has indicated no danger, long-term effects cannot be ruled out entirely.

Another disadvantage of MRI is high acquisition and operating costs. MRI systems use large magnets that are either permanent, resistive, or superconducting. Permanent magnets are very heavy (more than 100 tons) and thus require special attention to the structural bracing for the imaging facility. Resistive magnets require large amounts of electrical power and water cooling systems, whereas superconducting magnets require power only to start current flow but need cryogenic cooling with expensive liquid helium for proper operation. All magnet types require shielding, with the superconducting type needing extraordinary care in shielding to minimize the extent of the fringe field so that relatively innocuous objects are not turned into deadly projectiles. MRI installations tend cost more than do CT scanner installations, because of shielding and auxiliary equipment, and usually require more space. Also power and cooling requirements make MRI systems' operating cost higher than that of CT scanners.

Additionally, although one of MRI's advantages in brain imaging is the characteristic that bone is transparent, as a result MRI cannot be used for bone imaging, which CT scanners can readily perform. Further, MRI systems need long scan times (10 to 30 minutes is typical for a hydrogen scan) for an adequate signal to noise ratio (SNR), requiring that the patient be immobile for long periods. Present CT scanners are almost instantaneous with respect to scan time, and patient motion is less problematic. Because immobility is difficult for some patients, image blurring due to patient motion during MRI often occurs. Additionally some patients become claustrophobic from being confined to the inside of the magnet for long periods.

CT scanners also presently have an edge on MRI systems with respect to resolution (Crooks et al. 1984). Modern CT scanners have a limiting resolution of approximately 3 to 4 cycles per millimeter, whereas MRI systems have a limiting resolution of approximately 1 cycle per millimeter.

In view of other medical imaging systems' development, improvements in MRI systems will likely be made at least to mitigate some of their disadvantages. In the near future, however, it seems that MRI and CT will remain complementary clinical tools for those institutions that can afford their installation and operation.

Future Perspectives in MRI

Current research and development efforts are concentrated in the following areas: (1) the use of nuclear magnetic resonance to image other nuclear species besides hydrogen, (2) cardiac and blood flow MRI studies, and (3) determination of the optimal field strength for MRI (Portugal 1984, Stone and Taylor 1988b).

One of the most promising aspects of MRI is the potential to obtain information about body metabolism and biochemical reactions directly from MRI data. With respect to MRI the human body is an aqueous mixture of elements, some of the more important being carbon, iron, nitrogen, phosphorus, sodium, and sulfur. By manipulating these nuclei, MRI promises to enable the physician to identify which elements are present along with their chemical forms, concentrations, and locations. For example, iron deficiency causes anemia, which conceivably could be detected and measured noninvasively with MRI instead of the conventional technique of drawing a blood sample and performing a test for iron content. Iron is beyond the sensitivity of today's instrumentation, however, because only about 2 percent of the body's iron is present as the isotope iron 57, the odd-numbered form detectable by MRI. Also iron's gyromagnetic ratio is very low compared with that of hydrogen and requires a much higher field to generate a reasonable MRI signal. Sodium is another element of interest, which may be useful in localizing the site of infarction in strokes and heart attacks. Sodium concentrations increase dramatically at the site of infarcted tissues; therefore MRI, measuring sodium as the nuclear species, could readily locate the damaged area in the brain or the heart. Phosphorus is also of interest because it is a key element to a number of biochemical reactions. For example, it is contained within ATP, a major energy source for body cells, and it plays a role in muscle contraction and may also be useful in tumor detection.

Another nuclear species of interest, which occurs in the body only in trace amounts, is fluorine. MRI using fluorine is of interest in studying the effects of surgical anesthetics, many of which contain fluorine. MRI is helpful in studying the ways anesthetics are absorbed and distributed, the portions of the brain that are affected, and the length of time they take to produce anesthesia. Also fluorine has been used as an MRI contrast agent in blood flow studies. Biologically inert fluorinated hydrocarbons have been used as blood substitutes and theoretically could determine flow rates within small blood vessels deep within the body. One advantage of fluorine imaging is that fluorine-19 gives an MRI signal almost as large as that of hydrogen, as compared with the relatively weak signals of iron and sodium.

Cardiac imaging is another new area for MRI (Hamilton 1982). Presently this is accomplished by the gating technique described for CT scanners, whereby snapshots of the heart are taken at different points of the cardiac cycle. The image is reconstructed from data from the snapshots, with minimal blurring. Cardiac imaging offers the promise, in conjunction with sodium and phosphorus MRI, of providing new information about the heart's structure, tissue characteristics, and metabolism. This is potentially important for studying the alterations in cardiac metabolism in heart patients.

Blood flow can be measured without fluorine by taking MRI snapshots, 50 to 500 milliseconds apart, of the blood vessel of interest (Portugal 1984). The average cross-sectional area of the blood vessel can be calculated from the image, whereas flow velocity can be determined from the MRI data. This technique offers some advantages over other noninvasive flowmeters such as ultrasonic flowmeters in that it does not require laminar flow conditions to measure blood flow accurately.

Another area of interest is determining the optimal field strength for MRI. Most researchers concede that the SNR of MRI improves with field strength and that image resolution and SNR trade-offs can be made. Moreover higher fields allow imaging of more nuclear species. However, there is no consensus as to how strong a field is needed for imaging. For example, whole body images have been produced with permanent magnets producing a magnetic field of 3,000 gauss (G), and images have been produced by using superconducting magnets of 24,000 G and higher. Although higher field strengths do promise the imaging of nuclear species other than hydrogen, it is questionable whether the added capital and operating cost are justifiable for most clinical applications.

MRI provides valuable clinical information without ionizing radiation with resolution that approaches that of radiographic imaging. Therefore a new era has arrived in diagnostic imaging that shows great promise in redefining the clinician's expectations regarding the type of information available in the imaging process. Any discussion of the potential clinical uses of MRI should include several caveats (Stone and Taylor 1988b):

• The technology of MRI is still evolving rapidly; thus the various models currently available may differ substantially in their capabilities.
• More sophisticated users of MRI units may be able to get significantly better results than those who are less expert by making better use of methods to counteract motion artifact, the use of surface coils, more appropriate choices of pulse sequences, and so forth.
• Developments in MRI are so rapid that the clinical literature on specific applications, although an excellent source of general information about MRI, frequently cannot keep pace with technical innovations.
• New developments in MRI scanning will likely tip the balance toward MRI in many cases where it is currently considered equivalent or even inferior to other modalities.
• Economic considerations may make another technology preferable even where MRI is technically superior.

The vast majority of U.S. hospitals do not yet have MRI; radiologists, executives, and others with decision-making responsibility at these institutions have the difficult task of deciding whether to acquire this technology and, if so, which model to buy. Complicating this decision are the continuing development of low-field-strength units, the complexity of MRI techniques, and the fact that new ones are constantly being introduced. Thus, in addition to facing greater uncertainty about selection of a model, hospitals need to be concerned as to whether, if a unit were acquired, physicians with little or no training in MRI would be able to use it effectively.

Acquisition decisions are also complicated by the enormous growth of diagnostic imaging in recent years. Clinicians today frequently have two or more useful imaging tests available for a particular problem; while MRI is a potent tool, the way in which it should complement or replace these other modalities is not always clear. Since the mid-1970s CT has grown from a new technique to a fundamental of practice, ultrasound has become much more sophisticated, new ways of performing angiography have been developed, and nuclear medicine has added many new radiopharmaceuticals and has assumed an important role in cardiac diagnosis. The recent development of other revolutionary technologies such as positron emission tomography (PET) further complicates envisional uses of these imaging modalities in the future.

The Economics of Medical Imaging

The different imaging modalities examined in the previous section clearly provide some overlapping services in terms of the information they convey, and to that extent they are substitutes for each other in diagnostic procedures. CT scanners, diagnostic ultrasound, and nuclear imaging devices, for example, are all capable of providing images of specific organs of the body. Because many imaging modalities also provide different services, to that extent those different modalities are also complementary to one another. For example, the CT scanner provides data on brain tissue, nuclear medicine can be used to examine microcirculation, and an important application of ultrasound is the measurement of the velocity of blood flow. Consequently contemporary medical practitioners require a "portfolio" of imaging devices and techniques to ensure the provision of comprehensive high-quality data on which they can then base appropriate health care therapies.

However, there is a paradox to be explained. Even when a modality such as nuclear medicine is (virtually) technologically dominated by other modalities (for example, CT scanners) as a vehicle for obtaining images of the body, usually it remains a part of the hospital's imaging modality portfolio. The answer to this technological paradox is an economic one; such a strategy often helps hospitals to minimize the costs of providing health care services to their patients. In contrast other medical institutions do not have direct access to state-of-the-art imaging techniques using relatively expensive equipment such as CT scanners and MRI devices. Consequently in many regions both doctors and patients have complained about the lack of access to state-of-the-art imaging modalities. Elsewhere, however, some physicians have been accused of using high-cost imaging techniques excessively to increase their profits or to cover the costs of unwise investments on unnecessary equipment.

The Economics of Lumpy Investment Decisions

In this section we examine all of these issues, but, before proceeding, we need to develop some insights about the economics of investment in "lumpy" capital equipment. A lumpy piece of capital equipment is one that comes only in large sizes and in many situations is used at a rate lower than the one at which unit costs are minimized. Suppose that a certain type of medical imaging machine comes in only one size. If the machine carries out 1000 imaging tasks per month, the total cost of operating the machine is $20,000. Thus the unit cost of each image equals $20 if the machine is operated at this rate. However, if fewer images are made, the cost of each image will exceed $20 because the machine and the people who operate it are still necessary expenses. Alternatively, if more than a thousand images per month are made, unit costs will rise because of excessively heavy wear and tear on the machine and the need to pay operators overtime wages. The cost-minimizing level of use of the machine is thus 1000 images per month. Suppose also that under no circumstances would the hospital need more than one such machine. However, operating rates might be such that it could have massive excess capacity. Thus the imaging device represents a lumpy piece of capital for the facility. Both CT scanners and MRI devices fall into this category for most hospitals.

Two types of cost are associated with the use of any type of machinery: the fixed costs of acquiring and installing the machine and the variable costs associated with using it to carry out tests on an individual (to provide a unit of service). Fixed costs are often referred to as *overhead*. Once the machine has been acquired, such costs become *historical costs*; that is, they cannot be avoided. This aspect of fixed costs is extremely important in relation to the use of equipment that has already been acquired because once such costs have been incurred, they play no role in the decision about whether a machine should be used or left idle. What counts is whether the price of a unit of service provided by the equipment exceeds the variable cost of carrying out the service. If such is the case, the equipment that has been obtained already can be operated by the health care institution without additional losses being incurred. But such concerns are relevant only in the short run. If a technology is to remain economically viable over the long haul (that is, when the current machine wears out it will be replaced), all costs (overhead costs and variable costs) must be recovered by the institution. Therefore the way in which total and average (or unit) costs change as the use of the equipment changes is critically important in determining whether an institution can afford to obtain the equipment and whether (given the price it receives for the services provided by the equipment) ownership results in losses or profits.

The nature of the costs associated with lumpy equipment is perhaps best illustrated by an example: Suppose the acquisition and installation of an imaging machine (say a CT scanner) will cost the medical institution $250,000. To make the problem simple (in particular, to avoid concerns about the role of interest rates), let us pretend that the machine has a working life of one year. Using the machine to carry out a typical CT scan can also require the use of variable resources, the time of various health care professionals, power for the machine, administrative costs associated with billing, and so forth. These variable costs are likely to be constant on a unit basis for a wide range of output levels. Suppose, for example, that if the machine is operated at normal rates of use, during an eight-hour work day 10 CT scans can be carried out. Allowing for ten general vacation days and for festive occasions such as Christmas and Independence Day and assuming a five-day work week, the machine will operate in a normal mode for 250 days during its one-year life, permitting 2500 CT scans in normal working hours. Let us assume that each of these CT scans requires variable resources that cost the institution $60. Additional CT scans would have to take place at increasingly inconvenient times, perhaps overstressing the equipment and the space allocated for the process, resulting in higher and higher amounts of variable resources and correspondingly higher costs for each additional CT scan. Perhaps, for example, after the 2500th CT scan, each additional scan would cost 20 cents more than the previous one. Thus CT scan number 2501 would have an additional variable cost of $60.20, CT scan number 2502 would have an additional variable cost of $60.40, and so on, and, for example, CT scan number 2700 would have an additional scan cost of $100 ($60 + $0.2 X 200).* Finally, assume that the maximum (annual) capacity of the machine is 4500 CT scans. The total cost associated with the use of the machine at any level of output of CT scans is the sum of total variable costs and total fixed costs. Given this assumption, one can estimated the costs for any feasible level of use of the CT scanner. Such cost estimates are presented in table 7.3 for representative levels of use.

Notice that when the variable costs of additional CT scans begin to increase, total variable costs and (eventually) total costs begin to increase at increasing rates. The importance of this commonsense observation is reflected in the behavior of unit or average costs (total costs divided by total output). Just as there are three total cost concepts relevant to the issue (total variable costs, total fixed costs and total costs, the sum of the other two costs) so there are three unit or average costs to be examined: average fixed costs (AFC), average variable costs (AVC) and average total costs (ATC). These average or unit costs are computed for each output level in table 7.3 and are presented in table 7.4. Each of the average cost curves

*Readers familiar with economics will recognize immediately that we have introduced the concept of marginal cost, and we have assumed that the marginal cost of a CT scan is constant and equal to $60 up to an output of 2500 units. Thereafter the behavior of marginal cost is characterized by $MC = 60 + 0.2 (Q - 2500)$, for $Q > 2500$; that is, it is increasing, once output Q exceeds 2500, at the rate of $0.2 per unit. The total cost function for CT scans is the spline function $TC = 250,000 + 60Q$, for $0 < Q < 2500$, and $TC = 250,000 + 60Q + 0.1(Q - 2500)^2$, for $2500 < Q < 4500$.

Table 7.3
Total Costs of Operating a CT Scanner

No. of CT Scans	Total Variable Costs ($)	Total Fixed Costs ($)	Total Costs ($)
0	0	250,000	250,000
500	30,000	250,000	280,000
1000	60,000	250,000	310,000
1500	90,000	250,000	340,000
2000	120,000	250,000	340,000
2500	150,000	250,000	400,000
3000	182,500	250,000	432,500
3500	310,000	250,000	560,000
4000	465,000	250,000	715,000
4500	670,000	250,000	920,000

Table 7.4
Average Costs of Operating a CT Scanner

No. of CT Scans	Average Variable Costs ($)	Average Fixed Costs ($)	Total Costs ($)
0	—	—	—
500	60	500	560
100	60	250	310
1500	60	166.67	226.67
2000	60	125	185
2500	60	100	160
3000	60.84	83.33	144.17
3500	88.58	71.42	160
4000	116.25	62.50	178.75
4500	148.88	55.56	204.44

(AFC, AVC, and ATC) presented in figure 7.15 shows that the number of CT scans increases.

Several characteristics of the average costs curves are important. First, consider the behavior of AFCs. As the number of CT scans increases, AFC steadily declines, though at a decreasing rate. The reason is quite simple. The same amount of total fixed costs ($250,000) is being spread over a larger and larger number of CT

Figure 7.15
Average cost curves for the production of CT scans

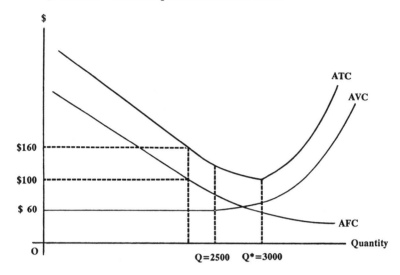

scans; that is, the overhead associated with the CT scanner is being speread over more and more units of output and therefore becomes smaller on a unit cost basis. Second, note that AVCs remain constant over the output range 0 to 2500 CT scans at $60 per scan. As a result, ATCs (the sum of AFCs and AVCs) steadily decline over that output range. For example, when Q is 500, ATC equals $560 (because AFC equals $500 and AVC equals $60). When Q rises to 2500, ATC falls to $160 because although AVC still equals $60, AFC has fallen to $60 as a result of overhead spreading. Third, note that once output exceeds 2500, AVCs begin to increase immediately. Again the reason is simple. Once the number of CT scans rises above 2500, the variable cost of performing an additional CT scan rises above $60 and continues to increase as Q increases. However, ATC continues to fall until an output level of approximately 3000 is reached. The reason for the continued decline in ATC after AVC has begun to rise is that AFC is still falling as Q increases. Over the output range of from 2500 to 3000, the effects on ATC of the fall in AFC dominate those of the increase in AVC; once output exceeds 3000 CT scans, the effects of increases in AVC dominate and ATC begins to increase. Thus the ATC curve associated with the CT scanner is U-shaped, reaching its minimum at 3000 scans, denoted by Q* in figure 7.15. Thus if the machine is used at that rate, average or unit costs will be at their lowest attainable levels.

The Use of Imaging Devices and the Issue of Hospital Profits
Frequently, perhaps even typically, a lumpy capital investment such as a CT scanner will not be used at the rate that minimizes unit costs (Q*). Instead, because of limited demand for services the equipment provides, utilization will be at some lower level. Several questions arise with respect to this fact. The first concerns whether a hospital can afford to have such a machine even though its presence would result in the provision of sophisticated imaging modalities that would improve the quality of health care. The answer to that question is quite simple. It depends on the price the hospital may charge for the services the machine provides.

Suppose, for example, that the hospital owning a CT scanner carries out 2000 CT scans over a one-year period. From table 7.4, it follows that the hospital will incur an average or unit total cost of $185 per CT scan. Its total costs of owning and operating the CT scanner at that level will be $370,000. Now suppose that insurance companies and Medicare/Medicaid are willing to reimburse the hospital at the rate of $185 per scan and that uninsured patients can pay their bills and are also charged $185 per scan. Obviously the hospital will break even on the activity because the price it charges for the service, which is also the average revenue it receives per unit of service, just matches its unit costs. The situation is illustrated in figure 7.16, which replicates the ATC curve presented in figure 7.15. In figure 7.16 the horizontal line PP represents the price received by the hospital for the CT scan. This line cuts the ATC at the point A, associated with an output level of Q (equal to 2000) CT scans. At Q unit cost equals price, and the hospital breaks even with respect to its provision of CT scans.

Also note, however, that a hospital could lose money. Suppose, for example, that the reimbursement rate (based on DRG criteria) remains at $185, but that the hospital finds that only 1500 CT scans are needed. This situation is illustrated in figure 7.16. The hospital is being reimbursed by the amount OP ($185) for each scan, but because of the relatively low rate of use ($Q' = 1500$), unit costs are OC ($226.67). Thus the hospital loses an average of PC dollars ($41.67) on each CT scan, a total loss amounting to the rectangle PCBE, or $62,500 ($41.67 X 1500).

The hospital could also make a profit on its CT scan operation if usage increased above Q (2000 scans) to some higher level, say, Q'' (2500 scans). At that level of use unit costs for each scan would fall to C' ($160), and the hospital would make a unit profit of C'P ($25) on each scan and a total profit amounting to the rectangle C'PFD, or $25 X 2000 on its CT scan operations.

In this discussion the hospital is assumed to be a price taker; that is, it has to accept the predetermined DRG payment of $185 per CT scan for each scan. It cannot freely adjust prices to match unit costs. The former assumption describes institutional arrangements for

Figure 7.16
Hospital CT scan profits and losses under different rates of use

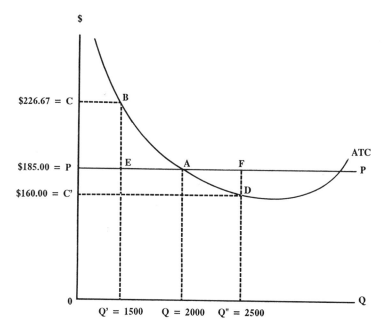

payments for health care of this type quite well and obviously, in such a setting, up to a point it is in the hospital's interest to increase the use of its CT scanner. The hospital can accomplish this goal by (1) attracting patients from other hospitals to its facilities and (2) allocating more patients to DRG categories that justify the performance of a CT scan and fewer patients to other DRGs that would justify less costly imaging modalities.

The first approach can be carried out in a legally and ethically acceptable manner. For example, two medical facilities in the same town could agree to use only one machine and to share jointly the burdens or benefits of the resulting losses and profits. The disadvantage of such a strategy is that some, perhaps many, patients will suffer additional health care costs because of longer journeys to and from the health care facility that provides the service they need. (This issue is in fact quite important in the context of decisions about the regional distribution of MRI devices and equity considerations and is discussed in more detail below). However, such actions by the health care providers are certainly not illegal or unethical and may represent good social as well as private economy by guaranteeing the provision of the service in the area.* It is also possible that a hospital might use unethical methods to attract patients by, for example, advertising and offering under-the-table payments to nonaffiliated physicians. Whether such strategies would be pursued vigorously is subject to some debate because they involve additional costs for the institution that quite conceivably could exceed the benefits associated with them.

The second method of increasing the number of CT scans is clearly unethical and probably illegal. Allocating patients to improper DRG categories is a violation of the contract the hospital has with third party health care payers such as insurance companies and Medicaid and represents fraud. Of course, because diagnosis is an

---

*In fact this type of behavior might be regarded as a violation of anti-trust laws, but could be defended if the two health care institutions were able to establish that they are price takers, not price setters (monopolists), in the health care market. If they operate under DRG schemes, certainly they would appear to be price takers operating as quasi-regulated natural monopolies.

an imprecise science, physicians often operate in a gray area with respect to the determination of what types of tests are necessary for any specific patient. However, if hospital administrators are faced with underused and unprofitable equipment, as the analysis suggests may sometimes be the case, they are likely to have strong incentives to encourage doctors to carry out economically lucrative but medically questionable tests on every possible occasion.

Can such behavior on the part of the medical institution be identified? The answer is probably No in the case of any specific patient (except when blatant abuse occurs). But the response is a firm Yes if there is a systematic bias toward the unnecessary use of such devices. Each hospital is required to maintain detailed records on its patient load and mix of diseases. Any hospital in a specific region having increases in its case load because of improper DRG allocations would soon stand out statistically like the proverbial sore thumb. Whether or not such behavior actually would be identified and curbed depends on the resources allocated by insurance companies and Medicare/Medicaid to the investigation of fraud. On the other hand a considerable body of anecdotal evidence suggests that the practice is not unknown.

The Optimal Distribution of Imaging Devices

The lumpy nature of many medical imaging devices also raises questions about what constitutes the optimal number and location of such devices in the health care system. the question is particularly relevant in the case of MRI devices that each cost in the order of $.5 million. At first blush it might seem that the discussion of the average cost curve for such devices provides the answer. Would it not be ideal if, for example, MRIs were allocated to regions and specific hospitals within each region to ensure that each machine was being used to minimize the average cost of a scan or test; that is, in terms of the example in figure 7.17, surely we would like to guarantee that each MRI device perform Q* scans, wouldn't we? The answer is that it is not necessarily so! Two factors mitigate against such a conclusion. First, the fact that patients incur

Figure 7.17
Average patient transportation and average total patient social cost of CT scans

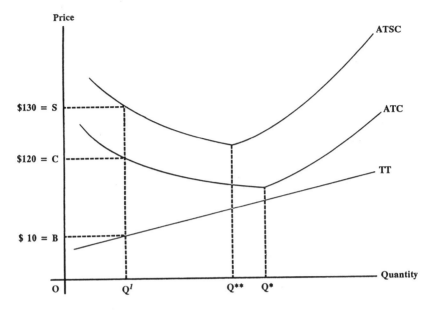

transport costs in acquiring health care services means that the ATC curve presented in figure 7.17 reflects only the cost to the hospital of the health care, not the total costs incurred by society as a whole. Second, there is an equity issue also related to transport costs. If, as is often claimed, health care is a special good, then health care services should be available to each individual at approximately the same total price, which includes both the payment to the hospital and transportation and other inconvenience costs that have to be borne by the patient. Both of these factors suggest that a simple allocation rule that ensures that each machine will be used to minimize the hospital's unit cost is inappropriate. Socially ideal usage rates should be lower, or, to put it another way, society would benefit from a larger number of devices than such an allocation rule would imply.

The allocation issue is important because in many states (for example, Connecticut) hospitals must obtain permits from the state before they can acquire large-scale and expensive equipment such as MRI devices. State regulatory agencies and commissions are concerned with controlling health care costs and, as part of the process, preventing the unnecessary spread and significant underuse of such devices that in turn would lead to unjustified increases in health care costs. However, the total cost of health care to society includes the transportation costs patients incur in travelling from their home to the health care center. If cachement areas for an MRI device are increased to increase use of the machine, travel costs for each additional patient are likely to rise as each such patient has to travel a farther distance from the home to the MRI location. When such travel costs are taken into account, the total social average cost curve associated with treatment (which also includes travel and inconvenience costs) will lie above the average cost curve associated only with treatment. Most significantly the level of use that minimizes average total social costs (patient travel costs plus treatment costs) will be lower than one that minimizes average treatment costs.

This situation is illustrated in figure 7.17, in which the curve labelled ATC represents average treatment costs and the curve labelled TT represents average travel costs. If the two curves are added together vertically, the average total social cost (ATSC) curve is obtained. For example, in figure 7.17, at Q´ the unit or average cost of a scan is represented by the vertical distance OC ($120). The average transportation cost for each of the patients receiving the Q´ scans is OB ($10) and the ATSC is thus OS ($130), equal to OC + OB. Note that because the unit transportation cost curve slopes upward, the ATSC curve achieves its minimum level at Q**, a lower utilization level than Q*, the number of scans that minimizes unit treatment costs. The assumption that transportation costs steadily increase with increased utilization of the machine, critical to the above conclusion, is quite reasonable. For the number of scans to increase, an MRI device must draw on patients from increasingly large and (in general) increasingly distant populations.

The second factor is equity. In relation to imaging devices such as MRIs, equity considerations have two dimensions: First, it is often argued that different consumers should not be charged different prices for the same product, whether that product is toothpaste, a new car, or a CT scan. Second, it is also argued that health care is a special commodity because it directly effects the availability of life to an individual and to that individual's inherent capacity to enjoy that life. The corollary of this assertion is that all citizens should therefore have reasonable access to acceptable health care, including high-quality techniques for the rapid diagnosis if their health status. This perspective implies that regulators should provide permits for MRI and other imaging devices in a manner that ensures that this goal is attained.

Is it Economical for Hospitals to Use Many Diagnostic Techniques?

Bearing in mind the importance of costs in hospital decision making, we are now ready to examine another of the major issues associated with the use of imaging modalities—that is, why, when certain medical technologies become outdated, are they still used extensively? The most obvious example is conventional X-ray imaging, which has been superseded by CT scanning from a technological perspective. The answer, in this as in many other cases, is cost. In many situations (for example, a simple fracture of the tibia) an X-ray image will provide all the information necessary for the physician to make an accurate diagnosis. Although a CT scan would provide even more information about the problem, the increased degree of precision it provides would be obtained only at a significantly higher cost and would not be required for a proper diagnosis. Moreover in such situations the physician is fully aware of that fact before she carries out the X-ray imaging.

Other situations may be less clear cut. Before any tests have been carried out, the physician may not be sure whether the patient's problem requires a CT scan. However, a low-cost X-ray image often will resolve the issue. Once it has been taken, the physician may find she has all the needed information, and the consequence is a satisfactory diagnosis made at a low cost. Alternatively the X-ray results may be ambiguous, and significantly higher-quality images may be required. The physician will then turn to CT scanning. And the costs incurred in diagnosing the specific patient's condition will be higher than if the physician had immediately used CT scanning. In these situations some patients will enjoy lower costs of treatment, and others incur higher costs of treatments if the X-ray image is used as a screening device. Should the physician ever have used X-ray imaging in the first place? The answer is ambiguous. In fact it all depends on the probability that a CT scan will be required once the X-ray data have been obtained and on the relative costs of the two procedures.

Suppose, for example, that a patient enters an emergency room with a complex fracture. Previous experience suggests that 30 percent of the time X-ray images will provide the physician with all the information she requires to make her diagnosis. Seventy percent of the time a CT scan will be required. Further suppose that a CT scan will cost $200, whereas the X-ray images will cost only $40. If a CT scan is performed automatically, the average cost of monitoring each patient will be $200. But now suppose that X-ray images are taken. Thirty percent of the time imaging will cost only $40 per patient, whereas 70 percent of the time it will cost $240 ($40 for the X-ray images and $200 for the CT scan). Thus on average the per-patient cost for this type of case will be only $180 if X-ray images are taken first ($40 X 0.3 + $240 X 0.7). In this case it is good economic practice, as well as good medical practice, for the physician to start with X-ray imaging.

Such is not always the case. Suppose, for example, that if a patient enters the emergency room with a complex fracture of the skull, there is a 95-percent chance that CT scans will be required once the X-ray images have been taken and that the X-ray images will be more costly than in the previous example (say, $80). In this case, if a CT scan is performed immediately, average imaging costs will be $200. If, however, X-ray images are taken first in 95 percent of the cases, total imaging costs would amount to $280 ($80 for X-ray imaging and $200 for CT scanning). Only 5 percent of the time would they amount to $80. As a result, if the X-ray images are taken, average imaging costs would amount to $270 ($80 X 0.5 + $280 X 0.95). Clearly, in this second example, from an economic perspective the sensible action is for the physician to require immediate CT scanning.

The Economics of Prevention as Opposed to Cure

Our discussion raises a more general issue that is troublesome to many people associated with or concerned about health care and the health care system, experts and nonexperts alike. When a state-of-the-art technology becomes available, should it not be used to enhance the quality of diagnosis and therapy provided to the patient? Moreover, at an extremely fundamental level, should society take actions to prevent the incidence of all diseases or let some take their course and treat those patients who end up sick? These issues span many areas of health care, of course, and are not relevant just to imaging modalities. However, this is an appropriate place to provide an economist's perspective on the problem because imaging modalities have a clear role in the prevention as well as diagnosis of health problems.

Louise Russell (1986) examined the issue in a controversial but important monograph published at the Brookings Institute, entitled

"Is Prevention Better than Cure?" The focus of her study is whether it is economically more efficient for society or individuals to prevent the occurrence of a disease through passive methods (for example, screening for hypertension or vaccination against flu viruses) or to let the disease occur and then cure the individuals who consequently suffer from it. Imaging modalities are not considered explicitly in her analysis, but much of what Russell has to say can be readily applied to the use of these technologies.

The argument put forward by Russell is quite simple: Preventive measures, such as vaccination to prevent measles, yield significant and valuable benefits in the form of reduction in the number of cases and savings in terms of the costs of treatment, reduction in lost work time (at school, in the home, or at work), and the avoidance of other unpleasantnesses associated with being sick. At the same time these benefits are obtained only at some considerable cost. The general vaccination of an entire population that has measurable exposure to the risk of disease can be expensive. Costs include expenditures for the vaccine itself, the time and transport of health care professionals who administer the prevention treatment, and the value of the time lost and cost of transport to the individuals who receive the treatment and the family members or friends who accompany them to the vaccination center. If these alone constituted the costs associated with prevention, then they would be large. Russell points out that between 1963 and 1968, over $205 million was spent to administer the measles vaccine to the target population. However, preventive measures themselves are not risk free. Often the preventive action creates health risks for the target population, and while these risks may be smaller than those associated with no prevention, they represent a part of the cost of the prevention program. In the case of the measles vaccination program examined by Russell, between 1963 and 1968 adverse reactions to the preventive therapy (the vaccine) caused occasional severe illness and millions of lost work days and school days due to mild reactions.

Frequently the benefits from a prevention program far exceed the costs of the program to society as a whole. In fact Russell carefully demonstrated that was almost certainly one reason for the measles and smallpox vaccination programs carried out in the United States and elsewhere in the 1950s through 1970s. But what was true for measles does not necessarily hold for every medical problem. Influenza is a case in point: Because repeat vaccination is required, preventive measures to protect the population against the flu are expensive.* Moreover the adverse effects of the disease on the health of most individuals is relatively modest. A widespread vaccination program designed to prevent the flu may therefore yield less benefits than it costs. Indeed the fact that such a program has not been instituted suggests that such may be the case.

Note that the decision not to implement a prevention program flies in the face of the ethical imperative requiring health care professionals to take actions that prevent harm to their patients; especially when there is some possibility that death could be the consequence of no action. After all, people can and do die of the flu, particularly if they are elderly or already in poor health. Nevertheless society does not do everything in its technological power to help people avoid all risks of infection or to detect the early onset of every adverse health condition to design effective preventive and early treatment therapies. As communities and nations we have decided, for example, that it is too expensive to annually provide everyone a CT scan or even a check-up carried out with low-tech equipment by family practitioners in local clinics.

Is our society, or any other, failing in its duty to each individual by choosing not to do everything in its medical power to guarantee the best possible treatment of disease? The answer is No. All of society's resources could be allocated solely to preventive and curative medicine, not only at the expense of perhaps frivolous commodities like compact disks, private swimming pools, and imported champagne, but also at the expense of products required to meet basic needs (food, clothing, housing, and the like). A more realistic appraisal suggests that the appropriate approach to preven-

*Influenza vaccinations are required annually whereas measles vaccinations are required only once in a lifetime.

tive medicine is to examine each program on a case-by-case basis. The question to be asked yet again is whether the benefits of a preventive program (for example, CT scans for all members of the population with some degree of demonstrated risk of cancer*) exceed its costs.

In fact the reluctance of our society as a whole to implement many preventive medical programs is reflected and even becomes more prevalent at the level of the individual. An individual bears at least some of the costs of any preventive medical program. For example, the state may pay for the costs of vaccination, but the individual has to face the terror of the needle, organize his schedule to visit the vaccination center, and bear the risk (even though typically it is very small) of an adverse reaction to the preventive treatment. Because the likelihood of actually catching the disease for which preventive measures are being taken is frequently very small, the individual often chooses not to take a "preventive" course of action. The benefits to the individual are small, or (perhaps more significant) perceived to be small, and the costs, though not great, are real and very present.

Of course it is the perception of benefits that determines whether an individual decides to undertake preventive actions. If an individual is aware, because of the family's medical history, that she is particularly susceptible to breast cancer, then she is much more likely to have biopsies on a regular basis. Similarly a man whose father and grandfather suffered from alcoholism is much more likely to avoid alcohol or excessive drinking. In both of these examples, information changes the individual's perception of the benefits to be derived from preventive medicine and increases the likelihood that those reactions will be adopted, irrespective of whether the individual bears all or only part of the cost.

In fact the point is quite general. One of the most important aspects of public policy directed toward the prevention of disease

---

*Such a target population might include all people over thirty who live in high-risk population regions, such as the Northeast and Los Angeles, and people whose families have a history of the disease.

is the provision of information. For example, public policy directed toward the dissemination of accurate information is one of the major components of the policy program that United States Surgeon General C. Everett Koop has advocated to deal with the AIDS (acquired immunodeficiency syndrome) epidemic. President Reagan's special advisory committee also viewed the provision of accurate information on AIDS as a strategy that would encourage people to modify their behavior to reduce their risk of exposure to the virus. These conclusions are not founded on speculation. People are known to adjust their behavior in the face of risk. For example, dozens of economic studies have shown that farmers adjust their production decisions to reduce risks of loss of income from natural disasters such as adverse weather and infestations. Environmental economists have shown that individuals with adverse pulmonary conditions are more likely to stay indoors and use air conditioners and dehumidifiers on days when the pollution index is high. However, obtaining accurate information about disease is often costly for the individual, often too costly relative to the benefits the individual expects to derive from acquiring it.

It is partly because of the costliness to the individual of developing such information that public administrators often step in to develop and disseminate that knowledge. There are economies of scale in information collection and dissemination that reduce the average burden of the associated costs to the individual (through taxes) to manageable levels. However, policymakers are also concerned with preventive measures to avoid contracting a communicable disease such as AIDS. Certainly when an individual acts to prevent AIDS, he also reduces the likelihood that it will be transmitted to the people with whom he associates. Thus the benefits to society of the actions he takes to prevent infection exceed those that accrue only to him. These considerations also justify policies that subsidize access to preventive medicine. They also validate the institution of controls over individuals who contract a contagious disease; for example, placing people discovered to have yellow fever or small pox in quarantine or, more controversially, passing laws prohibiting individuals with AIDS from sexual contact with other persons.* The important point is that preventive medicine often is a public good in nature, one that generates benefits to more people than just the individual practicing it, and thus government intervention to ensure the practice of preventive medicine is frequently justified.

In this section we have examined some of the critical economic issues associated with technological innovations in the imaging modalities used in medicine. These issues are closely linked to the fact that imaging devices often represent lumpy investments for medical facilities and therefore often appear to be underused. We have examined whether or not institutions face incentives to prescribe excessive use of such machines and what determines the optimal allocation of high-priced items such as MRI devices among hospitals. The latter concern is particularly important because in many states the distribution of MRI devices is regulated by state authorities concerned with ensuring adequate access to health care for all members of the population while controlling health care costs through the prevention of waste. We also have discovered that even technologically outmoded imaging modalities such as conventional X-ray imaging continue to be used because they play an important role in ensuring the provision of high-quality health care while minimizing resource use and health care costs. Finally, we have examined some of the broader economic issues associated with the role of preventive medicine in the health care system that are also associated with technologies that have diagnostic as well as therapeutic functions.

Technology Assessment: Safety and Efficacy

The medical imaging technologies discussed in this chapter are examples of the sophisticated, complex, and often costly new tech-

---

*We are not advocating such a law. Indeed, leaving aside extremely important questions concerning civil liberties that such a law would raise, as a practical matter it would probably be unenforceable. Society has found that morals laws in general may look good on paper, but are honored vastly more in the unpunished breach than in any other way.

nologies that are being integrated into the health care system at an increasingly rapid pace. In some instances these technological innovations have replaced previously existing technologies and have altered the nature of diagnosis and therapy. In other instances they have simply been added to the array of devices within the clinician's arsenal without substantially affecting the nature of medical care. In both cases it is appropriate to ask of any new technology's integration into the health care system whether the quality of care is thereby improved. The only reasonable way to answer this is by attempting to deliberately assess and validate new technologies well before their use becomes widespread. Otherwise adoption of new technologies may fail to benefit patients and may, in some instances, even subject them to harm. The traditional medical ethical obligations of beneficence, the duty to benefit patients, and nonmaleficence, the duty to avoid harming patients, as well as a general utilitarian concern with the general welfare, require that such assessment takes place. Indeed Congress, in recognition of the importance of such assessment, established its own Office of Technology Assessment.

Walsh McDermott, a physician, offered the following remarks on the importance of technology assessment in medicine, even in this era of science-based medicine (McDermott 1977):

At any one time . . . the body of knowledge that forms the practice, especially the therapeutic practice, of medicine is a curious mixture of a highly effective technology interspersed with islands of dogma, empiricism, conventional wisdom, and, at times, superstition. With the exponential growth of "interventions," however, this situation can no longer be tolerated. The persistence of unvalidated technologies leads not only to serious diagnostic error but to waste of skilled services and of money; it also contributes to the increasing load of medically induced, i.e., iatrogenic, disease and, by perpetuating untruths about serious chronic diseases, can give rise to untold human anguish and misery.

Two vivid examples, well worth quoting, of the harm that can be done by unvalidated technology are provided by McDermott (1977):

Chest X-rays were introduced early in this century, became the standard procedure in the twenties, and had come to be considered a most exacting diagnostic technique for tuberculosis by the early 1930s. Solely on the finding of an abnormal density on the X-ray or a change in the appearance of a density in serial films, momentous decisions were made that profoundly altered the lives of individuals. A young wife living in Brooklyn would be made to leave her husband and small children and be hospitalized in the Adirondacks for periods of a year or more; medical students would have to quit; young physicians, to change careers; school teachers, to abandon teaching. Moreover, these things happened frequently because, until the end of World War II, tuberculosis was the greatest cause of death and invalidism in the 15-to-35 age group. After the war, in the 1940s, in one of the first attempts at "validating" a technology of medicine, Yershalmy et al. found that, in making this X-ray interpretation, in one out of three cases the physician would not only disagree with a second or third "reader," but in 20 percent of the cases would not even agree with himself. That is to say, when confronted on two different occasions with the same pair of X-ray films, he would give diametrically opposing answers. Yet it was on this supposedly "decisive" technology that decisions radically affecting the lives of people were made.

During roughly the same time a serologic diagnostic test for syphilis was introduced by Wasserman. The test proved extraordinarily sensitive in that few patients with untreated active disease yielded negative reactions. What was not realized was that the test was overly sensitive. Of all those people yielding positive reactions, only about one-half were actually syphilitic. But this "validation," so to speak, of the test, this characterization of its inadequacies, was performed decades after large-scale public health campaigns (including laws governing premarital examinations) had brought thousands of people under treatment. Quite aside from the mental anguish brought on by a diagnosis of syphilis, the antisyphilitic treatment of the time carried considerable risk for those thousands of people, not the least of them the fact (discovered even later) that the treatment was an important source of hepatitis. These four or five decades, during which thousands of patients who did not have syphilis were subjected to the shame and dangers of antisyphilitic therapy, are not from the medical era of bleedings and leeches, but from the era of modern interventionist technology. It was science-based technology. The physician would choose and carefully administer the science-based technology, an arsphenamine derivative known to have a high degree of effectiveness in definable circumstances, specifically the presence of the microbes of syphilis. But those definable circumstances—the presence of the spirochete—were not actually there, or rather were not always there, or even there with a high degree of probability. Yet the particular bit of unvalidated technology that led to this massive 40-year-long unfortunate mistake represented the practical application of basic principles of the science of immunology.

The examples in this rather long quote illustrate quite well the importance of attempting to validate technologies before their full integration into the delivery of medical care. In the remainder of this chapter we focus on some of the concerns that efforts at medical technology assessment seek to address, as well as some of the problems these efforts raise. In this respect this section is somewhat different from the preceding sections of this chapter. Although those sections focused specifically on medical imaging technologies, the focus of this section is on the more general topic of the assessment of diagnostic technologies, and because the ultimate clinical value of such technologies lies in their effect on therapy, the general issue of the assessment of therapeutic technologies is taken up here as well.

No two concerns are more important to the assessment of medical technology than efficacy and safety (Banta et al. 1981):

Efficacy and safety are the basic starting points in evaluating the overall utility of a medical technology. If a technology is not efficacious, it should not be used, and if its efficacy is unknown, statements about its overall value cannot be made. . . . In addition, efficacy and safety data are needed to evaluate the cost effectiveness of a technology. Neither the need for a technology nor its appropriate use in medical care can be established without reliable and valid information on efficacy and safety. Federal programs to regulate and to provide medical care and medical technology depend on efficacy and safety information to assure wise decisions.

Information obtained from assessments of the efficacy and safety of new and existing medical technologies might help to ensure that new technologies demonstrated to have potential benefits with acceptable risks are made available rapidly, might constrain the diffusion and use of technologies which either lack efficacy or cause excessive harm, and might guide the appropriate use of all technologies.

Accordingly the concept of efficacy is the logical place to begin discussing the concerns addressed by efforts to assess medical technologies. Perhaps the best definition of efficacy is provided in *Toward Rational Technology in Medicine* (Banta et al. 1981):

Efficacy: The probability of benefit to individuals in a defined population from a medical technology applied for a given medical problem under ideal conditions of use.

This definition specifies four factors as especially crucial to the efficacy of a medical technology: "Benefit to be achieved, medical problem giving rise to the use of the technology, population affected, and conditions under which the technology is applied" (Banta et al. 1981). Because benefit to be achieved is the most complex of these factors, it is discussed first.

Benefit to Be Achieved: Diagnostic Technology

What constitutes benefit when the technology at issue is a diagnostic technology? One might think that specifying what counts as benefit for a diagnostic technology is rather straightforward. After all the point of medical diagnosis is to distinguish diseased from undiseased persons and to determine the severity of disease in the former. Accordingly a diagnostic technology is beneficial to the extent that it facilitates the attainment of this end. The medical literature on the assessment of diagnostic technology indicates that the matter of diagnostic benefit is not so simple. Several commentators have set out a hierarchy of several levels as constitutive of diagnostic benefit (Fineberg et al. 1977, Banta 1985, Guyatt et al. 1986, Banta et al. 1981). J. Sanford Schwartz provides the following version of the hierarchy (Assessing Medical Technology 1985):

1. *Technical capacity*   Does the device or procedure perform reliably and deliver accurate information?
2. *Diagnostic accuracy*   Does the test contribute to making an accurate diagnosis?
3. *Diagnostic impact*   Does the test result influence the pattern of subsequent diagnostic testing? Does it replace other diagnostic tests or procedures?
4. *Therapeutic impact*   Does the test result influence the selection and delivery of therapy? Is more appropriate therapy used after application of the diagnostic test than would be used if the test was not available?
5. *Patient outcome*   Does the performance of the test contribute to improved health of the patient?

Clearly, if diagnostic technology fails utterly at any step in this chain, then it cannot be successful at any later stage. If it succeeds at some stage, this implies success in the prior stages (even if they have not been explicitly tested), but does not tell what success may be attached to later stages. Thus an accurate test may or may not lead to a more accurate diagnosis, which in turn may or may not lead to a better therapy, and that in turn may or may not eventuate in better health of the patient.

Each level of this hierarchy of diagnostic benefit is significant and is examined in turn.

1. Technical Capacity
Two factors are crucial to a diagnostic technology's technical capacity: replicability and reliability (Assessing Medical Technology 1985):

Replicability (i.e., precision) reflects the variance in a test result that occurs when the test is repeated on the same specimen. A highly precise test exhibits little variance among repeated meas-

urements, an imprecise test exhibits great variance. The greater
this variation, the less faith one may have in a single test's
results. However, a precise test is not necessarily a good test. A
test may exhibit a high level of replicability yet be in error. A
good test must be reliable (i.e., unbiased); that is, it must exhibit
agreement between the mean test result and the true value of the
biologic variable being measured in the sample being tested.
Evaluations of clinical tests should consider both the replicability and reliability of the technology.

If the results of the use of a diagnostic technology are not replicable
or reliable, the technology will be unable to meet any of the
remaining four criteria of diagnostic benefit listed in our hierarchy.
Accordingly a test that fails at this level fails altogether.

2. Diagnostic Accuracy

The purpose of a diagnostic test or procedure is to discriminate
between patients with a particular disease and those who do not
have the disease. However, most diagnostic tests measure some
disease marker or surrogate (e.g., a metabolic abnormality that is
variably associated with the disease) rather than the presence or
absence of the disease itself. The performance level of a diagnostic test depends on the distribution of the marker being
measured in diseased and nondiseased patients and on the
technical performance characteristics of the test itself (its precision and reliability).
    Each disease marker has a distribution in populations of
diseased and nondiseased patients. Unfortunately these distribu-
tions frequently overlap so that measurement of the marker does
not permit complete separation of the diseased and nondiseased
populations. In these circumstances no matter what cutoff value
k is, it is not possible to ensure that all patients on one side have
the disease and all those on the other are free of the disease. We
are instead left with false positives and some false negatives.
(Assessing Medical Technology 1985)

    Four characteristics are crucial to a diagnostic test's capacity to
accurately discriminate between patients with the medical condition of interest and patients in whom it is absent: sensitivity,

specificity, predictive value positive, and predictive value negative (Assessing Medical Technology 1985):

Sensitivity measures the ability of a test to detect disease when it is present. It measures the proportions of diseased patients with a positive test. This can be expressed by the ratio

$$\frac{\text{true positives}}{\text{true positives} + \text{false negatives}}$$

Specificity measures the ability of a test to correctly exclude disease in nondiseased patients. It measures the proportions of non-diseased patients with a negative test. This can be expressed as

$$\frac{\text{true negatives}}{\text{true negatives} + \text{false positives}}$$

Although knowing the sensitivity and specificity of a diagnostic technology is necessary to make an assessment of its accuracy, such knowledge is not sufficient for arriving at a complete assessment (Assessing Medical Technology 1985):

Test sensitivity and specificity as measures of diagnostic test performance taken alone do not reveal how likely it is that a given patient really has the condition in question if the test is positive, or the probability that a given patient does not have the disease if the test is negative. The fraction of those patients with a positive test result who actually have the disease is called the predictive value positive of a test. It is calculated by the ratio of

$$\frac{\text{true positives}}{\text{true positives} + \text{false positives}}$$

The fraction of patients with a negative test result who are actually free of the disease is called the predictive value negative and is determined by the ratio of

$$\frac{\text{true negatives}}{\text{true negatives} + \text{false negatives}}$$

The predictive value positive and predictive value negative of a diagnostic test measure respectively how likely it is that a positive or negative test result actually represents the presence or absence of disease in a given population of patients with a given prevalence of disease. The positive and negative predictive values of a diagnostic test, however, are not stable characteristics of that test. Rather, they depend strongly on the prevalence of the condition being tested. As the disease prevalence (pretest likelihood of disease) decreases, the proportion of individuals with a positive test result who actually are diseased falls, and the proportion of nondiseased patients falsely identified as being diseased rises. Conversely, as the prevalence of disease increases, the proportion of patients with a positive test result who are in fact diseased increases, while the proportion of patients with a negative test who are not suffering from the disease falls.

These four aspects of diagnostic accuracy are of obvious importance. A test that performs poorly in any one of them may result in harm to patients. Indeed the example of the test for syphilis used in the early decades of this century, described by McDermott (1977) and quoted previously, is a case of harm to thousands of patients caused by a diagnostic technology that was too sensitive. A technology that performs poorly in any of these four aspects of diagnostic accuracy could result in unnecessary exposure of well patients to the risks entailed by therapy or in failure to treat those patients who would be likely to benefit from therapy.

An important difficulty facing efforts to assess diagnostic accuracy should be noted in this context (Assessing Medical Technology 1985):

A major difficulty in determining the performance characteristics of diagnostic tests is the lack of an appropriate reference standard (gold standard) against which to judge the test. The true state of nature generally is not known in clinical medicine. For most diseases even the best diagnostic test has some associated error rate. In practice one is forced to accept the best available, albeit imperfect, diagnostic test as a pseudo-reference

standard. Evaluating a diagnostic test against an imperfect reference standard obviously results in an inaccurate measurement of
test performance.

If this is correct, then because it is not possible to have perfect
knowledge of who has a given disease and who does not, assessment of diagnostic accuracy is always somewhat imperfect. A new
diagnostic technology can only be measured against existing technologies designed to differentiate between those patients who do
have a given disease and those who do not. But because those technologies are always prone to some measure of error, they constitute
a less-than-perfect basis on which to evaluate the new technology.

3. Diagnostic Impact

A new diagnostic technology may affect existing diagnostic procedures in several ways. Often a given diagnostic test is but one
among an array of diagnostic procedures that a clinician uses in
attempting to identify and understand the condition of the patient.
A new technology may alter which kinds of other diagnostic tests
are performed. It may increase or reduce the total number of tests
performed in identifying a given condition. Tests that were formerly regarded as irrelevant to identifying a particular condition
may come to be regarded as crucial because of the new technology.
Test that were formerly regarded as crucial may become irrelevant
because of the new technology. The new technology may even
replace some test or tests altogether. This is especially important
when the diagnostic procedures that are replaced or forgone are far
more invasive than the new technology. Indeed this is one the most
important benefits of the new imaging technologies discussed
previously in this chapter. They allow clinicians to obtain crucial
diagnostic information that formerly could be obtained only through
particularly invasive procedures such as exploratory surgery or
biopsy. Thus in many instances they radically reduce the risk and
discomfort that a patient must face to have his condition adequately
diagnosed.

4. Therapeutic Impact

A test result may have diagnostic impact and still not affect therapy. A health worker may be unaware of the significance of a test result or unfamiliar with the available treatment; the change in probability of disease may be insufficient to alter therapy; the patient may refuse treatment; there may be no therapy available; or the patient may already be receiving the best therapy. To change morbidity or mortality or the quality of life a diagnostic test must provide information that changes therapy. If test results lead to institution of an intervention of known effectiveness, patient benefit follows. If unproven therapy is instituted, change in health status as a result of the diagnostic technology remains a possibility.

If no therapeutic impact is found, it is extremely unlikely that the technology is of benefit. (Guyatt et al. 1986)

A new diagnostic technology may have therapeutic impact in a number of ways: Its use may be less taxing to the patient than older procedures. As a result the patient may not need as much time to recover from the effects of the diagnostic procedure, so therapy can be instituted sooner and perhaps with better results as a consequence. The new technology may alter the clinician's understanding of the nature and severity of the patient's condition and so cause him to make therapeutic choices that he would not have made otherwise. And this too may bode well for the patient. The new technology may reduce the amount of time needed to come to confident conclusions about the patient's condition and the need for therapy. This may allow therapy to be begun earlier in the development of the patient's ailment and thereby may mean that less drastic therapy is needed or that the course of therapy can be shortened.

5. Patient Outcome

Does one really need to go beyond the therapeutic impact, or even diagnostic impact, before concluding that technology warrants dissemination? There have been many instances in which a diagnostic technology provided information, but failed to change clinically relevant outcomes. In one case, application of a diagnostic test (serum cholesterol), when followed by a specific therapy (clofibrate), actually increased mortality! Emergency endoscopy in patients with upper gastrointestinal bleeding provides increased diagnostic information, but without lowering rates of surgery, hospital stay, or mortality. These examples illustrate the wisdom of demonstrating improvement in patient outcome before a diagnostic technology becomes widely disseminated. (Guyatt et al. 1986)

From the point of view of traditional medical ethics, the bottom line in medical care is avoidance of harm to patients and providing patients with benefits. A diagnostic technology helps the clinician in her efforts to discharge these traditional duties only to the extent that it has a positive effect on the patient's outcome. In other words a diagnostic technology must ultimately make a difference to therapeutic benefit, to either the medical or personal benefit that the patient receives from therapy. So all the other aspects of diagnostic benefit are rightly viewed as subordinate to patient outcome, and their value lies primarily, if not wholly, in promoting better health for the patient.

Benefit to Be Achieved: Therapeutic Technology

Now the question of what constitutes the benefit that is sought by the use of a given therapeutic technology can be addressed. First, there are improvements in morbidity, mortality, and life expectancy deriving from the technology's ability to correct physiology and control pathology. In short, therapeutic technologies are developed with the aim of relieving illness, saving lives, and increasing lifespans. These relatively quantifiable aspects of the performance of a medical technology can be referred to as the *medical benefit* of that technology. Although medical benefit is extremely important, it is not all that should be considered when the benefits of a medical technology are at issue. As Laurence Tancredi (1982) has pointed out, the scientific training of physicians often inclines them to emphasize the measurable aspects of disease processes and thus to think of benefit solely in terms of what is relatively amenable to measurement. But to limit assessment to the medical benefit of a technology is to omit consideration of aspects of therapeutic benefit that are especially significant to patients.

This brings us to the second element of the benefit that is sought by use of therapeutic technologies, the effect of the technology on the patient's quality of life, what can be referred to as the *personal benefit* of that technology. Tancredi (1982) has identified five aspects of this element of therapeutic benefit:

The first consideration regarding the quality of life is whether the patient will be able to return to a baseline of functioning, particularly in cases requiring dramatic therapeutic technologies such as renal dialysis or the proposed nuclear-powered heart. The medical treatment may be effective in keeping the patient alive and in correcting the biochemical and physiological abnormalities of disease or defect, but may at the same time create significant difficulties which change the quality of his or her life.

Although a technology may rate highly in terms of medical benefit because of its ability to relieve illness or save lives, it may leave the patient unable to return to work, unable to pursue favorite hobbies, unable to have a normal sexual life, or otherwise unable to engage in activities that are crucial to the quality of his life. That a technology has a high measure of medical benefit does not mean that it has an equally high measure of personal benefit or even that it offers much personal benefit at all.

According to Tancredi (1982), "The second consideration involving the quality of the patient's life would be the degree of relief from expected symptoms. The patient would be most concerned with relative freedom from pain or discomfort, including those that may be induced by the treatment itself." A therapeutic technology may be very effective at preventing a patient's death and at keeping him alive, yet it may fail to relieve him of the pain, discomfort, disorientation, or other unpleasant effects of his ailment. Indeed in some instances the technology itself may be the source of much suffering for the patient. In either situation, because of its effect on the patient's quality of life, the technology may rank very poorly with respect to personal benefit even if it is an excellent source of medical benefit.

"The third consideration, the requirements of the treatment, would also be particularly important to patients suffering from chronic illnesses, such as end-stage renal disease. The hemodialysis patient, for example, may have to spend as much as six hours, three times a week, attached to an artificial kidney machine in an institutional setting" (Tancredi 1982). A therapeutic technology may excel at relieving the symptoms of the patient's illness and at maintaining his life yet place demands on his time, energy, or other personal resources to a degree that severely undermines the quality of his life. This is especially important if the patient's condition is chronic and requires long-term use of the technology. In this sort of case the medical benefit of the technology may be more than overridden by the personal burdens it places on the patient.

"The fourth consideration would be the psychological effects of the technology—the short-term as well as the more permanent long-term changes that may ensue. In the case of patients on hemodialysis, studies have demonstrated a significant increase in the suicide rate, estimated in one study to be 100 times higher than the normal rate, which is about 10 in 100,000 individuals in the United States " (Tancredi 1982). A therapeutic technology can rate very highly with respect to medical benefit yet be a source of intense psychological distress to the patient, distress great enough to make the patient reluctant to use the technology even if it means renewed physical health and even if reliance on the technology will be required for only a relatively short time. In some patients a technology with great medical benefit may cause relatively permanent dysfunctional changes in the psychology of the patient—changes that substantially undermine the quality of his life. Again medical benefit and personal benefit do not always go hand in hand.

"The fifth consideration in understanding quality of life issues is the effect of the new technology on the self-image of the patient. This issue has particular relevance to mutilative surgery such as radical mastectomies, jaw resections, and other extreme surgical procedures for seriously ill cancer patients. Although radical procedures are usually employed only when dictated by severe necessity, and although they entail the potential of curing the disease, they can create immense difficulties for patients by causing distortions in body image and self-evaluation" (Tancredi 1982). Therapeutic technologies that mutilate the patient's body, even when such mutilation saves the patient's life or permanently relieves his illness, may have their medical benefit outweighed by the personal cost they force the patient to endure. A mutilated person may feel repugnant to himself and others. He may feel inferior to normal persons. He may believe that although he is alive, his life is less valuable, less meaningful, than the life of a "whole" person. Here the extremely poor performance of the technology with respect to personal benefit may make its use untenable for some patients, even if it offers them the prospect of avoiding premature death.

So there are two dimensions to assessing the benefit of a given therapeutic technology: medical benefit and personal benefit. The former consists of the technology's ability to relieve illness, avert premature death, and extend life expectancy as a result of its capacity to correct physiology and control pathology. The latter consists of the technology's effect on the patient's quality of life. Because medical benefit is relatively amenable to quantification and so to measurement, it may therefore seem more "real" to scientifically minded medical care practitioners. Yet it is not the only relevant aspect of therapeutic benefit. What the technology does to the patient's quality of life is no less real and no less important to the patient simply because it is not amenable to precise measurement. Indeed it is probably true that for very many patients, the medical benefit of a therapeutic technology is meaningful only insofar as it leads to an acceptable quality of life, an acceptable level of personal benefit. Accordingly assessment of the benefit to be attained by use of a technology must give attention to both these dimensions.

Now the three remaining factors that must be attended to in validating a medical technology can be discussed.

Medical Problem Giving Rise to Use of the Technology
"A technology's efficacy can be evaluated only in relation to the disease or medical conditions for which it is applied" (Banta et al.

1981). That a technology offers benefits for a particular medical condition means nothing concerning its efficacy with respect to other conditions. That must be established by empirical means. It may turn out that a particular technology, although highly efficacious with respect to a particular condition, does not merit support because this condition is exceedingly rare or lacks urgency. In any case knowing what medical condition a technology is suited for is crucial to being able to use it in a maximally beneficial manner and to avoiding causing undue harm by inappropriate use.

Population Affected

The effect of a medical technology varies, depending on the individual treated. Sometimes, however, enough uniformity of effect exists to permit careful generalizations. These generalizations, or extrapolations, apply to the specific population type within which the original observations were made and should be supported by valid and reliable statistical techniques. For example, in the late 1960s the Veterans Administration (VA) conducted a multi-institutional controlled clinical trial of treatment for hypertension using the drug hydrochlorothiazide, reserpine, and hydralazine. The treatment was shown to be efficacious for patients with diastolic blood pressure above 105 mm of mercury. However, all the patients in the trial were males. Thus, the treatment could be considered to be efficacious (based on that trial and other evidence) for the population studied—males—but no automatic assumptions can be made concerning its efficacy for females. (Banta et al. 1981)

Even after a technology has been determined to be efficacious when utilized with respect to a given condition or range of conditions, it is important to determine whether its efficacy is stable for all populations with that condition or within that range. A technology that is highly efficacious for some populations may be less so for others. Just as knowing for which conditions a technology is efficacious is central to maximizing beneficial use and avoiding harmful use, so is knowing how its efficacy differs across different populations of patients.

Conditions under which the Technology Is Applied
The outcome of the application of a medical technology is partially determined by the skills, knowledge, and abilities of physicians, nurses, and other health personnel; by the quality of the drugs, equipment, and institutional settings; and by support systems used by those personnel during the application. Cardiac surgery, for example, may result in better outcome when conducted by skillful, well-trained surgeons who frequently perform such operations than when conducted by surgeons who rarely use that technology. A situation where the physician is skillful and experienced, medication is administered carefully, and the patient receives the best care possible must be described as ideal. By definition, not all physicians are the most skillful, and not all conditions of use are of the highest possible quality. (Banta et al. 1981)

A distinction must be drawn here between a technology's efficacy and its effectiveness. *Efficacy* concerns the benefit likely to be generated by a technology under ideal conditions of use. As Jennett (1986) has noted, "[i]t is natural that early exercises in evaluation [of new medical technologies] are carried out in centres of excellence where the technical performance is unusually high, resulting in maximum benefit and minimal iatrogenic mishaps." Yet although such centers of excellence are the natural sites for the first assessments of a new medical technology, they are not likely to be the sites where most patients will come into contact with the technology if it is actually integrated into the health care system, especially if its dissemination is widespread. The conditions under which the technology will be used in these sites are likely to fall short of the more ideal conditions of Jennett's centers of excellence. Accordingly evaluation must continue beyond these centers of excellence to the less than perfect conditions under which many, if not most, patients receive medical care. A technology's *effectiveness* is its likelihood of benefit under these conditions. Complete assessment of a technology must include evaluation of its effectiveness as well as its efficacy. Obviously a technology does not merit dissemination if it is not efficacious. But if it is efficacious and cannot be made effective, that is, made to offer reasonable benefits under "average" conditions of use, its dissemination should be severely limited, that is, to the more ideal conditions of Jennett's centers of excellence. And it is quite possible that such limitation may make funding the technology an unwise use of medical resources, even though the technology has been shown to be genuinely efficacious.

Safety

"Safety, like efficacy, is a relative concept: no technology is ever completely safe or completely efficacious" (Banta et al. 1981). Every medical technology carries some degree of risk with its use. Therefore to judge a technology safe is not to judge it to be risk-free. Rather it is to conclude that the level of risk it poses is acceptable. The definition of risk provided by Banta and colleagues (1981) is useful here:

Risk: The probability of an adverse or untoward outcome's occurring and the severity of the resultant harm to health of individuals in a defined population, associated with use of a medical technology applied for a given medical problem under specified conditions of use.

This definition shows that assessment of risk must look to the likelihood that a given harm will occur as well as to the severity of that harm. A technology may be deemed safe if the risk it poses is serious but unlikely or if the risk it poses, although likely, is not very serious. Of course the seriousness and likelihood of the risk will be affected by the population of patients on whom the technology is used and the conditions of such use.

This discussion of risk reveals something important about the traditional medical ethical duty of nonmaleficence, the duty to do no harm. If the patient is subjected to diagnosis or therapy at all, some risk of harm is inevitable. So realistically the duty of nonmaleficence should be regarded as a duty to do no undue harm, a duty not to subject the patient to unacceptable risk. Now what constitutes unacceptable risk is a contentious matter and may be judged differently from the different perspectives of the physician, the patient, and society at large. No effort will be made here to specify which perspective, if any, should prevail or to specify what an acceptable risk is. Suffice it to say that acceptable risk is inescapably a matter of judgment.

Methods of Technological Assessment

How are the efficacy and safety of a medical technology to be determined? According to McDermott (1977),

Broadly speaking, there are two ways to validate a technology: one approach provides only a portion of a carefully studied group with the technological intervention, the individuals in the total group having been first selected randomly; the other approach provides the technological intervention to a consecutive series of persons with the disorder and the nature and extent of change in the disorder from its previous characteristic behavior serves as the basis for evaluation.

The first of these two methods is the method of random clinical trials (RCTs), and the second is the method of natural history or before-after case study. Each poses difficulties, and so each is discussed in the following. We consider the method of natural history first because it is generally believed to face serious difficulties that do not plague random clinical trials.

We should note, though, that before large-scale studies of either sort are undertaken, preliminary studies must be undertaken (McDermott 1977)

to delineate the specific conditions, and classes of patients, for which the treatment is likely to provide benefit. Except under extraordinary circumstances (such as striking reduction in the mortality of what was formerly a uniformly fatal disease) these types of studies cannot provide answers about whether a technology is effective, but they can tell us the conditions and the patients in which the treatment is promising enough to make further studies worthwhile. In addition, preliminary studies may delineate potential toxicity which would make further studies pointless or unethical.

The Method of Natural History

By this method the post-intervention and pre-intervention illnesses are compared in a significant number of people all having the same disease. Everyone receives the same treatment; there are no concurrent controls; theoretically each person serves as his own control. Imperfect and relatively primitive forms of this process represent the way virtually all our present interventions have been evaluated. Drug A is administered to patients with a particular disease. . . . If the results are truly without precedent, the case is strong that they represent effectiveness of the intervention, particularly if they occur with some consistency and within a relatively short time after the intervention. (McDermott 1977)

The basic idea here is to apply the technology to a number of individuals with the same illness and to determine whether their condition improves. If they do, then the technology is judged efficacious. But, as Guyatt and colleagues (1986) have noted, there is good reason to be very wary of such a judgment:

Unfortunately improvement in such a study cannot be necessarily attributed to the intervention. Other factors can explain improvement in a before-after study; these include the placebo effect associated with institution of treatment, the attention the patients received, other concomitant changes in therapy, bias in measurement or the natural history of the illness (i.e., they would have improved anyway).

The central deficiency here is the absence of a control group in this type of assessment effort. A control group is a group of patients who are as similar as feasible to the experimental group, the group to whom the new technology is being applied. This group, the control group, is not diagnosed or treated by means of the technology being assessed (Guyatt et al. 1986):

The absence of a control group leads, in general, to overly sanguine assessment of a technology's potential benefit. This is not really very surprising; the biases (by which we mean the factors which lead to findings systematically different from the truth) inherent in uncontrolled studies tend to favor the experimental treatment. Investigators tend to choose subjects to receive an as yet untested treatment at the low ebb of their illness. This means that the patients chosen are likely to get better on their own, even without the new technology. Any placebo effect (the improvement that is almost universally seen after institution of a new treatment, even if it has no biologic effect on the underlying disease) will be attributed to the technology. Both the patients themselves and the investigators would like to see the new technology work, and will therefore tend to err on the side of improvement when they evaluate whether any change with therapy has occurred.

It is precisely because of this tendency of natural history studies to overstate the benefits provided by a new technology that most commentators urge that RCTs be performed on any technology before it becomes widely used. If a technology is integrated into medical care on the basis of an overly optimistic assessment of its efficacy, methods of diagnosis and therapeutic procedures that are genuinely more efficacious than the new technology are likely to be used less frequently or displaced altogether with the result that patients receive inferior care. The health care professional's traditional duty of beneficence as well as society's concern with maximizing the utility to be achieved by use of a technology both require that overly optimistic assessments be avoided as scrupulously as possible.

Randomized Clinical Trials
Two features of RCTs are especially important: the presence of a control group and randomization as the means of choosing which patients are assigned to the control group and which are assigned to the experimental group. The presence of a control group is important because it reduces the likelihood of bias systematically infecting the assessment of the new technology. The control group

provides a set of responses with which the effects of the use of specific technology can be compared. The control group may receive no treatment, a different treatment, the same treatment administered differently, or a placebo. Whatever treatment the control group receives, when the treatment given to the experimental group is judged to be more efficacious by comparison, it is extremely important to know precisely why the treatment is effective. For even if the presence of a control group is sufficient to protect assessment of the technology from being affected by other biases, it may not be sufficient to protect it from being rendered overly optimistic by the placebo effect (Guyatt et al. 1986):

A placebo can be defined as any treatment or aspect of a treatment that does not have a specific action on the patient's symptoms or disease. The placebo effect is involved whenever patients are offered therapy of any sort. Even if the therapy does not have a specific action on the disease, nonspecific factors will influence the patient's response. These include the form of the intervention (whether a pill, surgical procedure, or a stroke rehabilitation program), the interaction with the health care provider, and the therapeutic setting. The placebo effect is extremely powerful; for most interventions one can expect that about one-third of patients will improve on the basis of the placebo alone. When an enthusiastic and trusted physician presents the interventions, the placebo response rate may increase to 70 percent. Perhaps the most effective demonstration of the placebo effect in medical history occurred in two controlled trials of internal mammary artery ligation for angina secondary to coronary artery disease. In these studies, the investigators subjected their placebo group to a sham operation, and found that symptom relief, reduction in nitroglycerine use, and increased activity occurred in over 75 percent of both active treatment and control groups.

The importance of avoiding contamination of the results of efforts at technology assessment by the placebo effect has led to virtually universal acceptance of the need to "administer an identical placebo to the control group" when the technology at issue is a drug (Guyatt et al. 1986). Because both groups will be prone to the placebo effect in the same measure, any improvement in the experimental group that is not also present in the control group can be attributed to the new intervention. Unfortunately for nondrug technologies it is often impossible to find a suitable placebo treatment. Take the case of a technology whose use requires performing an operation on the patient. Although it has been done in the past, contemporary medical professionals and commentators on medical practice would surely judge it immoral to perform sham operations on patients. Such operations would subject the patients to risk of harm with no possibility of benefit that could not have been achieved at lower risk. The objectionable nature of doing this sort of thing is captured in the notion that people should not be treated as guinea pigs. To treat a person as a guinea pig is to treat him as something of less dignity and value than a normal human being. It is to exploit and dehumanize him. Our society, at least in recent years, has manifested a reluctance to countenance such treatment even when it promises great benefits for society as a whole. Accordingly when the control group in an RCT cannot be provided with a suitable placebo, efforts of some sort must be made to determine to what degree the placebo effect is at work and to prevent the ultimate assessment from being contaminated by it.

Using randomization to determine which patients are assigned to which group, control or experimental, is important for two reasons: First, it prevents treatment comparisons from being rendered invalid by biased assignment of patients to one group or the other. Even the most conscientious efforts to make deliberate selections in an unbiased fashion may fail, and this failure may go undetected until the harm has been done. To ensure that this will not happen, assignment of patients to the two groups should be done randomly. Second, randomization tends to ensure that the two groups are indeed identical with respect to all relevant factors except treatment, even when it is not clearly known what some of those factors are. Randomization thus tends to ensure that the experimental group and the control group are genuinely comparable.

The main difficulty facing RCTs is an ethical one (Fried 1974):

The traditional concept of the physician's relation to his patient is one of unqualified fidelity to the patient's health. He may certainly not do anything which would impair the patient's health, and he must do everything in his ability to further it. The conduct of a patient's doctor in an RCT appears to conflict with these traditional norms.

In the RCT a physician (or group of physicians) determines each individual's precise therapy by considering not only the patient's needs, but also by considering the needs of experimental design, that is, the needs of the wider social group that will be benefited by a randomizing scheme. Does this not then clearly pose the dilemma of the physician's duty to the individual and his interest in serving a wider group that would be benefited by the results of the trial?

The whole point of conducting RCTs is to determine the efficacy of a technology and thereby to determine to what degree, if any, it offers benefits to patients. The aim is to serve the collective interests of prospective patients in getting the best medical care possible. Yet the physician is ethically bound to focus exclusively on the actual individuals who are under his care and to pursue for each that course of treatment that his professional expertise leads him to believe offers the greatest benefit. This moral obligation to serve the interests of the individual patient is at odds with determining his treatment, whether he is to be in the experimental group or the control group, by means of randomization. Yet randomization is crucial to the ends served by assessment. This conflict is especially clear when during the course of an RCT, solid evidence is obtained that clearly indicates that one course of treatment is superior to the other. The physician's duty to his patients as individuals requires that those who are not receiving the better treatment be given it as soon as possible. Yet this would surely spoil the RCT and undermine the attempt to arrive at a rigorous assessment of the technology at issue (Fried 1974):

Insofar as the RCT involves deliberately assigning some persons to a treatment which, overall, or in the particular circumstances of these persons—if those circumstances were inquired into—is not the most favorable treatment, to that extent those persons are being sacrificed in the name of greater knowledge which will be used to benefit future sufferers from that disease, presumably in large numbers. Intuitively it seems unfair to impose the burdens of experimentation on some who do not fully share in the benefits; a violation of their right not to be treated as means alone, not to be treated as a resource available to other people.

We are left with a dilemma that cannot be solved in these pages. Medical technology assessment is necessary if patients are to get the best medical care possible or at least if they are not to be unnecessarily subjected to undue harm. But both of the most promising methods of attempting to validate medical technology face serious ethical problems that cannot simply be ignored.

The Societal Effects of Medical Technology

Even if the problems of assessing the efficacy and safety of medical technologies are overcome, the task of assessment should not stop there. So far only the patient and the physician have been the primary focuses of consideration. What about the larger society of which the physician and the patient are members? New medical technologies will have effects on that society, and even technologies that are efficacious and safe may not deserve support in light of those effects.

Allocational Effects

Once a new medical technology has been determined to be safe and efficacious, its effect on macroallocational concerns must be assessed. Every society must make decisions about which social goods it will seek to supply its population from the resources under its collective control. Every society must determine what portion of its resources will constitute a health care budget, what portion will constitute an education budget, what portion will go to national defense, and so on. And it must decide what portion of its health care budget will be devoted to various kinds of care. Integration of a new technology into the health care system may be possible only if either the share of resources given to health care is increased or the share of resources given to the kind of care associated with the technology is increased. If the absolute size of society's total budget is not increased, then the former option would require taking resources from the provision of some other socially desired good and providing them to health care, whereas the latter option would require taking resources from the provision of one sort of health care to support another. Because it might not be feasible for society to increase its overall expenditures, one or the other of these situations may be unavoidable if the new technology is to be supported. But in that case society will have to decide if it wants to and if it is fair to sacrifice the interests of those of its members who prefer the social good or form of medical care that is given less support in order to support the new technology. It may be undesirable or unfair to support a new medical technology at the price of reducing the resources devoted to education or national defense, for instance. It may be undesirable or unfair to support the new medical technology at the price of reducing the resources devoted to, for example, neonatal care, stroke rehabilitation, or cancer prevention.

Supporting the new technology may also raise microallocational issues that undermine the case for integrating it into the health care system. Suppose that it is not feasible to provide a level of support that will allow the new technology to be produced in quantities sufficient to meet the demand for it. Decisions would have to be made about which institutions and which patients are to be provided with the technology. If the new technology is one that promises very important medical or personal benefit, some means of making these sorts of decisions fairly will have to be found. This problem may prove too onerous to make support of the technology reasonable.

Population Effects

Finally, integration of a new, safe, efficacious technology into the delivery of medical care can have (Tancredi 1982)

crucial effects on the size and composition of the population. An increase in heart transplantation for patients suffering from end-stage cardiac disease [for example] could result in increasing the elderly population in this country and thereby creating a deflection of medical resources not only for the cardiac care of these patients but for the numerous chronic and acute conditions that are concomitants of advanced age.

Technology can also affect the composition of the population through the development of [for example] techniques to minimize the childhood fatalities of genetic disease. If children with various genetic diseases survive the adolescent years and reach young adulthood, their procreating inevitably will bring about significant expansions of the gene pools for these conditions and a concomitant increase in the number of patients with genetic disease. Such an increase would add to the health care system's burden of caring for these patients.

These kinds of changes in either the size or the composition of the population may make it unreasonable to support the new technology. Indeed our society is presently concerned about the burden that taking care of its increasingly aging population will place on its younger members. Will these younger people be able or willing to support their older compatriots? Even if they are able to do so, would it be fair to compel them to do so? Even now many people are complaining that resources are being unfairly transferred from the young to the old. In large part this situation of alleged generational inequity is due to the success of medical care at keeping the members of the baby boom generation alive beyond the years of the generation of their parents. In turn this success of medical care is attributable largely to the efficacious and effective new technologies that have been integrated into contemporary medical care.

Three things should be noted by way of summary and conclusion. First, to ensure that patients are benefited, or at least not subjected to undue harm, efforts must be made to rigorously assess the efficacy and safety of medical technologies, therapeutic and diagnostic. Second, the two basic methods of conducting such assessment, the natural history method and RCTs, are both highly problematic. Third, assessment of medical technologies must look beyond efficacy and safety to their important societal effects.

Summary

In this chapter we have described and traced the development of several modern medical imaging technologies, including

• Nuclear medicine—the application of radioactive materials to medical diagnosis and treatment
• Diagnostic ultrasonography—the use of sound, beyond the range of normal hearing, to view internal anatomical features
• Computed tomography—the use of the computer to provide a picture of the distribution of X-ray absorption or transmission in a cross section of the internal anatomy of the patient
• Magnetic resonance imaging—the use of magnetic fields to obtain three-dimensional images of internal organs

Perhaps the most important feature of these technologies is their largely noninvasive nature. Each of these technologies allows clinicians to view and thereby assess internal physiological structures and processes without subjecting patients to the risks of exploratory operations. Although these technologies do invade the patient's body with ultrasound, with radiation, and with magnetic fields, these invasions appear to be far less radical and stressful to the patient than those involved in operations. This ability to obtain clinically useful information concerning the inner workings of the patient's body without inflicting the gross damage intrinsic to operations is a considerable boon to patient well-being. In many instances clinicians can now sort out those patients who need care from those who do not without subjecting any of them to the dangers and pains of an operation. For those patients who are shown not to require further care, the result is that unnecessary operations are avoided. In the case of those patients who need further care, contemporary medical imaging provides more accurate data at a much lower risk to the patient. Moreover the fact that the patient does not have to endure an exploratory operation enables her to cope better physically with subsequent therapy and the stresses it may involve.

We have also examined in this chapter a number of economic issues posed by the fact that many of these imaging modalities are examples of lumpy capital equipment. A lumpy piece of capital equipment is one that comes only in large sizes and in many situations is used at a rate lower than the rate at which unit costs are minimized. The fact that medical imaging equipment is often lumpy can result in low rates of use and high unit costs of operation. Fixed third-party payments for the use of these medical imaging devices (mandated under diagnosis-related group reimbursement schemes) can lead hospitals to experience losses associated with ownership of these devices. High levels of use generate low unit costs and can result in considerable profits. Given that this is the case, hospitals face incentives to find ways of increasing the use of such techniques. Some have been tempted to adopt ethically dubious practices (for example, incorrectly prescribing CT scans and the like for patients who do not need them). Some hospitals have entered into legally and ethically acceptable arrangements under which they share access to these technologies with other health care institutions. Such sharing arrangements sometimes cause considerable difficulties for patients who find themselves faced with long journeys for medical tests. From a societal perspective it has been shown that in fact when patient transportation costs are taken into account, the optimal distribution of high-cost lumpy medical imaging technologies is more extensive than might otherwise be recognized.

This chapter has further shown that there is an economic explanation for the paradox that new technologically superior medical imaging procedures have not completely displaced older and less precise ones. In many cases the older techniques (for example, X-ray scanning) are less costly and can be used as screening procedures to determine which problems require more sophisticated but more expensive diagnostic techniques. Using older technologies as screening procedures often, although not always, reduces the average cost per patient of diagnosis. Thus hospitals are often acting rationally when they use multiple diagnostic techniques.

Medical imaging techniques are important tools of preventive medicine. The economic issue of whether preventive therapies are always to be preferred to curative therapies from a benefit-cost perspective has also been examined in this chapter. It has been shown that prevention is not always better than cure from a benefit-cost perspective, although frequently it is the preferred approach from both personal and societal perspectives.

The final sections of this chapter turned from examination of particular modern medical imaging technologies to the more general topic of technology assessment, including both diagnostic and therapeutic technologies. The two most important issues for the assessment of medical technology are efficacy and safety. With respect to the efficacy of a medical technology, four factors are especially significant: the benefit to be achieved by use of the technology, the medical problem to which it is applied, the patient population for which it is used, and the conditions under which its use occurs. Although the issue of the benefit to be achieved by a technology might appear to be quite straightforward, the discussion in this chapter has shown that diagnostic benefit consists of at least seven dimensions. These include the technical capacity of the technology, its diagnostic accuracy, its effect on subsequent diagnostic efforts, its effect on selection and delivery of therapy, and, finally, its contribution to improvement in the patient's health. Similarly it was shown that a technology's therapeutic benefit has at least two dimensions: the technology's ability to effect improvements in morbidity, mortality, and life expectancy—that is, its medical benefit—and its ability to enhance the patient's quality of life—that is, its personal benefit. Although a technology's medical benefit is often obvious to scientifically minded medical practitioners because it is relatively quantifiable, from a patient's point of view the value of that medical benefit is crucially tied to its personal benefit, its effect on his quality of life.

This portion of the chapter also dealt with the two main methods of technology assessment—natural history and randomized clinical trials. We have pointed out that although randomized clinical trials avoid the problem of overstating the benefits of a technology being assessed (the central problem posed by the method of natural history), they pose an important ethical conflict between what is traditionally regarded as the physician's central obligation (the well-being of his individual patients) and the very feature that makes them a reliable means of assessing technologies. Doing what is best for the individual patient in a randomized clinical trial may well compromise the trial and thus ruin its value as a means of assessing the technology at issue.

Finally, this chapter has presented arguments that, whatever method of asessment is regarded as least problematic, technology assessment must look beyond the individual patient to encompass a technology's likely impact on society at large. At the very least this entails assessing its effects on the size of society's collective health care budget and the allocations made within that budget as well as its effects on the size and composition of the population. Even when a technology is determined to be safe and efficacious for individual patients, its overall societal effects may nonetheless militate against its widespread dissemination and use.

References

Andrew, E. R. 1979. Nuclear magnetic resonance imaging: Zeugmatography. In Kreel, L., and Steiner, R. E. (eds.): *Medical Imaging*. HMM Publ. Co., pp. 38-43.

Assessing Medical Technology. 1985. National Academy Press, Washington, D.C.

Baker, D. 1974. Physical and technical principles. In *Diagnostic Ultrasound*. C. V. Mosby.

Banta, H. D., Clyde, J. B., and Willems, J. S. 1981. *Toward Rational Technology in Medicine: Considerations for Health Policy*. Springer.

Barrett, A. S. 1978 (May). Image restoration. *Industr. Res. Devel.*, pp. 111-114.

Beck, R. N. 1975. Instruments: basic principles. In *Nuclear Medicine*. HP Publishing.

Bronzino, J. D. 1976. *Technology for Patient Care*. C. V. Mosby.

Bronzino, J. D. 1982. *Computer Applications for Patient Care*. Addison Wesley.

Brown, B., and Gordon, D. (eds.) 1967. *Ultrasonic Techniques in Biology and Medicine.* Iliffe & Sons.

Crooks, L. E., Kaufman, L., and Hoenninger. 1984. Spatial resolution in NMR imaging. *IEEE Trans. Medical Imaging* 3:51-53.

Damadian, R. 1977. Fonar image of the live human body. *Physiol. Chem. Phys.* 9:97.

Early, P. J., Razzak, M. A., and Sodee, D. B. 1975. *Textbook of Nuclear Medicine Technology.* C. V. Mosby.

Fineberg, H. V., Bauman, R., and Sosman, M. 1977 (18 July). Computerized cranial tomography: Effect on diagnostic and therapeutic plans. *JAMA* 238(3): 224-227.

Fleischer, A. C., and James, A. E. 1984. *Real Time Sonography Application.* Appleton-Century-Crofts.

French, L. A., Wild, J. J., and Heal, D. 1950. Detection of cerebral tumors by ultrasonic pulses: Pilot studies on post-mortem material. *Cancer* 3:705-708.

Fried, C. 1974. *Medical Experimentation: Personal Integrity and Social Policy.* North-Holland Publishing.

Gilbert, B. K., Storma, M. T., and Ballard, K. C. 1979. A programmable dynamic memory allocation system for input/output of digital data into standard computer memories at 40 megasamples/sec. *IEEE Trans. Comput.* 25:1101-1109.

Goldberg, B. B. 1973. *Diagnostic Ultrasound in Clinical Medicine.* Medcom Press.

Guyatt, G., et al. 1986. Guidelines for the clinical and economic evaluation of health care technologies. *Social Science and Medicine* 22(4):393-408.

Gordon, R., Herman, G. T., and Johnson, S. A. 1975 (Oct). Image reconstruction from projections. *Scientific American* 233:56-68.

Hamilton, B. (ed.). 1982. *Medical Diagnostic Imaging Systems.* F and S Press.

Hodges, P.C., and Moseley, R. D. 1974. Radiology. In Ray, C. D. (ed.): *Medical Engineering.* Year Book, pp. 702-723.

Holmes, J. H. 1974. Historical perspective. In: *Diagnostic Ultrasound.* C. V. Mosby.

Hounsfield, G. N. 1973. Computerized transverse axial scanning. Part I. *British Journal of Radiology* 46:1016-1022.

Hounsfield, G. N. 1980. Computerized medical imaging. *Science* 210:22-28.

Howry, D. H. 1957. Techniques used in ultrasonic visualization of soft tissues. In Kelly, E. (ed.): *Ultrasound in Biology and Medicine.* Waverly Press.

Howry, D. H., Scott, D. A., Bliss, W. R. 1954. The ultrasonic visualization of carcinoma of the breast and other soft tissues. *Cancer* 7:354-358.

Jennett, B. 1986. *High Technology Medicine: Benefits and Burdens.* Oxford University Press.

Lauterbur, P. C. 1973. Image formation by induced local interactions: Examples using nuclear magnetic resonance. *Nature* (London) 242:190.

Ledley, R. S. 1976. Introduction to computerized tomography. *Comput. Biol. Med.* 6:239-246.

Lerch, I. A. 1980. Beyond the CT scanner: In search for a point of light. *The Sciences* 20:6-9.

Macovski, A. 1983. *Medical Imaging Systems.* Prentice Hall.

Marvilla, K., and Pastel, M. 1978. Technical aspects of CT scanning. *Comput. Tomog.* 2:137-144.

Marx, J. L. 1980. NMR opens a new window into the body. *Science* 210:302-305.

McDermott, W. 1977. Evaluating the physician and his technology. *Daedelus* 10:135-157.

Meyer, D., and Gitter, K. A. 1967. History of ultrasound. In Goldberg, R. E., and Sarin, L. K., (eds.): *Ultrasonics in Ophthalmology.* W. B. Saunders.

Nagel, J. 1988. Beam/ray imaging. In Kline, J. (ed.): *Handbook of Biomedical Engineering.* Academic Press, pp. 243-314

Payne, J. T., and McCullough, E. C. 1976 (Mar–Apr). Basic principles of computer-assisted tomography. *Applied Radiology* 5:53-60, 103.

Portugal, F. R. 1984. NMR promises to keep. *High Technology* 4(8):66-78.

Puera, J. T. 1986. Principles of diagnostic ultrasound. In Bronzino, J. D. (ed.): *Biomedical Engineering and Introduction : Basic Concepts and Application.* PWS, pp. 347-386.

Pullen, B. R. 1979. Computed tomography limits and resolution. In Kreel, K., and Steiner, R. E. (eds.): *Medical Imaging.* HMM Publ. Co., pp. 10-15.

Reid, J. M. 1972. Ultrasonic echo ranging. In Rushmer, R. (ed.): *Medical Engineering.* Academic Press.

Russell, L. B. 1986. *Is Prevention Better Than Cure?* Brookings Institute.

Shaw, D. 1988. Magnetic resonance imaging. In Kline, J. (ed.): *Handbook of Biomedical Engineering*. Academic Press, pp. 315-354.

Spencer and Hosain. 1986. Basic science and practice of nuclear medicine. In Bronzino, J. D. (ed.): *Biomedical Engineering and Instrumentation*. PWS, pp. 279-346.

Steinberg, E.P. 1986. The status of MRI in 1986: Rates of adoption in the United States and worldwide. *AJR* 147:453-455.

Stone, S., and Taylor, P. L. 1988a. MRI: What are its applications? What will it replace? *Health Technology* 2(1):3-11.

Stone, S., and Taylor, P. L. 1988b. Future improvements in magnetic resonance imaging. *Health Technology* 2(1):12-19.

Tancredi, L. R. 1982. Social and ethical implications in technology assessment. In McNeil, B. J., and Cravalho, E. G. (eds.): *Critical Issues in Medical Technology*. Auburn House.

Taylor, K. D. 1986. Radiographic and nuclear magnetic resonance imaging. In Bronzino, J. D. (ed.): *Biomedical Engineering and Instrumentation*. PWS, pp. 387-432.

Taylor, K. J. W., Carpenter, D. A., and McCready, V. R. 1973 (Dec). Grey scale echography in the diagnosis of intrahepatic disease. *J ournal of Clinical Ultrasound* 1:284-287.

Wagner, H. N. 1968. *Principles of Nuclear Medicine*. W. B. Saunders.

Wagner, H. N., Jr. (ed.) 1975. *Nuclear Medicine*. HP Publishing.

Wagner, H. N., Jr., and Buddemeyer, E. U. 1974. Nuclear medicine. In Ray, C. D. (ed.): *Medical Engineering* . Year Book.

Walmsley, B. 1979. Computed tomography—Equipment. In Kreel, L., and Steiner, R. E. (eds.): *Medical Imaging*. HMM Publ. Co., pp. 15-22.

Wells, P. N. T. 1975. Ultrasonic diagnostics: A look into the future. *Biomedical Engineering* (Lond) 10:247-251.

Wells, P. N. T. 1977. *Biomedical Ultrasonics*. Academic Press.

Wild, J. J., and Reid, J. M. 1956. Diagnostic use of ultrasound. *British Journal of Physical Medicine* 19:248-257.

Winsberg, F. 1973. *Diagnostic Ultrasound: Echocardiography Manual*. Picker Corporation Publication.

Contemporary Medical Technology: Major Social and Ethical Issues

In the preceding four chapters the focus has been on specific medical technologies and with the economic and ethical issues they raise. This final chapter looks instead at some ethical and social issues posed by the highly technological nature of modern health care in general, issues that cut across all the specific technologies discussed in previous chapters. The first section of this chapter addresses the macroallocational question of whether acute care rescue medicine should receive substantially higher priority in society's health care budget than preventive medicine. The second section also considers a macroallocational question: Is basic medical research, a commodity whose benefits lie primarily in the future, being given an unduly low level of resources in favor of the development of so-called halfway medical technology, a commodity whose benefits lie much closer to the present? The third section turns to the complaint that the highly technological nature of modern medical care dehumanizes patients, and the fourth section deals with the issue of equitable access to medical technologies.

Acute Care Rescue Medicine versus Preventive Medicine

The technologies of acute care medicine (also known as crisis medicine) are the forms of medical technology that receive most discussion and generate most controversy in the popular news media. These big-ticket items include the supportive and resuscitative devices that are commonly used in modern intensive care units, medical imaging devices, and the artificial heart, to name only a few, all of which have been examined in preceding chapters.

The overriding goal that society aims to attain by devoting some of the resources under its collective control to health care is minimization (or at least significant reduction) of levels of morbidity and mortality in its population. Rescue medicine and preventive medicine are distinguished as different ways of attaining this goal. Rescue medicine includes all the types of medical care that attempt to restore health and normal functioning to individuals *after* illness or disability has already occurred. Preventive medicine is based on a completely different premise: It includes all the types of medical care that attempt to prevent or significantly delay the occurrence of illness and disability in the first place.

The ideal at which preventive medicine aims, and which it surely will never attain entirely, is to make rescue medicine unnecessary altogether. The extent to which this ideal is approximated, however, depends on the extent to which resources are devoted to development and use of preventive technologies rather than the technology of rescue. Because avoiding illness and disability in the first place is generally preferable to being restored to health and normal functioning, it is reasonable to assume that the medical technologies of greatest priority ought to be those with the greatest promise for medical prevention.* Yet our allocation of medical resources now substantially favors, and has favored for some time, rescue medical care. According to J. Michael McGinnis, Deputy Assistant Secretary for Health and Human Services, "more than 95 percent of the half-trillion dollars spent for medical care in the United States goes to treat rather than prevent disease" (McGinnis 1988-89). To many this implies that present medical priorities are badly misplaced!

Is this judgment correct? Are our medical priorities badly askew? Do we, as a society, devote far too much of our collective medical care budget to rescue medicine and its associated technologies? According to what may be the majority point of view on these matters, allocations within our collective health care budget should be based solely on the essentially utilitarian aim of getting the biggest "medical bang for our medical buck."† That is, in determining what portion of our total medical care budget should go to

*The popular image of preventive medicine falsely depicts it as a largely non-technological form of medical care. Although the technologies of effective preventive medicine are likely to be substantially different from those of acute care rescue medicine, technology will be crucial nonetheless.

†Freedman (1977) and Goodman and Goodman (1986) have surveyed the views of those who hold this opinion.

the development, production, or purchase of rescue medical care and associated technologies, the overriding and perhaps sole concern should be cost effectiveness.* Given that a certain portion of society's total resources are allocated to health care, our aim should be to use those resources to generate the mixture of rescue and preventive medical care that will most reduce levels of morbidity and mortality in our population. On this view to do less would be to waste health care resources and thereby to do less to maintain and restore health than can be done within the limits of our collective health budget. This in turn means that some persons would obtain less health care or lower-quality health care than otherwise could be provided within the constraints of this budget.

Proponents of this viewpoint claim that because acute care rescue medicine is very costly, its effect on morbidity and mortality statistics is marginal at best. In short, acute care rescue medicine is judged to have failed the test of cost effectiveness. (And the blame is frequently placed squarely on its technologies of acute care rescue medicine (Friedrich 1984). These technologies are alleged to be unable to address the underlying causes of illness and disability (Thomas 1971, 1977, 1983, 1986, 1987). Instead they only palliate, moderate, and compensate for some of their more important symptoms. Although this is certainly no trivial achievement, because it means the difference between premature death and extended life for many patients, critics adamantly insist that such technologies rarely extend life for long periods, are enormously costly, and in many cases may even significantly lower the patient's quality of life. The claim is made that when the disease and disabilities that are now treated by means of such halfway technologies are adequately understood, definitive technologies—technologies that are genuinely curative or, even more likely, genuinely preventive—will be forthcoming. On this view, until such understanding is attained, though, the lion's share of our collective health care budget should be devoted to basic medical research.)†

*See Caplan 1983 for an illuminating discussion of some of the ethical and conceptual difficulties involved in judgments of cost effectiveness.

†The debate over halfway technology is more fully examined later in this chapter (see pp. 519–534).

Suppose, for the sake of argument, that the biggest medical bang for society's medical buck is attainable only if substantial amounts of resources are shifted away from acute care rescue medicine to preventive medicine and consequently that efforts in the development of medical technologies would be best concentrated where they enhance the efficacy of prevention.* Would increased cost effectiveness alone be reason enough to substantially reduce our present efforts at rescue?†

Some argue that it would not. Proponents of this position claim that considerations other than cost effectiveness show that resources should not be shifted away from rescue to prevention, even if doing so would certainly purchase significant decreases in morbidity and mortality levels in our population. As one commentator wrote, "An ounce of prevention is not necessarily worth more than any amount of cure simply because it produces better mortality and morbidity statistics" (Freedman 1977). Arguments on behalf of this position often appeal to a distinction that concerns the respective beneficiaries of rescue and prevention. Although the former are identifiable persons, the latter are mere statistical persons. In other words rescue efforts, medical or otherwise, are undertaken on behalf of known persons who are presently in peril, identifiable persons whose health, normal functioning, or very lives are under seige here and now. Indeed how could it be other-

*According to McGinnis, ". . . for each of the leading causes of death and disability in this country, epidemiological and biomedical research have slow actions that can be taken to reduce risk. . . . such measures as early detection and intervention, immunization, and motivating changes in individual behavior could eliminate an estimated 45 percent of cardiovascular disease deaths, 23 percent of cancer deaths, and more than 50 percent of the disabling complications of diabetes. Bettter control of fewer than 10 risk factors—for example, poor diet, infrequent exercise, the use of tobacco and drugs, and the abuse of alcohol— could prevent between 40 percent and 70 percent of all premature deaths, a third of all cases of acute disability, and two-thirds of all cases of chronic disability. In contrast, technologically oriented medical treatment correctly promises to reduce premature morbidity and mortality by no more than perhaps 10 to 15 percent" (McGinnis 1988-89). For skepticism about claims on behalf of the efficacy of prevention, see Goodman and Goodman 1986.

†Presumably some level of expenditure on acute care rescue medicine would be included in any reasonable health care budget. Accordingly the issue is not whether this form of medical care should be eliminated but whether it should be deemphasized in favor of greater expenditure on preventive medicine.

wise? Rescue efforts can be made only when someone in need of rescue has been identified. When preventive efforts, medical or otherwise, are undertaken, although identification of the persons in a population who are at risk may be possible, in most instances one cannot determine which individuals are actual beneficiaries of those efforts. Furthermore individuals, even those at risk, may not fall victim to a particular ailment, even if no preventive measures are taken. Even if persons do remain healthy when preventive steps are actually undertaken, their good fortune cannot be proved to be due to those efforts. When rescue efforts succeed, actual individuals who benefit from the treatment can be identified. Yet when preventive efforts succeed, although statistical data may show that morbidity and mortality rates decreased in a given population, we usually cannot know which individuals in that population actually were prevented from becoming ill or disabled. Indeed some of those whose health or lives were thereby maintained might not have yet been born at the time the preventive measures were actually deployed.* Consequently advocates of preventive measures do not have the advantage possessed by advocates of acute care rescue medicine of being able to point to actual individuals who are better off because certain measures were taken.

This difference between identifiable and statistical lives gives acute care rescue a psychological/emotional impact that preventive medicine lacks. Refraining from attempting to help persons whose peril is immediate is extremely difficult. We actively empathize with their plight. Persons who have some chance of being in peril, but who are unknown to us, are likely to strike us as faceless abstractions. They are mere statistics. Their peril is not real to us in the immediate and visceral manner of persons who are sick or disabled here and now. The plight of the first recipients of the Jarvik 7 artificial heart, Barney Clarke and William Schroeder, for example, was more immediately meaningful to us than that of the many individuals who may contract severe heart disease if appro-

*Menzel (1983) marshals a number of arguments, suggesting that the alleged correlation between rescue medicine and identifiable persons, on the one hand, and preventive medicine and statistical lives, on the other, is not as strict as it might seem and even that the very distinction between identifiable and statistical persons may be untenable.

priate measures of prevention are not taken. Through the media millions of people came to know Clarke and Schroeder as fathers, husbands, brothers, neighbors—as people who mattered to other identifiable people. We came to know what their hopes and fears were, as well as what they and their families suffered because of their ailments. Consequently most of us were moved when William DeVries defended his implantation of a Jarvik 7 artificial heart in Murray Hayden with the following description of Hayden's condition before the operation: "Murray Hayden...had to wake up every morning with his head between his legs so he could breath. Every time he laid out flat in his bed he smothered and started coughing. I can't think of anything worse than not being able to catch your breath, not being able to breathe, and believing that every breath you take is your last" (DeVries 1986). This psychological fact— that we are far more easily and deeply moved by the plight of identifiable persons in known peril than by mere statistical persons facing the possibility of the same peril—may explain why some regard rescue medicine as meriting priority over prevention, regardless of cost effectiveness.

Yet from the fact that we are better able to empathize with the beneficiaries of rescue medicine than with those of preventive medicine because we know who they are, nothing follows concerning the allocative decisions that society ought to make. The persons whose lives and health are preserved by successful efforts at prevention are as real as those who benefit from successful rescue efforts. That we do not know who they are does not alter this fact. That we know their lives and health have been preserved, whoever they are, is what is important. We may not be as easily and deeply moved by their circumstances, but this is not a sound basis on which to make decisions within a society's health care budget. How we feel on this matter may not be how we ought to feel.* The fact that

*Focusing on the artificial heart, Albert Jonsen (1986) has argued that these kinds of feelings might lead to undesirable outcomes. He argued that the artificial heart, if successfully developed, could be a dangerous technology precisely because it would benefit identifiable persons. Specifically, because of our empathy for its beneficiaries, resources could be allocated to it that, if devoted instead to the prevention of heart disease, would benefit the lives and health of many more people (including some of society's most needy members).

we are more easily and deeply moved by the beneficiaries of rescue does not entail that our feelings are justified.

A somewhat more persuasive attempt to show that acute care medicine is special in some morally significant respect focuses on the alleged symbolic value of rescue efforts (whether medical or not). The symbolic value argument "suggests that rescue attempts show that individuals are 'priceless,'" that they are too important to be allowed to deteriorate or die prematurely without a fierce struggle on their behalf and that "society is 'too good' to let" its members "die without efforts to save them," that it is the kind of society that cannot "stand pat and let present victims die for the sake of future possibilities"(Childress 1981).

This argument is compelling only if acute care rescue medicine, in fact, is more effective and efficient at reducing morbidity and morality rates than is preventive medicine. If it is not, then giving priority to acute care rescue medicine in the allocations made within society's health care budget is to fail to do as much on behalf of human life and health as can be done within that budget's limits. It is to fail to value human life and well-being as much as the constraints of our health care budget will allow.* In this case rescue efforts can be regarded as having the symbolic value that is attributed to them only insofar as the members of society are unaware that such efforts in fact are not the most efficient means available of preserving health and lives. To believe that society better symbolizes its high regard for the well-being of its members by giving allocational priority to efforts at maintaining life and health that are not the most effective and efficient available within our budgetary limits is to be deluded. No society should base its health care priorities and decisions about what sorts of health care technology to develop and produce on a deluded symbolism. Charles Fried (1970) was quite correct when, in an ironic bit of understatement, he said, "... surely it is odd to symbolize our con-

---

*It is worth noting that what is symbolized concerning the value of human life and well-being by how a society utilizes the resources within its health care budget is tied to the size of that budget relative to expenditures on other socially desired goods. Efficient use of health care resources says little on behalf of the value of human life and well-being if a society chooses to devote an unduly small share of its total resources to health care.

cern for human life by actually doing less than we might to save life." If a society wishes to show that it values the lives and well-being of its members as highly as its collective health care budget will allow and wishes to engender in the members a sense that they are indeed valued, it would do well to place its resources where they are most effective and efficient in reducing morbidity and mortality and to make its population well aware of this.

Perhaps the most compelling consideration in favor of giving priority to rescue medicine is based on respect for personal liberty. According to many commentators, the vast majority of us are born healthy and made ill, disabled, or victims of premature death by our own voluntary behaviors. Each of the five leading causes of death in the United States includes risk factors posed by voluntary behavior (McGinnis 1988-89) (table 8.1). As John Knowles (1977) has pointed out, "Prevention of disease means forsaking the bad habits which many people enjoy—overeating, too much drinking, taking pills, staying up at night, engaging in promiscuous sex, driving too fast, and smoking cigarettes—or, put another way, it means doing things which require special effort—exercising regularly, going to the dentist, practicing contraception, ensuring harmonious family life, submitting to screening examinations." And the crucial fact is simply that many, perhaps most, individuals do not choose to give up their bad habits and make the special efforts that would significantly reduce their chances of needing medical rescue at some point.

For example, it is common knowledge that cigarette smoking plays a major causal role in heart disease, lung disease, and cancer. Cigarette smoking is still the single most destructive preventable health risk. It is estimated to account for nearly a third of all deaths that occur as a result of cardiovascular diseases, cancer, and respiratory diseases (McGinnis 1988-89). Far fewer people would need to be rescued from these ailments if they simply refrained from smoking. Given that many people continue to smoke and that many others choose to begin this habit, why doesn't society simply make the option to smoke unavailable by banning the production, sale, and use of cigarettes and related products? Likewise the

Table 8.1
The Five Leading Causes of Death in the United States and Associated Risk
Factors

Cause of Death	Risk Factors
Cardiovascular disease	Tobacco use
	Elevated serum cholesterol
	High blood pressure
	Obesity
	Diabetes
	Sedentary life-style
Cancer	Tobacco use
	Improper diet
	Alcohol use
	Occupational/environmental exposures
Accidental injuries	Safety belt noncompliance
	Alcohol/substance abuse
	Reckless driving
	Occupational hazards
	Stress/fatigue
Chronic lung diesease	Tobacco use
	Occupational/environmental exposures

From National Center for Health Statistics/U.S. Department of Health and
Human Services. Health United States: 1987. DHHS Pub. No. (PHS)
88-1232.

importance of regular exercise to maintenance of good health is
widely known. Yet many individuals choose not to make the extra
effort required to include regular exercise in their lives. Why
doesn't society simply force its members to exercise regularly?

Indeed it may be true in general that for preventive health care
and associated technologies to be effective enough to merit much
investment of society's health care resources, these sorts of coercive limits on individual choice would have to be enacted. But our
society has a longstanding and deeply held commitment to personal
autonomy based on the belief that a life worth living is a life based
on one's own freely made choices. Western ethicists, arguing from
widely disparate premises, have tended to agree with the Kantian

view that to be denied personal autonomy is to be degraded and reduced to the status of a mere thing. This commitment has led us to accord individuals a nearly absolute right to make their own choices as long as they do not impose harm on others, and this makes us reluctant to go very far in using coercive measures to alter individual behavior. This reluctance, coupled with the unhealthy choices that individuals make, may mean that acute care rescue medicine is our best bet, that big-ticket medical technologies like the artificial heart are precisely the sorts of technologies that our resources should be used to develop and deploy.* This is not because they are more cost effective—they may not be—nor because they are especially apt means by which society can express its regard for the well-being of its members, but because we greatly value our freedom to behave in ways that make it probable that we will eventually need rescue from life-threatening ailments. Unfortunately we are not at all reluctant to exercise that freedom. Society's respect for personal liberty means that we allow individuals to produce, advertise, and consume substances that harm them, and that we allow them to lead their lives unhealthily and to make their livings persuading others to engage in unhealthy behaviors. Even if it is not cost effective and hence a waste of health care resources, acute care rescue technology may merit priority in our collective health care allocations simply because it is compatible with our respect for personal liberty and the consequences of how we choose to exercise that liberty.

An irony worth noting is that the willingness of individuals to voluntarily jeopardize their health may itself be due, at least in part, to their faith in the ability of technology to rescue them if and when the harmful consequences of their choices eventuate. We may be

---

*Another argument against coercive attempts to regulate behaviors affecting health points to the high cost that such efforts often entail. One need only think of the example of Prohibition and the more recent experiences of this society in its attempts to stop the use of illegal recreational drugs. These examples show that enormous quantities of resources can be committed to coercive efforts to regulate behaviors that affect health without much success.

The argument based on respect for personal liberty cuts deeper though. It would hold true even if coercive regulation of behavior could be successfully achieved at very low cost.

victims of what ethicist Lance Stell (1985) has referred to as medical utopianism—"the view that medical technology will make possible ever-earlier diagnoses of killer diseases and provide sufficient back-up spare parts to extend progressively and perhaps indefinitely the population's life expectancy." As long as this faith induces individuals to exercise their rightly cherished freedom in unhealthy ways, any gains in cost effectiveness that might be gotten by devoting more of our collective health care budget to prevention may be inaccessible.

Halfway Technology versus Basic Research

Many of the technologies discussed in previous chapters, indeed the technologies of rescue medicine in general, are what Lewis Thomas (1971), former head of the Memorial Sloan-Kettering Cancer Center, has referred to as halfway technologies. And, according to Thomas and others, these technologies are being purchased at the cost of inadequate support of basic biomedical research, the only kind of research that can ultimately reveal the underlying physiological mechanisms involved in today's leading sources of disability, acute illness, and premature death.

According to Thomas, a halfway medical technology is one that can neither prevent nor cure disease. It can only compensate to some degree for a disease's effects on its victims. As James Maxwell (1986) put it, "Halfway by its very definition implies a technology that is . . . designed to make up for disease or postpone death." Included in this category are devices such as the mechanical ventilator, the artificial kidney, and the artificial heart. Halfway medical technologies are contrasted with what Thomas refers to as definitive medical technologies, technologies which successfully prevent disease or relieve the effects of disease by effecting genuine cure or prevention. Although, in the absence of an understanding of a disease's underlying mechanisms, a halfway technology may be the best that medical science can offer, proponents of Thomas's viewpoint argue that (Bennett 1977)

Much of the complexity and burgeoning cost of present medical care results from the use of halfway technologies—measures which merely palliate the manifestations of major diseases whose underlying mechanisms are not yet understood and for which no definitive prevention, control, or cure has yet been devised. The recent history of medicine is replete with evidence that each time a major disease has been controlled, the definitive technology has been cheaper and simpler than the technologies devised before the disease was understood.

If Thomas and his colleagues are correct, medical history indicates that halfway medical technologies are usually more expensive and complicated than definitive technologies and, for all their expense and complexity, do not provide the capacity to control disease provided by definitive technologies. Given that definitive technologies are superior in these respects to halfway technologies, why do we not forgo the former in favor of the latter? Quite simply because although a great deal of pressure is placed on medical professionals, both by the general public and by the medical profession itself, to do something that will maintain the lives and relieve the sufferings of patients, what actually can be done is limited by the state of medical knowledge at any given time. Development of a definitive technology for a given disease, according to Thomas, requires adequate knowledge of its underlying mechanisms. In the absence of such understanding, the only technologies that can be developed for therapeutic purposes are halfway technologies (Thomas 1971, 1986).

The pressure on medicine to do something for patients, something other than to soothe them as they suffer or die, makes it difficult for medical scientists and engineers to wait for the knowledge that will permit the development of definitive technologies to be obtained. Such knowledge can be obtained only by means of basic research. The process of scientific investigation, however, is often long and arduous and may take many years to yield results. During this search period patients continue to suffer and die prematurely if no intervention efforts are undertaken.

Consider what Thomas (1987) takes to be the last era of major progress in genuine prevention, control, and cure of disease—an era that began in the 1930s when "sulfanilamide turned up, penicillin, and then streptomycin and the other antibiotics." These drugs gave medicine power to control and cure infectious disease unlike anything known previously. This is not to say that medical practitioners failed to vigorously attempt to intervene in the course of disease before the development of effective anti-infectives. Indeed they did and in the process subjected their patients to a variety of "heroic" and "strenuous measures" (Thomas 1987):

By the end of the eighteenth century, getting ill and coming under the care of an energetic physician had become an athletic challenge. The dominant idea then was . . . that disease in general was caused by an imbalance in the distribution of fluids within the body. Bleeding, cupping, the application of blistering ointments to the skin over the affected organs, accompanied by violent purging with mercurical or plant cathartics, were the standard features of therapy for any serious illness. The bleeding often involved the removal of a quart or more of blood at one sitting, and the appearance of bluish pallor, a feeble pulse, and profound weakness were taken as signs that the treatment might be having the desired effect.

These vigorous efforts at medical therapy had little positive effect on the sufferers of infectious disease, or any other diseases for that matter, and often did grave harm to the patients unfortunate enough to be delivered into the hands of physicians. Consider the case of one their most famous (Thomas 1987):

. . . George Washington, by all accounts a hale and hearty man in his mid-sixties, developed a sore throat—probably a streptococcal infection—after a tiring horseback ride in the snow, on December 12, 1799. A blistering poultice of cantharides was applied to his neck; he was given repeated gargles of vinegar, molasses, and butter; he was bled 4 times for a total of about 5 pints of blood, and died 2 days later. Among his last words were, "I thank you for your attentions, but I pray you to take no more trouble about me. Let me go off quietly." And he died, quite possibly, in shock caused by blood loss.

The definitive treatments for infectious disease that began in the 1930s and continued to develop into the 1950s did not come into being overnight. Thomas (1987) wrote,

Behind the sulfonamides and penicillin, there lay a full 60 years of solid basic research in bacteriology and immunology. The microbial causes of infectious diseases had been identified and studied in detail: it was quite clearly known which microbes were responsible for diseases. The antibiotics did not simply drop into medicine's lap. If it had not been for those sixty odd years of research, penicillin could have come along in purified, stable form, and nobody would have known what to do with it.

What was true of the development of definitive treatments for infectious disease is taken by Thomas and others to be true of the development of definitive technologies in general. Although the desire to treat disease and thereby to relieve suffering and prolong life demands immediate action, technologies that can be used for genuine prevention, cure, and control of disease require knowledge of underlying disease processes, which in turn requires basic biomedical research that cannot be accomplished immediately. Although the research necessary to make definitive technologies available for a given disease or class of diseases may not always require 60 years, we can be sure that it will rarely, if ever, be completed soon enough to satisfy the urge for immediate action. And the intensity of this urge, coupled with the discomfort, pain, and death that are sure to occur if nothing is done on behalf of individuals suffering from diseases for which there is no definitive technology, is largely responsible for the development of various halfway technologies. After all, such technologies often do ameliorate suffering, compensate in some measure for impaired functioning, and avert premature death even when means of genuine prevention and cure are not available. In this respect many, perhaps even most, halfway technologies are not at all like the heroic measures characteristic of medical practice from the Middle Ages to the beginning of the twentieth century.

Why then are Thomas and others disturbed by the degree to which our health care budget is devoted to halfway technologies?

After all, halfway technologies are the best that can be done for individuals suffering from ailments for which there is no genuine means of prevention, control, or cure. Surely, even though palliation and postponement of death are not as good as prevention and cure, they are better than nothing. The concern of Thomas and his compatriots is not that halfway technology is not useful or not the best that can be had in the absence of adequate medical knowledge, but rather that such technologies, because they are often very costly, divert resources from the basic research that must be carried out if definitive technologies are ever to be developed. Their worry is that the greater our reliance on halfway technologies and the more this reliance undercuts basic research, the more such research will be slowed or put off altogether. This may well mean that the medical benefits generated in the short run by means of halfway technologies will be obtained at the price of delaying or rendering inaccessible far greater medical benefits in the long run. When we remember that the benefits in question are relief from suffering, restoration of normal functioning, and aversion of premature death, this price is high indeed, one that Thomas and others clearly believe is too high. The danger that halfway technologies will be allowed to consume too much of our collective health care budget should be apparent given the kind of impression they make on the public. Thomas (1987) wrote, "In the public mind, this kind of technology has come to seem like the equivalent of the high technologies of the physical sciences. The media tend to present each new procedure as though it represented a breakthrough and therapeutic triumph, instead of the makeshift that it really is." Are Thomas and the others correct? Is their concern about halfway technologies well founded?

We can obtain a somewhat fuller understanding of Thomas's perspective, and thus be better prepared to evaluate it, by examining a debate between Thomas and James Maxwell concerning the iron lung (Maxwell 1986, Thomas 1986). The iron lung is often cited by commentators who share Thomas's perspective as an excellent example of an expensive and complex halfway technology that, on the basis of fundamental medical research, was ultimately superseded by a simple, inexpensive, and easy-to-deliver definitive technology, the Salk polio vaccine (Bennett 1977, Thomas 1986). According to Maxwell (1986), a careful examination of the history of the iron lung reveals both that it was not a compelling example of a halfway technology and, more important, that the very concept of halfway technology is not very helpful for evaluating medical technologies.

Maxwell (1986) claims that a reexamination of the history of the iron lung establishes three important facts: First, the iron lung was a clinically effective technology. Its use saved the lives of many people who surely would have died otherwise, and it successfully began to perform this lifesaving function from its very first clinical application. Moreover, this success grew as experience with the device increased, as knowledge of optimal patient populations increased, and as important ancillary therapies were increasingly used in conjunction with the iron lung. Furthermore, contrary to what is widely assumed, very few patients became chronically dependent on the iron lung. Most patients either regained normal or nearly normal respiratory functioning or improved sufficiently to be placed on less radical respiratory-assist devices.

Second, the iron lung was not an especially expensive treatment modality (Maxwell 1986):

Data on the early costs of respiratory care are not readily available, but rough estimates can be formulated from different data sources. At the height of the large polio epidemics of the 1950s, average treatment costs for the acute respiratory patient, who spent less than four months on the iron lung, were modest even for that period, at about $2,500 ($10,800 in 1982 dollars or approximately 60 percent of the 1955 median family income). A large proportion of treatment costs were concentrated on the chronic patient, confined to the tank respirator for a year or more. The expense of maintaining a chronic patient equaled more than $12,000 per year or $52,000 in 1982 dollars. Total societal costs for respiratory treatment probably approached $10 million per year, a small figure when compared to the amounts currently spent on many life-sustaining medical technologies.

Although, by current standards, the costs of respiratory treatment with the iron lung were not extraordinary, they did place large financial burdens on patients at a time when there were few insurance or other third-party payment schemes available to ameliorate medical expenses. Maxwell has noted, though, that publicity concerning the plight of iron lung patients generated a variety of local, regional, and national efforts that did much to help the needy afford adequate respiratory care.

Third, "[t]hose who characterize the iron lung as an awesome antique that was replaced by polio vaccine fail to recognize the critical role the iron lung played in the development of modern-day respiratory care" (Maxwell 1986). The iron lung was crucial to that development in at least three ways: First, the clinical prowess of the device stimulated development of other forms of respiratory-assist equipment. Second, it motivated medical scientists to seek better understanding of human respiration and thereby stimulated increased research on respiratory physiology. Third, it stimulated development of novel methods of care for patients suffering from respiratory ailments (Maxwell 1986):

Today mechanical ventilation is one of the major lifesaving technologies relied on in the intensive care unit. Although some investigators question whether the widespread use of intensive care units has resulted in improved survival or quality of life for certain categories of patients, there is little dispute about its lifesaving potential. Many patients with respiratory failure resulting from trauma, illness, and as a consequence of surgery receive temporary mechanical support that sustains life until normal breathing is restored. The expansion of mechanical ventilation parallels the exponential growth of intensive care in the United States, and it is probable that out of a total of more than 66,000 adult intensive care unit beds, a large number receive some kind of mechanical respiratory assistance. Thus Drinker's "tinkering" in the Harvard machine shop back in the late 1920s [and the consequent development of the iron lung] initiated a process that culminated in one of the essential practices of modern-day critical care medicine.

Maxwell's (1986) view is that collectively these points establish that "[i]n retrospect, it is clear that the iron lung does not provide a compelling example of halfway technology nor does it provide the most appropriate example of the ultimate, palliative technology. The iron lung must be viewed as a simple respirator, similar in its physiologic effects to modern respiratory equipment." Thus, as Maxwell notes, it is incorrect to regard the iron lung as a halfway technology for treatment of polio replaced by the definitive technology of polio vaccine. Rather the iron lung should be seen as a primitive stage in the evolution of more modern respiratory equipment—an early and therefore necessarily primitive and inelegant point in the incremental evolution that has characterized the development of respiratory equipment and most other forms of medical technology as well. Although polio vaccine eliminated one source of acute respiratory distress and thereby one source of need for respiratory equipment, it did not eliminate acute respiratory distress altogether as a medical problem. The polio vaccine made the iron lung obsolete as a treatment for respiratory distress due to polio, but for other forms of respiratory distress to which the iron lung would have had clinical application, it was made obsolete by the more modern forms of respiratory equipment to which it led. Indeed medicine's current ability to save the lives and alleviate the suffering of individuals stricken by such distress is indebted significantly to the iron lung. Furthermore the individuals whose lives were saved by the iron lung before the development of an effective polio vaccine certainly would have died if the iron lung had not been available (Maxwell 1986):

The iron lung represents but one example of a technology unfairly characterized by the halfway technology concept. A large number of medical technologies effectively compensate or restore premorbid function without eliminating the underlying cause of disease. Thomas himself describes the cardiac pacemaker as a technological marvel that dominates the heart's electrical conduction system and regulates its rhythm with perfection. The pacemaker clearly maintains the function of the natural heart, both reliably and at a relatively modest cost. The practical success of the pacemaker depended not only on advances in cardiac physiology, but upon a number of crucial engineering developments in microelectronics and power sources.

Other so-called halfway technologies, such as insulin, diuretics, digitalis, antihypertensives, antidepressants, anti-inflammatory agents, narcotics, and a variety of surgical procedures, enhance patient quality of life, while offering cost-effective treatment.

Maxwell (1986) has argued that underlying Thomas's point of view is a specific conception of the relation between science and technology, a conception in which technology is merely the "handmaiden of science." According to this view, "a progression occurs from investment in basic research to an understanding of disease processes and then to applied science and technology, which results in a useful drug or vaccine." Thus science, particularly medical science, learns nothing from technology, and technology, particularly medical technology, is only as good as the level of basic scientific knowledge available at a given time. Thus "good" technology—technology that is more than merely makeshift—must wait until adequate scientific research has taken place. Maxwell (1986) challenges this view by noting that in medical as well as other sciences, the causal chain of events often leads from technology to science rather than from science to technology:

. . . major technologies, such as the telescope and the steam engine . . . not only evolved from other technologies but also led to significant scientific breakthroughs. Galileo was able to use the telescope, a device made possible through advances in the craft of lens making, to discover that the planets revolved around the sun rather than the earth. Similarly, Watt's invention of the steam engine worked and ultimately led to the development of thermodynamics.

The causal chain of events from technology to science appears no less true in medicine than in other fields. The X ray, the electron microscope, the electrocardiogram, mechanical heart valves, and cancer chemotherapies all provide instances of technologies that contributed greatly to the understanding of basic physiologic processes, which in turn resulted in improvements in medical care. By probing a disease with chemotherapy, cancer researchers uncovered the concept of the pharmacologic sanctuary in the central nervous system, which is referred to as the blood/brain barrier. Many medical discoveries in fact seem to be made by individuals playing with new techniques or instruments to see what would happen.

History shows that the technologies existing at a given time stimulate the development of new, better technologies by stimulating efforts to improve what is at hand. History also shows that existing technologies stimulate efforts in basic research that likely would not otherwise have occurred and that this research in turn often leads to the development of new and more efficacious technologies. Maxwell (1986) has argued that this lesson of history is as true of medicine as it is of other fields of human endeavor. Therefore,

. . . by categorizing technologies as halfway or definitive, Thomas disparages the benefits that technology makes to medical science and treatment. The halfway technology concept minimizes the role of technology as a source of ideas and understanding. Science and technology are not in conflict; knowledge from these fields flows along separate tracks, proceeding from different necessities, and enjoying different serendipities. Technical change in medicine is not a single linear process, but one in which science and technology interact in complex and largely unpredictable ways.

Thomas's concepts have the potential of creating issues and misunderstanding, especially in the increasingly cost-conscious health care environment. Instead of relying on preconceived notions, society must judge technologies on their individual merits. Society must ask not whether a technology results from basic research, but how it compares to existing treatments in terms of costs, benefits, and risks. Policy prescriptions based on careful analyses of the specific strengths and weaknesses of the technology are likely to lead to more balanced and discriminating policies toward both technology and the basic sciences in medicine.

If Maxwell is correct, Thomas's notion of halfway technology is dangerous. It can mislead us in our assessment of the merits of specific technologies, such as the iron lung, and it can lead us to misjudge a whole class of technologies, disposing us automatically to undervalue technologies that are developed for disease treatment in the absence of understanding of the relevant underlying physiologic processes.

In response to Maxwell, Thomas has conceded that the iron lung was an extremely useful technology that led to the development of the more effective and elegant respiratory equipment of today's intensive care unit, and he admits that this has been true and is true of many other halfway technologies as well. On Thomas's view technologies such as the iron lung may be undeniably of great utility when definitive technologies are unavailable. Nonetheless Thomas (1986) said of the iron lung,

... this kind of technology could not possible serve as a satisfactory answer to the problem of poliomyelitis. It was required at the time, but only because nothing was better. The real problem in polio, unsolved until the early 1950s, was the inner mechanism of the disease. This could only be got at by basic research, and that, in the end, is what luckily happened, although it did not happen quickly. Several decades of elegant research were needed before it could be known that there were three (only three) distinct types of polio virus, each possessing its own, unique antigen, and that each virus could be grown to abundance in tissue cultures. Once these two items of basic information were in hand, it was a sure thing that a vaccine could be made.

The pertinent question here is, Why is the inner mechanism of polio or any disease "the real problem" for medicine? The purpose of medicine is to provide individuals with those benefits that medicine alone can supply. Thus it would seem more appropriate to take restoring normal functioning and averting the premature deaths of people who are ill to be the real problem of medicine. Ideally the solution to this problem is to keep people from becoming ill in the first place or to cure their illnesses when prevention is not possible. But it does not follow from the unavailability of the ideal solution to medicine's real problem that nothing except basic research should be done until the ideal is reached. Thomas himself conceded as much when he acknowledged the clinical usefulness of the iron lung and many other so-called halfway technologies.

Given this, what is the source of Thomas's concerns about halfway technologies? Before offering an answer to this question, we should note that Thomas (1986) also conceded that technology is not always the mere "handmaiden of science," at least in the physical sciences. This much surely must be conceded for, as Maxwell noted, the steam engine, for instance, was developed and put to many important uses before anyone understood the principles of thermodynamics on which its successful development and use depended; indeed the steam engine itself stimulated efforts that led to such understanding. But, according to Thomas (1986), the situation is different for medical science and medical technology:

Medicine has nothing resembling a steam engine in its past history, waiting there to lead it to something like thermodynamics. Looking back, medicine's old technology consisted in large part of things that really never worked at all and usually did more harm than good: bleeding, cupping, huge doses of mercury, incantations, blistering ointments. If pressed, medicine can bring up a few useful old pieces of empiricism: digitalis, quinine, cowpox vaccination, a few others, but it is a very short list.

What Thomas has said here is true of much of Western medicine's history. In the past medicine has had little in the way of technologies with the potential to stimulate progress in medical science. But, as Maxwell (1986) has made clear, the iron lung did lead to greater understanding of respiratory physiology, and other more recent medical technologies have also led to increases in basic medical knowledge. Thomas has failed to appreciate an important difference between the technologies of the era of heroic medicine and those in use today. As he noted, whereas the technologies of heroic medicine were not subject to rigorous evaluation aimed at determining their clinical efficacy and safety, many modern technologies are not put into general use until they have been shown to be safe and to have the promise of significant clinical benefits (Thomas 1986). Such evaluation is often continued even after widespread dissemination has occurred. True, many contemporary technologies do manage to become widely disseminated without much rigorous evaluation, and others have been inadequately evaluated. Nonetheless the difference between the degree to which contemporary medical technologies are subject to rigorous evaluation and the degree to which the heroic measures of the past were evaluated at all is enormous. This means that contemporary medical technologies are far more likely to lead to progress in medical science than their historical counterparts. They are far more scientifically grounded than their historical predecessors. In short there is much reason not to accept Thomas's conclusion that, in medicine, technology should be secondary to basic research.

Thomas also has conceded that halfway technologies can be useful in yet another way: They can, and do, lead to even better technologies. Again the iron lung is an excellent example from recent history. This device was a crucial, and perhaps even necessary, stage in the development of the respirators on which the successes of modern acute care medicine depends. Yet Thomas (1986) attempted to soften the blow here as well, claiming that the respiratory technologies to which the iron lung led are themselves halfway technologies insofar as they are used in "situations where we still lack a deep understanding of the mechanisms of the diseases in question and have no other choice in treatment." (Included among the diseases in question are emphysema, lung cancer, heart disease, and other lung diseases.) But if, as we have argued, halfway technologies are useful in the two respects already noted—therapy and progress in medical knowledge—then it is difficult to see what is troublesome about the fact the new technologies engendered by one generation of halfway technologies are themselves halfway technologies.

Now we can return to the question of what fundamentally troubles Thomas and others about halfway technologies. Their main concern is with the costs of such technologies. Implicitly, and sometimes explicitly, the concern is that the portion of our collective health care budget consumed by halfway technologies will leave basic medical research with an inadequate level of resources.

Thus the benefits obtained in the short run by means of halfway technologies will cause us to delay, or even to forgo altogether, the far greater benefits that basic research will bring in the long run. Several points are worth noting: First, we cannot simply assume that the long-run benefits of basic research will always outweigh the short-run benefits of halfway technologies. Research programs often lead to dead ends altogether. Second, we cannot simply assume that the benefits of halfway technologies are short-run benefits. As Maxwell (1986) has shown, the halfway medical technologies of a given period often lead in the long run to better technologies, to new and better forms of patient care, as well as to new and more fruitful directions in basic research. Third, even if the benefits of halfway technologies are limited to the short run and are outweighed by the long-run benefits of basic research, the benefits of halfway technologies are typically certain, whereas those of basic research are uncertain. Benefits that occur in the here and now are certain, even when we are not sure who receives them, as is the case with successful preventive medicine. We know that they occur. Benefits that are expected to occur in the future may not occur. Even if medical history tells us that research always has led to definitive technologies, what has always occurred does not necessarily occur in the future. There may be limits to our medical knowledge that research cannot surmount. Fourth, even if the benefits of research were as certain as the benefits of halfway technologies, the interests of people existing now would have to be considered against the weight of the interests of those who are expected to exist in the future. Suppose we were certain that these future persons would exist and that basic research would benefit them. It does not follow that they have greater or equal claim on today's medical resources as do people who presently exist. Perhaps justice requires that the interests of existing people be given greater weight, or at least not less weight, than those of future people. It is certainly not obvious that the interests of future persons have greater moral weight than those of existing persons.

These considerations may not suffice to prove that we are not devoting too much our collective health care budget to halfway technologies, but they do show that we cannot easily conclude that basic research is superior and should be given greater budgetary priority. What these considerations do show is that there are difficult empirical and ethical issues that must be resolved before we can say with a reasonable degree of confidence what share of our health care budget should go to research or to halfway technologies. The confidence of Thomas and others in basic research is not clearly justified.

A further point about the concept of halfway technology is worth mentioning: Whether a particular technology is categorized as halfway or definitive depends on whether it is efficacious as a means of cure or prevention of disease. (Presumably Thomas and those who share his perspective would agree that a technology that cures or prevents disease is definitive even if it was developed without knowledge of the relevant underlying physiological processes.) To which category does a technology belong if its use is ancillary but crucial to some technology that is definitive? Suppose, for instance, that cure of some disease requires an operation under general anesthesia. Surely the success of such an operation would very much depend on the availability of efficacious means of mechanical ventilation. In this case respiratory technology could not be regarded as definitive for it neither prevents nor cures disease. Nor could it be regarded as halfway for in this case it is not simply something used to palliate the effects of disease in the absence of means of genuine prevention or cure. Yet it might be used as a halfway technology in other cases, as with emphysema. What these remarks add up to is that whether the halfway/definitive distinction applies to a technology is a matter of its use. In some uses this distinction simply may not be applicable and therefore not useful to evaluation of the technology, even if it is a halfway technology for other purposes.

Medical Technology and the Dehumanization of Patient Care

Although medicine is now generally regarded, by patients and practitioners alike, as having far greater diagnostic, preventive, and therapeutic efficacy than even a few decades ago, some commentators claim that modern medicine is in a state of crisis. Just as the new efficacy of medicine is attributed largely to modern medical technology, so is this alleged crisis. The crisis at issue here is not the one posed by escalating costs, a genuine crisis in its own right and a situation also often attributed largely to modern medical technology, but rather one that is relatively independent of matters of cost. The crisis of concern here is posed by the fact that, although they acknowledge the benefits that modern medicine offers, many patients feel alienated and dehumanized by contemporary medical care.* This widespread sense of dehumanization constitutes a crisis because it poses a threat to the health of many individuals. Precisely because they find modern medical care dehumanizing, many individuals are avoiding entry into the health care system until their illnesses are acute and perhaps beyond hope, or they are ignoring the advice of their physicians, or—perhaps the greatest threat to health posed by this sense of dehumanization—they are turning to the unproved, and sometimes proved harmful, methods offered by so-called holistic medicine. Loren Eisenberg (1977) put matters this way:

It is, at first glance, curious that dissatisfaction with medicine in America is at its most vociferous just at a time when doctors have at their disposal the most powerful medical technology the world has yet seen. The "old fashioned" general practitioner, with few drugs that really worked and not much surgery to recommend, is for some reason looking good to many people— in retrospect, at least. In the past, if the patient had a serious infection, all the doctor could do was wait for its resolution; now he has powerful antibiotics. In the past, if the patient had cancer, the recommendable surgery was limited and uncertain in its effects; now, potent antimetabolites, radiation, and many curative operations are available. Compared with the "good old

---

*Accordingly Wolpe (1985) was quite correct when he wrote, " . . . [this] 'crisis' of medicine is thus not addressed by protestations that physicians are curing our population better than ever before," and again when he wrote, "In medicine today, how well we cure sick patients has become only one competing standard of value by which the health care system is assessed."

days," modern treatment provides an effective range of drugs and procedures that are curative for a long list of diseases, and palliative for many more. How, then, can we explain the paradox that more effective medicine has resulted in nostalgia for a medically primitive past?*

The short answer to Eisenberg's question reiterates what has already been noted, namely, that many patients find modern medical care dehumanizing. To simply state this is not of course to explain how and why they find it so. Before turning to a more complete answer to Eisenberg's question, it should be noted that the complaint that modern medical care is dehumanizing has at least three interrelated aspects (Howard 1982): First, patients feel that they are "being perceived and treated as things—as unfeeling quantifiable objects with standardized parts and wholes." In particular, patients feel that they are treated as machines rather than as persons. Second, patients feel that they are often reduced to the diseases that bring them to seek medical care. That is, they feel that doctors and other medical practitioners are concerned exclusively with their physical ailments and not with the persons afflicted by these ailments. Third, patients feel that caregivers are too often indifferent, detached, and cold, too often lacking in empathy, warmth, and compassion.

Although many complain that modern society in general is dehumanizing and place the blame squarely on the degree to which contemporary life is dominated by technology, the charge that modern medical care is dehumanizing because of modern medical technology deserves attention itself. As Wolpe (1985) has noted,

Medicine provides us with the best opportunity to examine our relationship to our new machines, for it is in the medical realm that we have the most intimate contact with our technology. We do more than "interface" with medical instruments, we assimilate them, allowing them entry past the natural defining barriers of our physical bodies.

*See also Schwartz and Wiggins 1985, Boyle and Morriss 1981, as well as Wolpe 1985.

Our contact with medical technology is far more intimate than our contact with any other form of technology. Furthermore contact with medical technology typically occurs during times of illness and therefore times of great insecurity and vulnerability. Accordingly it is not surprising that complaints about the dehumanizing effects of technology should be most intense and give rise to the greatest concern in the context of medicine.

Perhaps no aspect of medical care is more important than the relation between the medical caregiver and the patient, and here the effect of technology is especially profound (Reiser 1984):

Technology has altered significantly the form and meaning of the medical relationship. It allows us to direct our vision and attention to variables singled out by it as significance. Thus stethoscopes increase the significant of chest sounds, X-rays of anatomic shadows, electrocardiograms of waves on a graph, computers of printouts, dialysis machines of chemical balances, and so forth. Such evidence is important for diagnosis and therapy, and the more precisely it can be stated, the more valuable it becomes. In comparison, evidence given by patients, altered by its passage through the prism of their experience and personality, has seemed to the technological age of the past two centuries less substantive, accurate, and meaningful as a basis for clinical decision making and actions. Increasingly, practitioners encounter patients for relatively brief and intermittent periods— such as the consultant visiting a hospitalized patient whom he or she has never before met. In such visits the technical aspects often dominate, for there is no time or prior relationship to determine much about who the patient is, or what the patient thinks about the illness or the needs it engenders. And even in medical relationships that are not so discontinuous, technological measurements and measures tend to crowd out other dimensions of evaluation and therapeutics.

To gain insight into why the highly technological nature of modern health care has come to be so widely regarded as dehumanizing, a brief examination of the history of medicine is necessary (Reiser 1984):

. . . some of . . . [the] key ideological and pragmatic goals [of contemporary health care] reach back in time to the Renaissance and the scientific revolution, which transformed the reigning view of medicine that had been shaped by the ancient Greeks. The Greek physicians, steeped in the learning of the Hippocratic school, saw the natural world as an environment to live with and adapt to, rather than to conquer and dominate. Their theory of illness, the humoral theory, concerned an equilibrium that existed among the four basic constituents of the body (humors), which in turn were connected with environmental elements (physical, social, and personal) that surrounded and interacted with the humors and determined health and illness.

According to this view, illness occurred when one or more of the four humors became excessive or deficient and upset the equilibrium in which they existed. The resulting imbalance produced symptoms related to the particular humoral dysfunction. The idea that disruption of one aspect of this biological system would affect all connecting parts implied that illness involved the whole person, not just a segment of the body.

This focus on the whole person, which typified the Hippocratic perspective on disease, began to erode around the sixteenth century with an important change in the way medical practitioners studied human anatomy. Until then anatomy had been studied primarily by reading the texts of traditionally revered medical authorities, especially the ancient Roman physician Galen. As Reiser (1984) noted,

Medievalists did not believe that they could essentially improve on the views of revered, seemingly authoritative ancient geniuses. Although the Hippocratic Greeks had freely acknowledged their ignorance, they were committed to continual explorations of diseases, mainly by bedside observation of the response of patients to illness. This spirit of inquiry became dulled in the Middle Ages, when scholars accepted as true what the ancients had written.

The change was a turn to empirical study of human anatomy through the dissection of cadavers. Reiser has noted that Andreas Vesalius, an Italian anatomist, in a text first published in 1543, showed that Galen's traditionally revered account of human anatomy contained over 200 errors. Vesalius thereby began a process that effectively undermined strict reliance on allegedly authoritative texts as the ultimate source of medical knowledge. He demonstrated the importance of empirical study of the body at an especially propitious time, when scientists outside medicine were also successfully challenging reliance on authority and insisting on empirical study of phenomena as a key to scientific progress. Vesalius's push to base the study of human anatomy on dissection of cadavers was timely also in that it took place as scientists in general were insisting that the analysis of complex entities and phenomena into simpler parts was a crucial, if not the critical, empirical means of obtaining genuine scientific understanding (Reiser 1984):

... the analytic, anatomic perspective generated a view of illness that segregated disease and disorder to specific places in the body. This concept replaced the ancient Greek notion of illness as a dynamic process involving place and lifestyle and affecting the whole person. It has become the prevailing concept of the nature of illness, and has influenced greatly the development and adoption of health care technology.

Whereas the Hippocratics viewed disease as a dynamic phenomenon, successful treatment of which required information about the body of the patient and his place in and relation to his natural, social, and personal environments, information about the whole person (Reiser 1984),

Anatomists developed a whole new perspective on Disease: It was a disorder of a bodily structure localized in a site. A small change in a vital part could result in serious damage to an otherwise normal body. Anatomists thus isolated diseases in places, and introduced into the discourse of health care what has become the principal question in the evaluation of patients: Where is the disease?

A second important change in medicine, a change largely made possible by the turn to empirical studies of anatomy and the resultant new conception of disease, was the rise and increased acceptance of medical specialization (Reiser 1984):

The rise of anatomical thinking and specialization are events inextricably bound. Specialization elevates the status of those who develop great knowledge about an aspect of things. Specialization of medical function did not begin with the acceptance of anatomy; it had existed since the earliest days of medicine. Specialists in the first half of the nineteenth century were largely informally trained people—midwives, bone setters, and so on. Those with a formal medical training, who so narrowed their practice, were looked down upon by their colleagues.

But the logical justification for specialism, offered by an anatomically influenced perspective that isolated illness in specific bodily places, and the practical justification for specialist based on a need to create and master technology in order to detect and treat anatomical disruptions, combined to initiate the growth and acceptance of specialization in orthodox medicine during the nineteenth century.

Two of the sources of the dehumanization felt by modern patients are now clear. The first is the historical shift of medicine from an art based in Hippocratic doctrine, and its perspective on the nature of disease, to an art based in the empirical and analytic science that today is generally recognized as the source of much of medicine's powers. This shift caused medical practitioners to divert their attention from the patient regarded as a whole person in dynamic interaction with a complex environment and to focus instead on the patient as a body. Disease then came to be viewed as something localized in specifiable bodily structures, and treatment required that the physician look only to these structures. The second is historical rise of medical specialization. This necessarily meant that practitioners increasingly focused on ever narrower aspects of the body. The almost inevitable result was that concern with the patient even as a whole living body, much less as a whole person, receded further and further into the background.

A third source of dehumanization also came into play with the rise of medical specialization during the nineteenth century as what had been a longstanding medical bias against technology was undermined (Reiser 1984):

This anti-technological bias in medicine can be traced to the thirteenth century, when the medieval universities, which were just beginning, became the locus of formal medical training. Learning their skills alongside theologians and lawyers in graduate faculties of the university, physicians came to accept and adopt the belief of these disciplines that the use of tools and manipulation lowered the esteem and social standing of scholars and physicians. Learning in the university was book learning. In pursuing their activities, scholars could not act like trades-men, with the whom the use of instruments was associated.

The practice of dissecting cadavers, a process increasingly accepted throughout the sixteenth and seventeenth centuries as a source of genuine scientific understanding of disease, ultimately dissipated the scholarly bias against the use of medical instruments and equipment. To perform dissection, it was necessary to use tools and to touch human bodies. In short it was necessary to abandon what might aptly be called "ivory tower medicine." Furthermore detecting the internal bodily changes that dissection revealed to be linked with diseases and their symptoms required development of more and better diagnostic tools that could give practitioners information about the inner workings of living persons. And to treat the inner workings of living patients so as to repair or excise the bodily structures affected by disease required development of more and better surgical tools and techniques.

The drive to develop and use reliable diagnostic technology, which began in earnest in the nineteenth century, is especially important for understanding the dehumanization experienced by patients of the modern health care system (Reiser 1984):

Microscopes, incubators, staining agents, colorimeters, and other such devices were gathered together to form the diagnostic labo-ratory. This institution depended increasingly on technicians with the time and skill to apply the new techniques to science. Physicians, impressed with the connection of these instruments and techniques to science, and thinking of themselves as practicing scientifically by using them, increasingly turned from the judgments of their own senses and the techniques of physical diagnosis to the impressive data of the laboratory.

This change was in keeping with the original emphasis of the scientific revolution of the seventeeth century. One of its tenets was to place inquiry on an objective basis. In the investigation of nature, it sought to eliminate evidence influenced by human values or bias. It attempted to establish a rigorous set of methodologies to establish facts, such as experimentation, and to describe the facts wherever possible in objective ways, such as through the use of numbers.

The development of technologies that allowed the inner workings of the living body of the patient to be revealed and the drive to use these technologies to gain objective, and therefore scientific, information on which to base diagnosis, had two important effects: First, it meant that the physician took an increasingly objective stance toward the patient, a stance in which the physicians own values and feelings were put aside. This necessarily meant increased detachment from the patient as a person and an increased tendency to view the patient as an object. Second, it meant that the information that the physician found crucial to arriving at a diagnosis no longer came from conversation and dialogue with the patient. This surely accounts for some of the feeling that patients today have of being treated by clinicians who lack empathy and warmth. Indeed with the objective information that diagnostic technologies could provide, the inherently subjective information that the patient could provide had to be deemphasized. The feeling of patients that physicians are concerned only with their diseases and not with them as whole persons must be attributable, at least in part, to this. And the effect of these two aspects of diagnostic technology was surely only intensified by the impact of surgical technologies, technologies that can treat only the patient's body or some part thereof and that have no capacity to address the subjective aspects of being an ill person.

So far three historical sources of the dehumanization that afflicts the highly technological care provided by modern medicine have been identified: acceptance of empirical study of human anatomy

and the resultant conception of disease, increased medical specialization, and increased reliance on technology as means of diagnosis
and treatment. A fourth historical source of dehumanization, one
especially relevant to understanding the sense that many patients
today have of being regarded as mere living machines by those who
dispense medical care, can be traced to the seventeeth century
French philosopher, Rene Descartes. Although Descartes, widely
regarded as the father of modern philosophy, is best known for his
views on the nature of and relation between mind and body, his
views on the nature of the body are most germane here.

In the most widely studied of his works, *Meditations on First
Philosophy*, Descartes argued that the essential feature of the
nonmental world, the world of physical entities, is extended matter
and that all extended matter is subject to mechanical laws of nature.
Because the body is a physical entity, it too must function in conformity to these laws. Boyle and Morriss (1981) wrote:

Having studied medicine as well as physics and mathematics
Descartes began with zeal to apply this mechanical model to
biology and physiology.

The human body was envisioned under the not-so-hypotheti-
cal metaphor of an "earthly machine," and this metaphor ushered in the vocabulary of cause and effect, mechanical laws of
operation, and the notion of the whole as the sum of its parts.
Because of its great explanatory efficacy the mechanical model
was destined to become a fundamental root metaphor of medical
science.

In his work Descartes gradually became convinced that it
would be in medicine that his most important and beneficial
discoveries would be made. For, if all bodily processes were
mechanistically caused, bodily diseases, once their causes were
determined, should be remediable with the same precision and
certainty as the disorders of a clock.

Descartes's conception of the body as a machine was an instance
of his general conception of all natural phenomena, a conception
that came to be shared by the majority of scientists. With its
implication that the body is a thing that can be analyzed into com-

ponent parts and, like everything else, be understood and subjected to repair in terms of mechanical laws of nature, Descartes's conception of the body as a machine was clearly quite compatible with the three historical sources of the dehumanizing effects of modern medicine. According to Boyle and Morriss (1981), this mechanistic conception of the body is current in medical thinking even today. Indeed, in their view, it is correctly regarded one of the more important sources of modern medicine's advances:

The mechanistic model of medicine has been exceedingly successful. In the biology and medicine of the nineteenth and twentieth centuries, it led to the development of laboratory medicine and technological medicine. The image of the person as a machine is retained. The machine is merely identified as more complex than before.

Not all of the sources of the dehumanization of patient care are rooted in medicine's distant past. A fifth source, one local to the late twentieth century, is often referred to as the technological imperative in medicine. This phenomenon consists of the identification of high-quality medicine with highly technological medicine, that is, the belief that the best health care is the care that uses as much sophisticated medical technology as possible. Several forces contribute to this phenomenon: First, the contemporary pervasiveness of third-party payment schemes insulates both physicians and patients from the full cost of using medical technology and thus allows them to overuse it without paying the full cost. Second, physicians are trained to pursue the medical interests of their patients aggressively and relentlessly. This process of medical socialization instills them with the desire to leave no therapeutic or diagnostic stone unturned. Accordingly they are motivated to use all the means of diagnosis and therapy available to them, including the full array of medical technologies. Third, clinicians are well aware that medical knowledge consists of generalizations and that no matter how well grounded in empirical fact these generalizations are, they cannot cover every case. Each patient is a unique individual and so may present an anomalous case, a case that simply does not fit generalizations. The desire to avoid misdiagnosing or mistreating a patient by failing to recognize anomalous cases also leads to a tendency to want to use every technological means available. Fourth, although patients complain about dehumanization and point to technology as the culprit, they nonetheless also identify high-quality care with highly technological care. Communication of this belief to their caregivers also motivates overuse of medical technology.

One of the most graphic and tragic forms of medical dehumanization most closely connected with the technological imperative in particular is modern medicine's apparent obsession with death prevention. For many in the modern health care system, every death, even when it is by no means premature, represents a failure. The understandable desire to avoid this failure motivates physicians to act aggressively to keep their patients alive. Clearly there are circumstances where this kind of vigorous struggle is appropriate and in keeping with medicine's longstanding humanitarian commitments. But when it becomes the predominant response to death, as many believe it has, genuine problems are posed. The most worrisome is the danger—which may already be a reality—that a whole class of patients, particularly elderly patients whose lives are at a natural and inescapable end, will be kept alive by mechanical means long beyond what can be of any benefit to them. (Indeed the death-with-dignity movement and the desire of many to obtain legal recognition for a right to die are largely motivated by this danger.) Technology is particularly well suited to keeping the body going when life is threatened. But it is not particularly well suited to the needs of persons who are not benefited by having their bodies kept alive. Those persons need their humanity attended to. They need to be helped to die humanely. This means attending to their subjective needs, needs that are especially acute when death is being faced.

Having identified the historical roots of modern medical dehumanization as well as a more recent source, the question of to what degree the crisis posed by this dehumanization is to be laid against medical technology can be posed. The first point is that technology clearly is a part of the problem. Technology constitutes a layer interposed between the medical caregiver and the patient and thus necessarily alters the nature of the human interaction. Through this ever-increasing layer of technology the physician comes into contact with the patient. Through this layer of technology the physician gains information crucial to her diagnostic and therapeutic choices. And by means of this layer of technology the physician undertakes to act on the patient. Thus the physician is less intimately in contact with the patient than would otherwise be the case and that her attention is often directed primarily toward the technology, even though her ultimate aim is to benefit the patient. But as Reiser (1984) has asserted,

To speak so of the . . . technological features of practice does not diminish their great significance and benefit. Rather, it points out that they do not encompass all critical aspects of diagnosis or treatment. In diagnosis, we seek to place a patient in a category with earlier patients who exhibited similar manifestations of illness. The similarity of the pattern or symptoms between one patient and another does not make them identical. That which is unique in a patient's illness can often be learned best by nontechnological inquiries based mainly on dialogue. Such inquiries deal with the patient's sensations and perceptions, with the values held by the patient that are relevant to treating the illness. An illness is not only a physical disturbance of the body's structure and function; it is an experience by the patient, who invests the illness with meaning. Knowing what this personal response involves is crucial information for therapy. Therapeutics is successful to the extent that it meets not only the needs of the disease of the patient, who modifies and feels the effects of disease in a unique way.

Dialogue, a means to retrieve and understand the experiential dimensions of illness, is critical. To the extent that inordinate fixation on the technical aspects of medicine diminishes the possibilities for or the importance of person-to-person dialogue, we become less effective in meeting human illness.

As Reiser's remarks show, acknowledgment that technology is somewhat responsible for the dehumanization experienced by today's patients must not be allowed to eclipse its benefits. Also, perhaps most important, it must be recognized that the key to overcoming this dehumanization is not to become medical Luddites who renounce technological medicine but to reintroduce warmth and human contact into the interaction between medical caregiver and the patient. Reiser thinks that this necessarily means ensuring that genuine dialogue in which the views and concerns of the patient are expressed and truly heard takes place. Whatever the necessary practical steps are, they must take the form returning medicine to what it was in the days of Hippocrates, a *humane* science. Medicine is surely a science, and the benefits it has to offer are grounded in that fact. The science of today, brought into existence by the scientific revolution of the sixteenth and seventeeth centuries, is vastly superior to the science of the Hippocratics. We should have no desire to return to the limited medical knowledge of classical civilization. But we should, given the number of people driven away from modern medicine by dehumanization, return to the humaneness of the Hippocratics. This means above all else making medicine attend to the whole person, especially his subjective aspects, precisely those aspects to which the scientific side of medicine is least able to attend. On the one hand this requires that medicine broaden its scientific base and learn from the social sciences—psychology, sociology, and anthropology, to name only those most obviously relevant. On the other hand it means that medicine must look beyond science to the humanities and what they can teach about the social, personal, and even spiritual dimensions of persons, especially persons who are made vulnerable by illness.

Can the necessary steps, whatever the specifics may be, be carried out? There are many reasons to believe that they can. One of the most compelling is that for some time some members of the medical profession have been calling for more attention to the social sciences and humanities in medical school curricula and in ongoing physician education. This interest is manifested in the traditional medical journals as well as in relatively new journals in which medical professionals are educating themselves and their colleagues in the medical merits of the social sciences and the humanities and their importance in clinical practice.

To conclude this section, let us address two matters: the turn of many alienated patients to so-called holistic medicine and the nostalgia many such patients feel for the old family practitioner of several decades ago. The first of these can be dealt with briefly. Although the holistic health movement has many sources, one clear source is the disaffection generated by the highly technological character of modern medical care. In some instances individuals are rejecting modern medicine on the basis of a general belief that it is part of a society that is far too technological. Persons of this persuasion often feel that the prominence of technology in contemporary society stems from an attitude that nature is to be conquered rather than cooperated with. On their view most of the ills facing modern humanity are results of this attitude. These persons would take heart in the Hippocratic conception of medical practice. As Reiser (1984) has noted, Hippocratic

[p]hysicians viewed their task as assisting the natural powers of recovery by intervening at strategic and critical times in the course of illness. The ability to detect these moments was a central feature of Greek medical learning.

But one need not reject medical technology because one agrees with the notion that nature should be cooperated with rather than dominated. Although much depends on what counts as cooperating with nature rather than attempting to overpower it, there is no reason to believe that cooperation with nature is inherently inconsistent with technology. Indeed many of the most efficacious drugs available to modern medicine are natural substances.

A more troublesome aspect of the turn to holistic medicine is that it is driven somewhat by the alienation produced by modern medical care. This is troublesome because some of the therapies of holistic medicine have been proved to have no medical value and even to be clear sources of harm. Furthermore many holistic practices have not been rigorously evaluated. Their value and the risks they pose are simply not known. Holistic practitioners often defend themselves by asserting that Western science is not capable of assessing holistic practices. In our view this amounts to a rejection of the only tool we have for assessing practices. At the very least holistic medicine owes its patients a rigorous account and persuasive argument on behalf of some alternative nonscientific means of assessing practices that promise to prevent, control, or cure disease. In any case much of the ability of holistic medicine to compete with orthodox medicine derives from its attention to the whole person. Holistic practitioners typically are concerned not only with what ails the patient but with the patient's way of life, worldview, emotions, and values, and they believe that such attention is crucial to effective therapy. This means that holistic practitioners typically must pay much more attention to who their patients are and to what their patients value. This challenge is one that orthodox medicine can meet if it is willing to determine and do what is necessary to make medicine both scientific and humane.

As for the nostalgia that many patients feel for the days of the general practitioner, this too is easily explained. Again the history of medicine will provide the needed understanding. Even as medical research and diagnosis became more scientific, for much of its history, indeed until the early twentieth century, medical practice often consisted of the heroic measures that we have discussed. Bleeding is a particularly familiar and apt example. In the early nineteenth century some practitioners began to have doubts about these measures and set out to study their effects. The results showed that patients more often were harmed than helped by these measures. Nonetheless, according to Thomas (1987),

It was a full century before this kind of medicine was given up. Sir William Osler led the way toward a rational medicine through his teaching at the University of Pennsylvania, at Johns Hopkins, and at Oxford. In effect, Osler ushered in a phase of medical practice later referred to as "therapeutic nihilism," in which almost all of the standard, accepted forms of treatment were simply given up. There were a few medications, but only a few, that could be trusted on the empirical record . . . Osler's students were instructed to look after the patient, to explain things when possible to the patient and the family, but not to meddle.

This era of therapeutic nihilism did not end until the development of the anti-infectives beginning in the 1930s. Clearly what many disaffected patients find attractive about the old family practitioner was the result of the therapeutic nihilism of his day. Because physicians could offer little in the way of therapeutic intervention, they had to rely on their skills as comforters of patients and their families. This meant talking to them, explaining things to them, and expressing warmth and compassion. Recapturing this humaneness and integrating it with the technological efficacy of today's medicine is a crucial challenge facing medicine and society. The growth of family practice, for example, indicates that this challenge is by no means insuperable. The pertinent question of course is, Are the very patients who complain of dehumanization in medicine willing to bear the costs of adding this form of medical care to the forms already available? If not, they may find obtaining what they most want—the warmth of the family doctor of the past along with the efficacy of today's high-tech medicine—is an impossibility.

Equitable Access to Medical Technologies

Access to quality health care, and thereby to validated medical technologies, is often quite unequal. Although some fortunate individuals have very high levels of access, others have vastly lower levels of access. Presently an individual's access to medical technology depends on a number of different variables. These include personal wealth, employment benefits, eligibility for federal and state social welfare programs, and geographic region. Needy individuals, whose circumstances leave them with low levels of access to validated medical technologies, often have no alternative but to appeal to the charitable impulses of their fellow citizens. (Such appeals have often been made in recent years on behalf of children needing liver transplants, for example.) Unfortunately many individuals are unable to make such appeals or to make them successfully. Is this kind of unequal access to medical technology and other elements of health care justifiable? Does society have an obligation to guarantee some specific level of access to validated medical technologies, or at least to certain kinds of validated medical technologies, to all its members? Questions of this nature challenge the notion that there are no morally significant differences between medical technologies and health care and all the other kinds of goods and services produced in this society. They challenge the notion that medical technologies are morally indistinguishable from food processors, television sets, and microwave ovens. Although these sorts of nonmedical technologies are highly valued by many people, little support exists for the view that society has an obligation to ensure any particular level of access to them to all its members. The types of microwave ovens, for instance, that are produced, their distribution to the population, the means by which they are delivered to those who get them, and the means by which they are financed are matters determined primarily by the market, that is, by voluntary exchanges between individuals using their own legitimately held resources. Although such exchanges take place within a framework of laws and regulations designed to protect the overall welfare of society and to protect the vulnerable from exploitation, no one is guaranteed any level of access to the majority of nonmedical technologies thereby generated. What is produced, who gets it, and how it is paid for are left up to the decisions individuals choose to make with their own resources. Such decisions may result in high levels of access to these technologies for all, low levels of access for all, or, more likely, varying degrees of unequal access. No particular pattern or level of access is ruled out as long as it is genuinely the outcome of voluntary exchanges between individuals using their own legitimately held resources in pursuit of their own aims and ends. Why shouldn't medical technology be treated in the same way? Why shouldn't the types and quantities of medical technologies produced, the distribution of those technologies, and the means of financing them be left up to the voluntary transactions individuals engage in with their own resources in light of their own preferences?

Commentators have noted several features of health care that give it special significance, but perhaps none is more important in the context of a society like the United States than its effect on opportunity. To begin to see this, it is necessary to focus on the nature and importance of health care needs. Of course people do have health care needs and thus needs for medical technology. To know this though is not to know what is special about the practices and technologies that are used to maintain and restore health. Indeed, given one very widely accepted sense of "need," virtually anything can be legitimately regarded as a need. As ethicist Norman Daniels (1987) has written,

Without abuse of language, we refer to the means to reach any of our goals as needs. To reawaken memories of Miller's, the neighborhood delicatessen of my childhood, I need only the smell of pickles in a barrel. To stay at Cape Cod next weekend, I need a motel reservation.

In this sense what an individual needs is a matter of that individual's wants and preferences, a matter of the particular aims and ends that individual seeks to realize. On this understanding needs are entirely relative to the individual and the importance of a need is primarily, and perhaps wholly, a subjective matter, a matter of the intensity of the individual's wants and preferences. If X is wanted/preferred more intensely than is Y, then those things that are required for the attainment of X are more important needs than those that are required for the attainment of Y. Accordingly if this conception of need is the only acceptable one, whether needs for medical technology and health care in general are of special importance is a question of whether the wants individuals can satisfy by means of medical technology and the other instrumentalities of health care are more intense than wants satisfiable by other means. If this is the only acceptable sense of the term *need*, then it is difficult to see how health care and the medical technologies used to provide that care can plausibly be regarded as importantly different from any of the other kinds of goods and services produced in a society. Indeed the wants that are satisfied by means of microwave ovens, alligator shoes, and portable stereos can be felt just as intensely and even more intensely than those that are satisfied by medical technology.

Although this sense of need is indeed legitimate and widely accepted, it is not the only legitimate sense. This can be seen by looking to the notion of need that in fact does function in attempts to determine when individuals have some responsibility to provide others with the things they need. When we attempt to determine which needs of others we have some responsibility to meet, we do in fact discriminate between kinds of needs. Some needs are regarded as making genuine claims on our resources, claims we have some duty to meet. Others are not. When this sort of distinction is made, it is not based on the intensity of the individual's wants. As Daniels (1987) has pointed out,

...not all the things we say we need are equally important. It is possible to pick out various things we say we need, including needs for health care, which play a special role and are given a special weight in a variety of moral contexts. If I appeal to my friend's duty of beneficence in requesting $100, I most likely will get a quite different reaction if I tell him I need the money to get a root-canal than if I tell him I need the money to go to the Brooklyn neighborhood of my childhood to smell pickles in a barrel. Indeed, it is not likely to matter in his assessment of his obligation that I strongly prefer to go to Brooklyn. Nor is it likely to matter that I insist I feel a great need to reawaken memories of my childhood—I am overcome by nostalgia.

In making this sort of distinction, according to Daniels, we are refusing to rely merely on the individual's own assessment of his well-being as the basis for deciding which of his needs may obligate us to come to his aid. But how then do we decide which of his needs can be the source of such obligations? What, other than the intensity of the individual's wants, could possibly be a basis for regarding some of his needs as important enough to place responsibilities on others?

Daniels answered these questions by appealing to an especially useful distinction between kinds of needs delineated by philosopher David Braybrooke, a distinction that is crucial to understanding how needs can be objectively sorted into those that can confer responsibilities on others and those that cannot. According to Braybrooke, all the things an individual can need are either "course-of-life needs" or "adventitious needs" (Daniels 1987):

Adventitious needs are the things we need because of the particular contingent projects (which may be long-term ones) on which we embark. Their relative importance will reflect many idiosyncratic facts about our personal choice of a plan of life. Course-of-life needs are those needs people "have all through their lives or at certain stages of life through which all must pass." They include food, shelter, clothing, exercise, rest, companionship, a mate (in one's prime), and so on. Such needs are not themselves deficiencies, for example, when they are anticipated. But a deficiency with respect to them "endangers the normal functioning of the subject of need, considered as a member of a natural species." Thus course-of-life needs are important when we abstract from many particular choices and preferences individuals might make or have.

Clearly the notion of a course-of-life need is a different sense of need from the sense in which an individual's needs are entirely relative to his particular wants and preferences, a sense that is captured by the notion of adventitious needs. A course-of-life need remains a need even when it cannot be connected to some particular aim or end an individual wants to realize. Such a need is identified not by looking to the individual's specific wants, to his particular goals, but rather by looking to a standard that Daniels has referred to as species-typical normal functioning. This standard identifies needs that individuals will have all through their lives or in particular phases of life through which all individuals pass (if life is not ended prematurely), whatever their particular aims and ends, wants and preferences happen to be.

What then is species-typical normal functioning? The fundamental idea is that for any given natural species, there are kinds and levels of functioning that are normal for a typical member of that species. What is normal for such an individual will be a matter of the species to which it belongs. Although an account of species-typical normal functioning can be given for an oak tree, a dalmatian, or a human being, clearly the correct account will be quite different for each of these kinds of entities. The task of determining exactly what is species-typical normal functioning falls on the sciences that study natural species and thereby determine their defining properties. In the case of the human species, this will include the science of human biology and all the biomedical sciences, and given that humans are intrinsically social and psychological creatures, at least some of the social sciences will also be relevant to defining species-typical normal functioning for human beings. As these sciences reveal more about human nature and thus more about species-typical normal functioning for human beings, which things are viewed as course-of-life needs will likely change. By whatever means the notion of species-typical normal functioning is specified, those things that are necessary to maintain and restore such functioning and to compensate for incurable losses of such functioning are course-of-life needs.

Species-typical normal functioning and the course-of-life needs that must be met to maintain or restore it and to compensate for its loss constitute a dimension of well-being that is distinguishable from the particular wants and preferences of the individual and how intensely those wants and preferences are felt, a dimension of well-being that is objectively specifiable. We intuitively recognize this objective dimension of well-being when we recognize that something a person is interested in may not be in his interest. For example, we recognize that although an individual may intensely desire to consume large quantities of alcohol, it would be to his genuine detriment to satisfy this want. Although he is truly interested in the pleasures attainable by such consumption, it is not in his interests to obtain them in the sense that he would thereby impair his ability to function in the ways and at the levels that are normal for a typical member of his species. Some such notion of well-being that can be identified apart from the individual's wants and their intensity seems to be presupposed whenever we judge an individual's wants to be contrary to his good.

Forms of health care and the technologies used in this care belong in the category of course-of-life needs. This follows from the very nature of health and disease. As Daniels (1985) has noted,

The basic idea is that health is the absence of disease, and diseases (I include here deformities and disabilities that result from trauma) are deviations from the natural functional organization of a typical member of a species.

Thus health care needs, by their very nature, are things that we must have "in order to maintain, restore, or provide functional equivalents (when possible) to normal species functioning" (Daniels 1987). Although this will include a diverse body of things, ranging from adequate nutrition to adequate rest to adequate sanitation, clearly it will include access to some validated medical technologies as well.

So far it has been argued that health care needs are course-of-life needs, needs for things that are necessary for maintaining and restoring species-typical normal functioning and compensating for its loss. If this is true, then medical technologies and health care in general are indeed different from microwave ovens, portable stereos, and similar nonmedical technologies. But how does this support the notion that medical technology and health care in general are special in some morally significant way? How is it relevant to the question of whether society has a duty to ensure some level of access to these goods and services to all its members rather than allowing who has what level of access to be determined entirely by the voluntary exchanges of individuals using their own legitimately held property, that is, to be determined by the market? At least some medical technologies are necessary for the maintenance of species-typical normal functioning. Many kinds of nonmedical technology are not. So what?

The answers lie in the effect that disease has on opportunity. To see this, it is necessary to turn to Daniels's (1987) notion of "a normal opportunity range":

The normal opportunity range for a given society is the array of life plans reasonable persons in it are likely to construct for themselves. The range is thus relative to key features of the society—its stage of historical development, its level of material wealth and technological development, and even important cultural facts about it.

Whereas many factors affect how the normal range of opportunity for a given society in fact is distributed across its population, access to health care is surely one of the most important. An individual who cannot function in the ways or at the levels that are normal for a typical human will not have access to as much of the range of opportunity that is normal for her society as unimpaired fellow citizens. One of the most highly regarded moral norms of contemporary American society is fair equality of opportunity. According to this norm, any given individual's share of his society's normal opportunity range should be determined as far as possible only by his talents and skills. Any other constraints on the individual's share of that range of opportunities are not necessarily equal in the sense that every individual has exactly the same share as every other individual. Rather they are equal in the sense that the share held by each person, no more or less than the share held by any other person, is determined by only his skills and talents. Because individual skills and talents differ, the shares of the normal opportunity range held by different individuals will differ also even when those shares are fair. Some will have larger shares than will others. Although opportunity will be unequal in this sense, it will be equal, and thus fair, in the sense that everyone is constrained only by their own individual skills and talents.

Whereas disease is clearly one constraint that can cause an individual's share of the normal opportunity range to be less than the share his talents and skills would otherwise make available to him, it is not the only constraint (Daniels 1987):

. . . skills and talents can be underdeveloped or misdeveloped because of social conditions, e.g., family background or racist educational practices. So, if we are interested in having individuals enjoy a fair share of the normal opportunity range, we will want to correct for special disadvantages here too, say, through compensatory educational or job-training programs. Still, restoring normal functioning through health care has a particular and limited effect on an individual share of the normal functioning range. It lets him enjoy that portion of the range to which his full array of skills and talents would give him access, assuming that these too are not impaired by special social disadvantages.

In recent years American society has gone to great lengths to identify and to attempt to eliminate social conditions that unfairly limit the individual's share of the normal opportunity range, and such efforts continue today. Although there has been and will continue to be much debate about which social conditions have this kind of effect and what responses are appropriate to these conditions, the commitment to ensuring that the individual's opportunities are, as far as possible, a function only of his talents and skills is widespread and is becoming deeply rooted. The kind of commitment to fair equality of opportunity that underlies efforts to mitigate social barriers to fair equality of opportunity for all individuals justifies similar efforts against those conditions that can be prevented, removed, or at least somewhat ameliorated by medical technology and health care in general. This ultimately is the reason for regarding health care needs as special and for wanting to ensure some level of access to medical technology to all individuals, the reason for not wanting who gets what kinds and quantities of medical technology and health care and how these are to be financed and delivered to be determined in the way that these matters are determined for microwave ovens, for instance.

Although the powers of modern health care and its technologies far exceed those of the health care available only a few decades ago, even now there are diseases and impairments that cannot be prevented, cannot be cured, and whose effects cannot be ameliorated to any significant degree. No matter how great a society's commitment to fair equality of opportunity may be, it cannot be expected to do the impossible. The fact that people do get ill, live impaired lives, and die prematurely does not show by itself that a society has failed to ensure that medicine is doing all it can to provide fair equality of opportunity. It may simply mean that what medicine can do is limited. Whatever equitable access to medical technologies may mean, it does not mean absolute protection against illness, impairment, and premature death.

What then does equitable access to medical technologies and health care in general mean? What level of access to medical technologies and health care in general must society guarantee to all its members to provide equitable access? Insofar as society's commitment to fair equality of opportunity is the basis for a social obligation to provide equitable access, society need not ensure access to all kinds of medical technology and care to all people. Not all kinds of medical technology and health care prevent loss of species-typical normal functioning, restore such functioning, or compensate for its loss. An individual may desperately want to have a beautiful nose, and the technology to satisfy this want does exist. Yet providing him with such a nose does not necessarily maintain, restore, or compensate for loss of species-typical normal functioning. If it does not, then society has no obligation to provide him with access to this cosmetic surgery, although he may very intensely desire to have it. Only if some genuine course-of-life need is met by a form of medical technology or health care does society have an obligation, based on its commitment to fair equality of opportunity, to provide that form of technology or care. Only then does that form of technology have the kind of effect on species-typical normal functioning that is relevant to fair equality of opportunity.

Ideally, insofar as the obligation to provide equitable access is founded on a commitment to fair equality of opportunity, equitable access should mean equal access for all to any and all forms of medical technology and health care that can prevent loss of species-typical normal functioning, restore such functioning, or compensate for the effects of impairment in such functioning. But given that society's resources are finite and that society has other ends it seeks to accomplish, equitable access in this sense is probably not possible.

If the ideal in equitable access to medical technology and other aspects of health care is simply not feasible, what then would be a realistic sense of equitable access? In answer to this question it should be noticed that not all forms of medical technology and health care in general will have the same effect on species-typical normal functioning. Among preventive technologies, some will prevent more serious impairments than will others. Among cura-tive and rehabilitative technologies, some will restore more impor-tant kinds and levels of functioning than will others. Among tech-nologies that compensate for incurable losses in functioning, some will compensate for more important losses than will others. In addition some mixes of preventive curative, rehabilitative, and compensatory technologies will have greater positive effect on spe-cies-typical normal functioning than will others. (This is why it is important that society deal with the debate about prevention versus rescue as well as the debate about halfway technology versus basic biomedical research.) If society, given its resource constraints and other ends, cannot provide access for everyone to everything that can prevent, cure, or compensate for the effects of disease, then it ought to do as much as it can to ensure that everyone has a reasonable level of access to those technologies that have the most significant effect on species-typical normal functioning, those technologies that meet the most important course-of-life health needs. By definition these will be the technologies that do the most toward realizing fair equality of opportunity. These technologies will do the most to remove impairments that prevent an individual's share of her society's normal opportunity range from being as large as it would be if it were determined by her skills and talents alone. Which technologies do in fact have the greatest positive effect on keeping individuals at the level of functioning normal for typical human beings is a matter to be determined by scientific judgment. What level of access to such technologies is a reasonable, and thus the level that ought to be guaranteed to all, is a matter for social decision in light of society's resource constraints. Although this question cannot be answered here, it is one that society is forced to face through its commitment to fair equality of opportunity.

What implications would a commitment to providing all indi-viduals with the level of access to validated medical technologies mandated by fair equality of opportunity have for efforts to contain the costs of medical care? First, it would mean that efforts to control the costs of medical care should initially be directed at those kinds of medical technology and those forms of health care that do not meet course-of-life medical needs, that do not function to maintain, restore, or compensate for loss of species-typical normal functioning. Second, it would mean that should those efforts prove insufficient and should savings are attainable only by turning to cost-containment measures with respect to technologies that do affect species-typical normal functioning, before measures are taken that affect access, efforts should be made to eliminate all inefficient use of such technologies. Only then should access be restricted. And insofar as these measures reduce access, such reductions should be directed at those technologies that are least important for maintaining, restoring, or compensating for loss of species-typical normal functioning. Third, it would mean that society must ask whether, to reduce health care costs, it is willing not to guarantee access for all to those medical technologies and forms of health care that are most important for keeping individuals functioning in the ways and at the levels that are typical of normal human beings. A decision of this nature would be a decision to be less committed to fair equality of opportunity for some than for others. For it would mean not ensuring that as much will be done to avert or neutralize the most severe limits on opportunity that disease and impairment can engender for some persons as would be done for others. Can society make that decision without implying that it values some of its members less than others? If it cannot, one must wonder what reason such less valued individuals could have for feeling much allegiance to society. Is the moral cost of treating certain of its members as second-class citizens with respect to health care a moral cost that society can afford to bear? This question cannot be answered here, but it is unavoidable for a society that is committed to both fair equality of opportunity for all and to containing the costs of health care.

Presently, although American society is committed to fair equality of opportunity, it has failed to ask what this means for access to medical technologies and other forms of health care. Failure to ask this question has left society unable to see that this commitment requires that a reasonable level of access to certain kinds of such technology and care be provided to all or that it must face up to whether the financial costs of providing such access are preferable to the moral costs of not doing so.

Summary

In this chapter we have examined four broad social and ethical issues that encompass a wide spectrum of medical technologies. The first of these issues is the question of whether enough health care resources are allocated to preventive medicine as opposed to rescue medicine. In chapter 7 this issue was considered solely from an economic perspective. Evidence was provided to show that preventive medicine is not *always* cost effective from the standpoint of either the society as a whole or the individual. Nevertheless, in many circumstances, preventive medicine *is* cost effective and *more attractive* in terms of its net benefits than many forms of rescue medicine. In this chapter, however, we have seen that our society often seems predisposed to allocate medical resources to rescue medicine because it meets the crisis needs of real and very present people (our families, our friends, our neighbors, and the little girl whose need for a liver transplant is so touchingly described on the evening news).

Crisis or rescue medicine is often high-tech and costly (as noted in chapter 5 in relation to intensive care units) because of heavy reliance on health care professionals rather than costs associated with the hardware. Preventive medicine, frequently much cheaper in relation to the benefits it generates, is often left on the shelf largely because it seems to benefit only "statistical" lives rather than identifiable human beings. This has been shown to be a poor reason for ignoring preventive medicine if the goal of a health care system is to provide its beneficiaries (those we know *and* those we do not know) with the maximum benefits from the resources available for health care. In fact greater emphasis should probably be placed on certain forms of prevention and less on certain forms of rescue. In either case medical technology will play an integral role. If there is a misallocation in favor of rescue, then medical technology is not at fault. Ultimately our choices in the marketplace and our political actions determine the structure and focus of health care policy and the distribution of resources between different forms of care.

The second issue considered in this chapter is the debate over the allocation of resources between halfway technologies and basic research. Halfway technologies, such as the iron lung and subsequent respiratory technologies (similar to those described in chapter 4), have been shown to provide substantial benefits to many patients. They may not cure disease, but they often ameliorate its effects and extend life. We disagree with the view that halfway technologies are necessarily bad. In fact it is by no means clear that the concept of a halfway technology, one that fails to get to the root of the patient's disease and only keeps him in a limbo of discomfort, has meaningful empirical content. We also reject the related view of medical technology as the handmaiden of basic science. Innovations in medical technology frequently lead to innovations in basic science. Rather we suggest that there is far more merit in the view that basic science and technology (including so-called halfway technologies) enjoy a synergistic relationship—medical technology improves basic research and basic research improves medical technology. Both work together to improve the overall welfare of the patient. This, when all is said and done, is what medicine is about.

In fact the halfway technology debate is something of a red herring. What is of real concern is whether enough resources are being allocated to basic science to allow our society to improve its health care at an optimal rate. Chapter 2 documented the huge cutbacks in federal funding for basic research that occurred under the Reagan administrations. These cutbacks decimated the academic medical research community and have done lasting damage to our society's capacity to develop new therapies in all fields of health care. The solution to the problem is not, however, to rob Peter (current expenditures on beneficial therapies) to pay Paul (support for basic science). It is to persuade policymakers and especially the current Bush administration to correct the wrongdoings of the previous administration and to increase support for basic research without undercutting other equally beneficial health programs.

The third issue is the dehumanization of medicine. Technological innovation has certainly played a part in this process by creating layers of hardware between the patient and the doctor and drawing the physician's attention away from the patient toward the information provided by the machines. However, the origins of the process of dehumanization are found in events that occurred long before twentieth century hardware became the focus of medical treatment. They can be found in the fifteenth century drawings of Leonardo da Vinci that reflected the increasing importance of the study of anatomy, in which the human body rather than the human soul becomes the object of interest; in medical specialization and the widespread acceptance of Descartes's conception of the body as a machine, both of which took hold in the eighteenth and nineteenth centruries; and in the medical technological imperative of the twentieth century, which leads patients to demand high-tech care because, rightly or wrongly, they perceive it to be the best possible care.

Medical technology is not *the* problem that has to be addressed to rehumanize medicine. Attitudes have to be altered. Health care professionals need to rediscover the patient as a thinking and feeling being. Patients need to adjust their expectations, recognizing that not all physicians can be family doctors. One solution to the problem may be the rebirth of family practice. Several medical schools' programs now concentrate on educating physicians who can act as a family's "black bag" physician and mediate on their behalf with the complex medical world of specialists whose services are needed sometimes to save and often to improve their lives. However, family practice adds another layer to the health care system and itself increases the health care bill. Compassion is costly, and if, as a society, we want to be sure it is available in the health care system, we must be willing to bear those costs.

The final issue examined here is in many ways the source of the deepest concerns about medical technology and the health care system: To what extent should access to different forms of health care be available to all members of our society? Should the rule be "to each according to his needs and from each according to his ability to pay," or should it be "to each according to his ability to pay"? Neither rule seems universally appropriate if only because it is difficult to sort out what constitutes a person's needs. We have suggested, however, that a helpful distinction can be made between types of need—course-of-life needs as opposed to adventitious needs. Course-of-life needs are those that must be met if a person is to enjoy fair equality of opportunity in a normal life. If there is a consensus in our society, it is probably that these types of need (basic shelter, adequate food, and the like) should be satisfied irrespective of an individual's ability to pay. Thus, through its governmental agencies, our society should step in to fill the gaps between such needs and access to the resources required to meet them. Many aspects of health care are targeted toward course-of-life needs (for example, trauma centers, emergency care facilities, vaccination programs, and prenatal care). Often they involve sophisticated technologies (for example, coronary bypass operations and pacemaker implantations). To the extent that our society succeeds in providing all of its members with access to such care (through existing programs such as Medicaid and Medicare or new programs such as catastrophic health insurance for the elderly), it is viewed as a just and fair system. To the extent that it fails, it stands condemned on its own terms because it then fails to provide its citizens with equality of opportunity for a productive and normal life.

The issues discussed in this chapter should not be taken to imply that medical technology is wholly a bane or wholly a boon. Medical technology does have the capacity to provide our society with massive benefits: longer and more comfortable lives for larger and larger segments of our population. Yet utilizing that capacity poses hard economic and ethical questions. If our society wants health care that is both efficient and humane, it cannot avoid the responsibility of tackling these questions.

References

Boyle, J. M., and Morriss, J. E. 1981 (July/Oct). The philosophical roots of the current medical crisis. *Metaphilosophy* 12(3 and 4).

Childress, J. 1981. Priorities in the allocation of health care resources. In Shelp, E. E. (ed.): *Justice and Health Care*. D. Reider.

Caplan, A. 1983. How should values count in the allocation of new technologies in health care? In Bayer, R., Caplan, A., and Daniels, N. (eds.): *In Search of Equity: Health Needs and the Health Care System*. Plenum Press.

Bennett, I. J. 1977. Technology as shaping force. In: *Doing Better and Feeling Worse: Health in the United States*. W. W. Norton.

Daniels, N. 1985. *Just Health Care*. Cambridge University Press.

Daniels, N. 1987. Justice and health care. In Van DeVeer, D., and Regan, T. (eds.): *Health Care Ethics: An Introduction*. Temple University Press.

DeVries, W. 1986. The artificial heart. In Bronzino, J., Smith, V., and Wade, M. (eds.): *Technology and Medicine*. Trinity College.

Eisenberg, L. 1977. The search for care. In: *Doing Better and Feeling Worse: Health in the United States*. W. W. Norton.

Freedman, B. 1977 (April). The case for medical care, inefficient or not. *Hasting Center Report* 7(2):31-39.

Fried, C. 1970. *An Anatomy of Values: Problems of Personal and Social Choice*. Harvard University Press.

Friedrich, O. 1984 (10 Dec). One miracle, many doubts. *Time* 124:70-77.

Goodman, L., and Goodman, M. 1986 (April). Prevention—How misuse of a concept undercuts its worth. *Hastings Center Report* 16(2):26-38.

Howard, J. 1978. Health care: Humanization and dehumanization of health care. In: *The Encylopedia of Bioethics*. Vol. II. The Free Press.

Jonsen, A. 1986 (Feb). The artifical heart's threat to others. *Hasting Center Report* 16(1):9-12.

Knowles, J. 1977. The responsibility of the individual. In Knowles, J. (ed.): *Doing Better and Feeling Worse*. W. W. Norton .

Maxwell, J. H. 1986. The iron lung: Halfway technology or necessary step? *The Milibank Quarterly* 64(1):3-29.

McGinnis, J. M. 1988-89 (Winter). National priorities in disease prevention. *Issues in Science and Technology*, pp. 46-52.

Menzel, P. 1983. *Medical Costs, Moral Choices: A Philosophy of Health Care Economics in America.* Yale University Press.

Reiser, S. J. 1984. The machine at the bedside: Technological transformations of practices and values. In Reiser, S. J., and Anbar, M. (eds.): *The Machine at the Bedside: Strategies for Using Technology in Patient Care.* Cambridge University Press.

Schwartz, M. A., and Wiggins, O. 1985 (Spring). Science, humanism, and the nature of medical practice: A phenomenological view. *Perspectives in Biology and Medicine* 28(3):331-361.

Stell, L. 1985. Medical technology, medical utopianism, and morality. In: *Bioengineering and Health Technologies: The Physics, Clinical Applications, Economics and Bioethics of Magnetic Resonance Imaging.* Davidson College-Sloan New Liberal Arts Program.

Thomas, L. 1971 (Dec). Notes of a biology-watcher: The technology of medicine. *New England Journal of Medicine* 285:1366-1368.

Thomas, L. 1977. On the science and technology of medicine. In Knowles, J. (ed.): *Doing Better and Feeling Worse.* W. W. Norton.

Thomas, L. 1983 (Feb). Who will be saved? Who will pay the cost? *Discover*, pp. 30-32.

Thomas, L. 1986. Response to Maxwell's essay, "The Iron Lung." *The Millbank Quarterly* 64(1):30-33.

Thomas, L. 1987. The new medicine: Something different. *International Journal of Technology Assessment in Health Care* 3:5-10.

Wolpe, P. R. 1985 (Winter). Medicine, technology, and lived relations. *Perspectives in Biology and Medicine* 28(2):314-322.

Index

Acute care medicine, 509-516, 518-519, 531, 534
Aesculapius, 3
AI (artificial intelligence). *See under* Patient care computing
Aid to families with dependent children (AFDC) program, 66
Allocation of scarce medical resources, 125-134
Allocational effects of medical technology, 499-500, 505
Alveolar ventilation, 220
Ancient Greek medicine, 4, 106, 537
Angiography, 12
Arrhythmia, 29, 248
Artificial heart, 192-212
complete artificial heart, 136-137, 161-165
devices, 136-137, 155-166, 175-180
FDA regulations of, 197-208
left ventricular assist devices (LVAD), 136, 156-166
temporary implantation of, 208-211
Artificial intelligence (AI). *See under* Patient care computing
Automated patient-monitoring systems, 28-29, 247-249

Beneficence, 105-108, 114, 125, 129, 192
Benefit-cost analysis of medical technology, 38, 39, 71-75
discount factor, 80
mechanics of, 78-85
of new cardiac technologies, 166-170
present value, 72-73, 78-85
Biomedical engineering, definitions, 13

Cardiac catheterization, 12
Cardiac monitoring system, 239
cardiac output, 239-240
Cardiac technology, 136-213
cardiopulmonary bypass, 156
fundamental principles of heart action, 138, 144
heart disease, 136
Clarke, Barney, 195, 208, 211, 513-514
Clinical chemistry, 22, 283
Clinical information system, 300-304
example of, PATIENT LOOKUP system, 301-304
Clinical microbiology, 22, 283
College of physicians, establishment of, 7
Computed tomography (CT), 30-31, 200, 416-473, 503
compared with MRI, 450-451
CT scanners, 31-33, 416-473
scanning gantry, 431-435, 503
Computer-based technologies, 304-364. *See also* Patient care computing
effect on employment in health care industry, 334-338
ethical issues, 338-362
liability for negligence, 340-352
privacy and confidentiality, 358-362
rate of adoption, 331-334
strict liability, 340-348, 352-357
tort liability, 339-348
Computer revolution, 20-23
Computerized axial tomography (CAT), 30, 416. *See also* Computed tomography
Computers in health care, 280-364
effect on costs, 322-331